Multiscale Scratch Mechanics of Laminated Shale

Liu Yang

Multiscale Scratch Mechanics of Laminated Shale

Springer

Liu Yang
State Key Laboratory for Tunnel Engineering
China University of Mining and
Technology (Beijing)
Beijing, China

Translated by
Kunheng Zhang
Department of General Education
Beijing Health Vocational College
Beijing, China

ISBN 978-981-95-7040-9 ISBN 978-981-95-7041-6 (eBook)
https://doi.org/10.1007/978-981-95-7041-6

Jointly published with China University of Petroleum Press

This Springer imprint is published by the registered company Springer Nature Singapore Pte Ltd.
The registered company address is: 152 Beach Road, #21-01/04 Gateway East, Singapore 189721, Singapore

Preface

Shale is a variety of rock widely existing in nature, which is closely related to human production activities. During the exploitation of coal and oil/gas and the sequestration of carbon dioxide and nuclear waste, the shale layer is one of the most common strata. At the same time, shale exists in projects, such as subway tunnels, water conservancy, and hydropower. In particular, the development of shale oil and gas resources will continue to be a pillar of energy for a long time. Therefore, how to ensure the resource development in shale formations and the use of underground space is a key issue that needs to be solved. This requires further research and an understanding of the physical and mechanical characteristics of shale.

However, the intensive research on laminated shale is full of challenges. The layer-like structure is widespread, including micron-millimeter-level, centimeter-level, and meter-level layers. The lengths of drilling cores and tunnels in shale formations reach several kilometers. Within such a large range, the heterogeneity of shale mechanical properties is more pronounced. The traditional experiments of rock mechanics that are based on standard cylindrical samples are insufficient to reflect the influence of cross-scale layered structures on mechanical properties. The processing difficulty of shale samples is high, and the specimens are completely destroyed after experimental testing, resulting in the significant consumption of rock cores. The test results have a large divergence, making it difficult to form a continuous trend curve. In view of this, the author used non-destructive or micro-damage scratch testing techniques to study the spatial continuity of mechanical properties on the surface of laminated shale, which can serve as an important supplement to traditional rock mechanics methods.

The work presented in this book has been funded by the National Natural Science Foundation of China (No. 52374014) and the Fundamental Research Funds for the Central Universities (No. 2024ZKPYSB03). During the writing of this book, I received guidance and assistance from Prof. He Liu and Prof. Manchao He. Thank you for the support of the postgraduates from my research team. In addition, as the author's level is limited, there are inevitably omissions in this book. I urge all colleagues and readers to provide criticisms and suggestions.

Beijing, China Liu Yang

Contents

Chapter 1
Introduction

Shale is widely distributed in the Earth's strata and has received extensive attention in the fields of mineral resource extraction (e.g., shale oil and gas extraction) and underground space resource utilization (e.g., soft-rock tunnel support, CO_2 geologic sequestration, radioactive nuclear waste, etc.). Shale is a rock formed by dehydration and cementation of clay, which is dominated by clay minerals and has an obvious laminated structure. Its mechanical strength tends to be low, and it is susceptible to softening and swelling when exposed to water, with poor stability. The study of the physical and mechanical properties of shale is of great significance to the safe mining and efficient utilization of minerals and underground space. However, due to the complexity of shale's own structure and the limitation of testing technology, the past has mostly focused on small-size samples' full load destructive measurements, which are difficult to use to respond to the influence of complex structure on mechanical properties. In addition, the specimen sampling requirements are higher, the test results are more discrete, and multiple experiments are needed to ensure the reliability of the results, which makes it difficult to form a continuous and trendy understanding. At the end of the experimental test, the specimen is completely destroyed, the core consumption is large, and the economic cost is high. In recent years, with the development of non-destructive and micro-destructive scratch test technologies, it has become possible to study the characteristics of spatial continuity changes in the mechanical properties of shale. This book focuses on some research results of macro- and micro-scratch mechanical tests and numerical simulations of shale, in order to provide a reference for understanding and recognizing the macroscopic mechanical properties of shale.

L. Yang, *Multiscale Scratch Mechanics of Laminated Shale*,
https://doi.org/10.1007/978-981-95-7041-6_1

1.1 Background

The lithosphere and hydrosphere are the foundations of human production and life. The lithosphere provides abundant mineral resources (e.g., coal, oil and gas, iron ore, and non-ferrous metals) and geospatial resources (such as for CO_2 geological sequestration, radioactive nuclear waste and toxic sewage treatment, etc., or for use as underground water reservoirs, underground oil reservoirs, underground nuclear test sites, etc.). Rock is a porous medium with abundant pores, fissures, and laminae, forming complex, non-homogeneous structural features. Both the development of mineral resources and the utilization of geospatial resources require the excavation or disturbance of rocks, which requires further research and understanding of the physical, mechanical, and chemical properties of rocks.

Shale laminated structural surfaces are widely developed, including micrometer-to-millimeter-scale laminae (Qingshankou Formation, Songliao Basin), centimeter-scale laminae (Longmaxi Formation, Sichuan Basin), and meter-scale thin interbedded layers (Luchaogou Formation, Jimusar Depression). Laminated structural surfaces are the smallest macroscopic layers in sedimentary sequences and usually comprise different mineral compositions with highly variable mechanical properties. Currently, horizontal drilling or tunneling in shale formations reaches several kilometers in length, and the non-homogeneity in shale mechanical properties is even more pronounced over such a large span. The low strength of shale makes sample processing difficult, and the specimen is completely destroyed at the end of the experimental test, with high core consumption and high economic costs. The specimen sampling requirements are higher, the test results are more discrete, and multiple experiments are needed to ensure the reliability of the results, which makes it difficult to form a continuous and trend-oriented understanding. It is found that the mechanical properties of the transition zone at the interface of grain layers have a great influence on the deformation and damage behavior of shale; however, there is a lack of intuitive understanding of the distribution of interfacial mechanical parameters. Multiscale scratch testing is a near-nondestructive surface mechanical testing method with only minor material loss. It can provide continuous spatial information on rock mechanical properties and capture their trend variations, thereby serving as an effective supplement to the discrete measurements obtained from conventional rock mechanics experiments. This technique supports the construction of a more rigorous grain-scale mechanical model of shale. It also improves the understanding of shale deformation and failure mechanisms. These insights provide a theoretical basis for mineral resource development and efficient underground-space utilization in shale formations [1].

1.2 History of Micro/Nano-Mechanics

1.2.1 Nanoindentation Method

The indentation experiment does not destroy the integrity of rock samples, has the characteristics of repeatable testing, short testing time, and simple specimen fabrication, and has been widely used to study the properties of biomaterials, polymers, and thin-film materials [2]. Sun et al. [3] carried out nanoindentation studies on clay minerals in mudstone and analyzed the effects of quartz and kaolinite on the mechanical properties of organics in mudstone. Yang et al. [4] carried out nanoindentation tests on the surface of shale to obtain its mechanical properties and proposed a new micro-brittleness index and explored the distribution characteristics of the brittleness index on the surface of shale as well as the influence law of the mineral composition on the brittleness index. Liu et al. [5] used nanoindentation to measure the local modulus of elasticity and hardness of shale and obtained the mechanical properties of the mineral phases, such as quartz, organics, and clays, in the shale, and studied the mechanical properties and indentation creep behavior of shale before and after CO_2 treatment [6]. Currently, most of the research results focus on analyzing the local elastic modulus, hardness, fracture toughness, and other characteristics of shale. However, relatively few studies have been conducted on the effects of brittle minerals and the mechanical properties of grain layers. This limits the wider and more reasonable application of nanoindentation in characterizing the mechanical properties of shale.

Nanoindentation has the advantage of small sample size requirement, which greatly improves the utilization of precious cores. In addition, due to the non-homogeneous nature of shale, conventional mechanical tests cannot fully elucidate the macroscopic mechanical property changes caused by brittle minerals and grain layers. Nanoindentation can analyze the complex mineral compositions of the cores and can also make up for the shortcomings of the conventional experimental methods that cannot identify the weak surfaces of shale, which can be used as an important supplement to the macroscopic rock mechanics experiments.

1.2.2 Scratching Method

Considering the limitations of the nanoindentation test, the scratch test can realize the measurement of mechanical parameters on different scales by setting a certain loading speed, which can effectively establish the connection between microscopic phases and the overall mechanical properties of shale and greatly improve the use of valuable cores [7]. Liu et al. [8] carried out nanoindentation tests and scratch tests on shale samples with different mineralogical compositions and compared the fracture toughness values obtained by the two methods. Hernandez [9] and others proposed a method to quantify the brittleness of mud shale based on the micromechanical parameters collected by the scratch test, defining the brittleness index as the ratio of the energy associated with brittle damage to the total energy generated by the scratch test process. Scratch tests can be reorganized and planned within the framework of inline elastic fracture mechanics. The airy stress function method is used in the analysis to determine the stresses and displacements near the scratch-blade-material interface, which serves as input to evaluate the energy release rate by integration. The energy release rate was thus found to be proportional to the sum of the squares of the applied forces compared to the previous model. Utilizing the linear relationship between scratch width and scratch depth, the method can be shown to provide an easy way to determine fracture toughness for scratch tests with different scratch widths and depths by validating the experimental data on cement paste and sandstone scratches [10].

Due to the non-homogeneous nature of shale, conventional mechanical tests and indentation experiments cannot fully elucidate the macroscopic mechanical property changes caused by brittle minerals and grain layers. Compared to nanoindentation experiments, scratch experiments can capture continuous mechanical property parameters, which greatly improve the utilization of precious cores.

1.2.3 Indentation and Scratching Theory

Shi et al. [11] investigated the distribution area of the grain layer in Longmaxi shale samples from the Youyang area, Chongqing, on the mechanical properties of the grain layer, such as Young's modulus, hardness, and fracture toughness, using the dotted nanoindentation method. Through the field-emission scanning electron microscope (FESEM) and the energy spectrum spectrometer (EDS), the morphology and mineral composition of the indentation zone near the grain layer were quantitatively analyzed. Yuan et al. [12] introduced a ternary micromechanical model based on the classification of mechanical strengths and used the Mori-Tanaka model to enhance the mechanical parameters from the nanometer scale to the centimeter scale and further compared the results with those of uniaxial compression. Due to the small thickness of the grain layer, it is difficult to determine its elastic modulus by conventional standard mechanical methods. Li et al. [13] investigated the differences in elastic modulus of terrestrial-phase shale grain layers using microindentation techniques and scanning electron microscopy (SEM) equipped with automated mineral identification and characterization system (AMICS). The results showed that the elastic modulus of shale grain layers is highly heterogeneous and can be categorized into three types: harder grain layers, intermediate grain layers, and softer grain layers.

Interfacial transition zone (ITZ) is the boundary region between different materials in a non-homogeneous sample, which usually exhibits different properties from a single material, including low density, high porosity, and complex structural features [14]. Vignoli et al. [15] assumed that the ITZ thickness of the fiber-matrix tends to a constant value. Then, by comparing finite element simulations with experimental data, it was verified that the interfacial transition zone of fibers has nonlinear damage characteristics and plays a key role in damage occurrence and crack propagation. Kato et al. [16] evaluated the mechanical properties of the interfacial transition zone of cement pastes doped with different additives and then quantified the thickness of the ITZ by using a wall effect model and compared the analytical results with experimental results. Wei et al. [17] conducted nanoscratching for concrete materials with different vertical loading rate experiments and analyzed the effects of friction coefficient and fracture toughness on the quantitative results of ITZ thickness of cement paste materials. Nie et al. [18] introduced the concept of the interfacial transition zone between mineral aggregates, such as pyrite, calcite, and apatite lumps, and the grain layer in shale and proposed a method of quantifying ITZ fracture toughness in conjunction with nanoscratch tests.

Combining numerical analysis and finite element simulation can further analyze the relationship between the indentation energy and the mechanical parameters of the material. Cheng et al. [19] found that there was a direct proportionality function between indentation energy recovery rate and dimensionless indentation hardness through nanoindentation dimensional analysis in combination with dimensionless energy. Yang et al. [20] chose 20 typical materials to conduct nanoindentation experiments, finding that both metallic and nonmetallic materials conform to an approximately linear relationship. The scale factor of the positive correlation is about 0.2, and the experiment verifies the accuracy of the theory. Finite element method (FEM) and molecular dynamics (MD) are widely used to simulate nanoscratching and nanoindentation processes. Both methods have limitations and advantages. MD simulation can describe the nanoscratching process at the atomic level, including material changes such as dislocation motion, grain boundary effects, laminar dislocations, and twinning under the indenter during the simulation process. Shi et al. [21] investigated the effect of indenter deflection angle on the collected mechanical parameters of the material in nanoscratch tests using the FEM method. Chatterjee et al. [22] simulated the scratch test process of a rigid indenter on a new crystalline material of HfB2 by using the FEM method and found that the finite element calculations agree well with the experimental results for the same indenter scale. Wagih et al. [23] investigated the nanoindentation of nanocomposites by using a two-dimensional finite element model. The results showed that the load-displacement curves obtained from finite element simulations were in good agreement with the experimental data.

1.2.4 Brittleness Index

There is no unified standard and evaluation method for shale compressibility evaluation at home and abroad. At present, scholars at home and abroad have established many compressibility evaluation models based on shale experimental research. Compressibility characterizes the degree of difficulty for tight oil and gas reservoirs to be effectively transformed, and it has become an important evaluation criterion for the development potential of tight unconventional natural gas. Compressibility refers to the ability of a rock reservoir to be effectively fractured and form fractures during hydraulic fracturing. Jarvie et al. [24] proposed a computational method for evaluating shale brittleness based on a brittleness index, which quantitatively characterizes the tendency of shale to undergo brittle failure. The size of brittleness is quantified by calculating the relative weight of quartz-dominated brittle minerals in the total rock minerals. The greater the proportion of quartz in the total rock minerals, the greater the brittleness of the rock and the better the compressibility.

With the deepening of research, more and more researchers have realized that the compressibility of rocks is affected by other factors besides brittleness. Wang [25] conducted a follow-up in-depth study, which also used the specific gravity of brittle minerals to measure the brittleness size, but considered that brittle minerals included dolomite in addition to quartz, and the content of quartz and dolomite in the reservoir accounted for a higher proportion of the total rock minerals. The higher the proportion of quartz and dolomite in the rock reservoir to the total amount of rock minerals, the higher the brittleness of the shale reservoir, and the better the compressibility. This research result opens up a new research path for future scholars on the study of brittle mineral evaluation method. Gong et al. [26] utilized microstructural knowledge extracted from rock images for geologic facies analysis and developed a machine learning workflow for predicting Young's modulus and assessing rock compressibility, taking into account mineralogical composition, geology, and microstructure. The results show that the use of microscale thin-section analysis in conjunction with fracture bonding models enhances our understanding of fracture geometry and understanding and contributes to compressibility assessment. Zeng et al. [27] established a brittleness index calculation method to obtain parameters such as Young's modulus and Poisson's ratio of the reservoir, taking into account factors, such as shale depositional environment and confining pressure, and combining logging data and in situ rock mechanical test results. Based on the fractal method, an in situ rock bifurcation/irregular fracture toughness calculation method was established. The hydraulic fracture induced index is introduced to respond to the influence of parameters, such as stimulation-induced displacement, reservoir seepage, fluid viscosity, and horizontal principal stress on compressibility. A method for evaluating compressibility with the coupling of reservoir rock brittleness, fracture toughness, and hydraulic fracture induced index is developed.

References

1. Yang L, Yang D, Zhang M, et al. Application of nano-scratch technology to identify continental shale mineral composition and distribution length of bedding interfacial transition zone—a case study of cretaceous Qingshankou formation in Gulong depression, Songliao Basin, NE China. Geoenergy Sci Eng. 2024;234:212674.
2. Zhang M, Zhang G, Yan P. Characterizing the micro-fracture in quasi-brittle rock using nanoindentation. Eng Fract Mech. 2024;308:110378.

3. Sun C, Li G, Gomah ME, et al. Meso-scale mechanical properties of mudstone investigated by nanoindentation. Eng Fract Mech. 2020;238:107245.
4. Yang L, Mao Y, Yang D, et al. The characteristic and distribution of shale micro-brittleness based on nanoindentation. Materials. 2022;15(20):7143.
5. Liu Y, Liu S, Liu A, et al. Determination of mechanical property evolutions of shales by nanoindentation and high-pressure CO2 and water treatments: a nano-to-micron scale experimental study. Rock Mech Rock Eng. 2022;55(12):7629–55.
6. Manjunath GL, Jha B. Nanoscale fracture mechanics of Gondwana coal. Int J Coal Geol. 2019;204:102–12.
7. Bandini A, Berry P, Bemporad E, et al. Role of grain boundaries and micro-defects on the mechanical response of a crystalline rock at multiscale. Int J Rock Mech Min Sci. 2014;71:429–41.
8. Liu K, Jin Z, Zakharova N, et al. Comparison of shale fracture toughness obtained from scratch test and nanoindentation test. Int J Rock Mech Min Sci. 2023;162:105282.
9. Hernandez-Uribe LA, Aman M, Espinoza DN. Assessment of mudrock brittleness with micro-scratch testing. Rock Mech Rock Eng. 2017;50:2849–60.
10. Manjunath GL, Nair RR. Microscale assessment of 3D geomechanical structural characterization of Gondawana shales. Int J Coal Geol. 2017;181:60–74.
11. Shi X, Jiang S, Lu S, et al. Investigation of mechanical properties of bedded shale by nanoindentation tests: a case study on lower Silurian Longmaxi formation of Youyang area in Southeast Chongqing, China. Pet Explor Dev. 2019;46(1):163–72.
12. Yuan SS, Zhu QZ, Zhao LY, et al. Micromechanical modelling of short-and long-term behavior of saturated quasi-brittle rocks. Mech Mater. 2020;142:103298.
13. Li L, Huang B, Tan Y, et al. Using micro-indentation to determine the elastic modulus of shale laminae and its implication: cross-scale correlation of elastic modulus of mineral and rock. Mar Pet Geol. 2022;143:105740.
14. Tang BX, Xie RF. Micromechanical properties of concrete under freezing-thawing condition. Russ J Nondestruct Test. 2020;56:527–39.
15. Vignoli LL, Savi MA, Pacheco PMCL, et al. Micromechanical fiber-matrix interface model for in-plane shear in unidirectional laminae. Mech Adv Mater Struct. 2023;31:1–13.
16. Kato Y, Tsukahara E, Uomoto T. Evaluation of characteristics of transition zone existing at aggregate-cement paste interface. In: 2nd International Conference on Engineering Materials, San Jose, CA; 2001. p. 621–8.
17. Wei Y, Kong W, Wang Y, et al. Multifunctional application of nanoscratch technique to characterize cementitious materials. Cem Concr Res. 2021;140:106318.
18. Nie Y, Wu B, Zhang G, et al. Determination of the thickness and fracture toughness of shale interfacial transition zone near various micromineral aggregates using nano-scratch tests. SPE J. 2024;29:1–18.
19. Cheng YT, Cheng CM. What is indentation hardness? Surf Coat Technol. 2000;133:417–24.
20. Yang R, Zhang T, Jiang P, et al. Experimental verification and theoretical analysis of the relationships between hardness, elastic modulus, and the work of indentation. Appl Phys Lett. 2008;92(23):231906.
21. Shi C, Zhao H, Huang H, et al. Effects of probe tilt on nanoscratch results: an investigation by finite element analysis. Tribol Int. 2013;60:64–9.
22. Chatterjee A, Polycarpou AA, Abelson JR, et al. Nanoscratch study of hard HfB2 thin films using experimental and finite element techniques. Wear. 2010;268(5–6):677–85.
23. Wagih A, Fathy A. Experimental investigation and FE simulation of nano-indentation on Al–Al2O3 nanocomposites. Adv Powder Technol. 2016;27(2):403–10.
24. Jarvie DM, Hill RJ, Ruble TE, et al. Unconventional shale-gas systems: the Mississippian Barnett shale of north-Central Texas as one model for thermogenic shale-gas assessment. AAPG Bull. 2007;91(4):475–99.
25. Wang FP, Gale JF. Screening criteria for shale-gas systems. Gulf Coast Assoc Geol Soc Trans. 2009;59:779–93.

26. Gong Y, Ilham EM, Mohamed MM. Learning and data fusion approach for elastic rock properties estimation and fracturability evaluation. Energy AI. 2024;16:100335.
27. Zeng F, Gong G, Zhang Y, et al. Fracability evaluation of shale reservoirs considering rock brittleness, fracture toughness, and hydraulic fracturing-induced effects. Geoenergy Sci Eng. 2023;229:212069.

Chapter 2
Nanoindentation Experiments

Unlike conventional rocks, the complexity of the internal structure of shale determines the diversity of shale macro- and micro-mechanical behaviors, i.e., the macroscopic deformation of shale under external loading often exhibits obvious non-uniformity, discontinuity, anisotropy, and inelasticity, while the micro- and fine-scale mechanical behaviors are even more complex and not yet adequately researched [1], and how to accurately characterize the micromechanical characteristics of terrestrial shale is not yet clear. Therefore, in this chapter, the nanoindentation test is carried out with Qingshankou Formation striated shale as the research object, the hardness and elastic modulus parameters are tested, and the damage mode and fracture characteristics in the indentation process are analyzed, and finally, the energy dissipation relationship in the indentation process of striated shale is investigated by introducing the dimensionless parameter.

2.1 Nanoindentation Method

In this experiment, Cretaceous Qingshankou Formation laminae shale of Songliao Basin is selected as the research object for analysis. Its geographical location map is shown in Fig. 2.1. The Songliao Basin, spanning Heilongjiang, Jilin, and Liaoning provinces in northeast China and surrounded by mountains and hills, is a typical continental sedimentary basin. Daqing Oilfield, located in Heilongjiang Province, is one of the most important oil producing areas in recent years. The laminated shale of Qingshankou Formation is an important reservoir in unconventional oil and gas exploitation in this area, and its rock composition and geological characteristics are of great significance to petroleum exploration and development. In order to make a better comparative study, clay and carbonate laminated shales with different brittle mineral contents of quartz were selected for analysis. The processing of rock samples was carried out in the National Key Laboratory of Disaster Prevention and Control and Intelligent Construction of Tunnel Engineering, China University of

L. Yang, *Multiscale Scratch Mechanics of Laminated Shale*,
https://doi.org/10.1007/978-981-95-7041-6_2

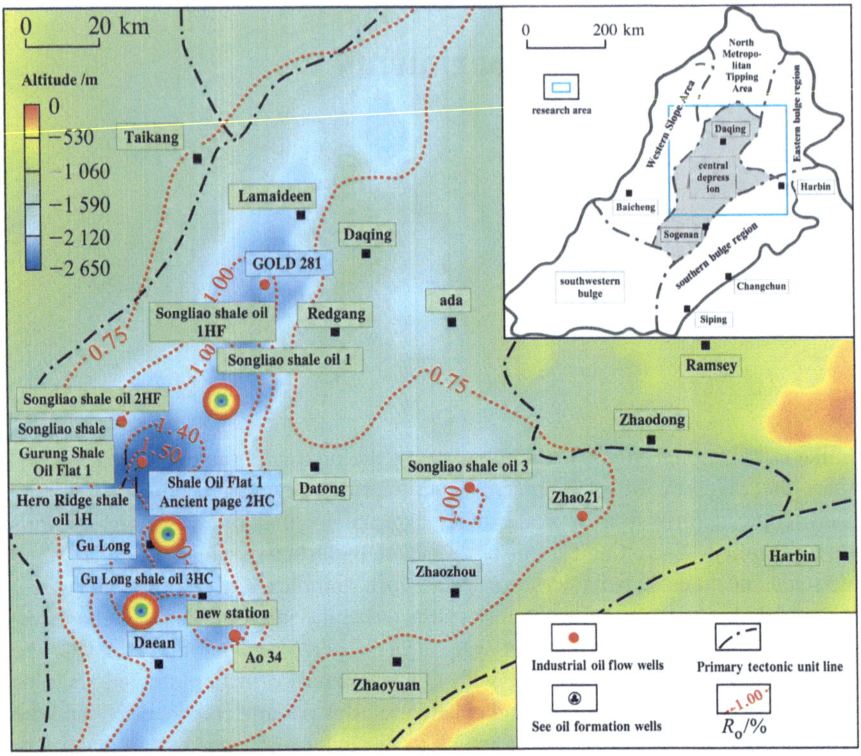

Fig. 2.1 Geological map of Songliao Basin

Mining and Technology (Beijing), and the samples were prepared in strict accordance with the shale and mudstone thin-section identification standard (GB/T 35206-2017).

2.1.1 Test Principle and Procedure

The basic principle of nanoindentation technology is to apply the test indenter to a small volume at the nanoscale to study the mechanical properties of the material according to the obtained load-displacement curve. The nanoindentation process consists of loading and unloading stages, as shown in Fig. 2.2. When the load applied by the test head to the surface of the material is small enough, with the increase in the applied load, the material will first undergo elastic deformation and then plastic deformation. When the load is large enough, the recessed area enters the full plastic stage. When the indenter exits, the elasticity of the material in the previously depressed area of the indenter recovers, while the residual depression

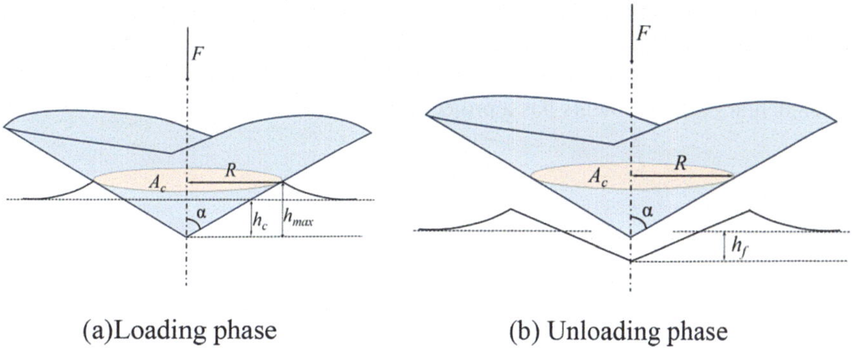

Fig. 2.2 Principle of nanoindentation. (**a**) Loading phase. (**b**) Unloading phase

Fig. 2.3 Typical load-displacement curve

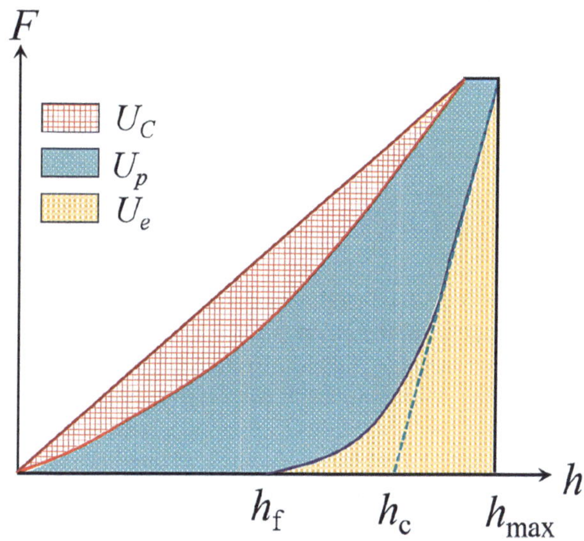

caused by plastic deformation remains. The sensor records the indentation depth and load in real time during the nanoindentation test and outputs the load-displacement curve, as shown in Fig. 2.3, from which the contact area can be obtained in real time.

Mechanical parameters such as elastic modulus and hardness can be calculated based on changes in load and indentation depth recorded by the indenter sensor. The Oliver-Pharr method is widely used to analyze the relationship between the load obtained from the experiment and the indentation depth [2]. The hardness H of the shale can be obtained from the ratio of the maximum load to the projected contact area.

$$H = \frac{P_{\max}}{A_c} \tag{2.1}$$

where H is hardness, GPa; F_{max} is the maximum load, mN; A_c is the projected contact area, nm^2. For a particular indenter geometry, the projected contact area is a function of the contact depth [3]. The ideal Berkovich indenter used in shale nanoindentation experiments has a projected contact area:

$$A_c = 24.5h_c^2 \tag{2.2}$$

Contact depth can be obtained using the following equation:

$$h_c = h_{max} - \varepsilon \frac{F}{S} \tag{2.3}$$

where ε is a constant dependent on the indenter geometry and, for Berkovich indenters, $\varepsilon = 0.75$, S is the contact stiffness and h_{max} is the maximum indentation depth. Once the contact depth and contact area are obtained, the elastic modulus can be calculated using the following equation:

$$E_r = \frac{\sqrt{\pi} S}{2\beta\sqrt{A_c}} \tag{2.4}$$

$$\frac{1}{E_r} = \frac{1 - \nu^2}{E} + \frac{1 - \nu_i^2}{E_i} \tag{2.5}$$

where E is the elastic modulus of shale, GPa; ν is the Poisson ratio of shale; E_i is the elastic modulus of the indenter, GPa; ν_i is the Poisson ratio of the indenter; and E_r is the reduced modulus, GPa, for Berkovich indenter, $\beta = 1.304$.

The physical quantities involved in the indentation problem are the material parameters (elastic modulus E, Poisson's ratio ν, and yield strength Y), the indenter geometry (equivalent half-cone angle θ), and the control variables (usually selected h as the indentation depth, which contains the corresponding maximum indentation depth h_m). E and h are selected as basic quantities from which the remaining mechanical parameters can be derived.

During the loading phase, the load F can be expressed as:

$$F = f_F(E, \nu, Y, \theta, h) \tag{2.6}$$

H defined the equation as follows:

$$H = f_H(E, \nu, Y, \theta, h) \tag{2.7}$$

Apply the π-theorem, where the dimensionless relation is:

$$F = Eh^2 \prod_F \left(\frac{Y}{E}, \nu, \theta\right) \tag{2.8}$$

$$H = E_r \prod_F \left(\frac{Y}{E}, \nu, \theta\right) \tag{2.9}$$

Using the above equation, the total energy can be expressed as:

$$U_t = \int_0^{h_m} F\,dh = \frac{Eh_m^3}{3} \prod_t \left(\frac{Y}{E}, \nu, \theta\right) \tag{2.10}$$

$$U_e = \frac{Eh_m^3}{3} \prod_e \left(\frac{Y}{E}, \nu, \theta\right) \tag{2.11}$$

$(U_t - U_e)/U_t$ has the following relationships:

$$\frac{U_t - U_e}{U_t} = 1 - \frac{\prod_e \left(\frac{Y}{E}, \nu, \theta\right)}{\prod_t \left(\frac{Y}{E}, \nu, \theta\right)} \tag{2.12}$$

There is a strong linear relationship between the energy recovery rate U_e/U_t and H/E_r as shown in the following equation:

$$\frac{U_e}{U_t} = \beta_c \frac{H}{E_r} \tag{2.13}$$

Parameter β_c represents the brittleness index related to the release of elastic energy, according to which the expansion ability of shale fracture network can be further evaluated.

2.1.2 *Sample Preparation*

The experimental samples were taken from the Cretaceous Qingshankou Formation terrestrial shale in Songliao Basin, buried about 1600 m. Figure 2.4a shows viscous shale and Fig. 2.4b shows carbonate shale. In the nanoindentation experiment, the local pores or the sudden change of roughness at different phase interfaces can affect the accuracy of the test results. The surface of the sample is mechanically polished several times using silicon carbide sandpaper, followed by a secondary polishing using a wide beam argon ion device to obtain a smooth surface. The SLATE surface was randomly scanned using the Zeta-20 3D morphometer to ensure that the roughness requirements were met. The sample is then ultrasonically cleaned, treated with acetone to remove rust spots and other contaminants, and finally cured with epoxy resin.

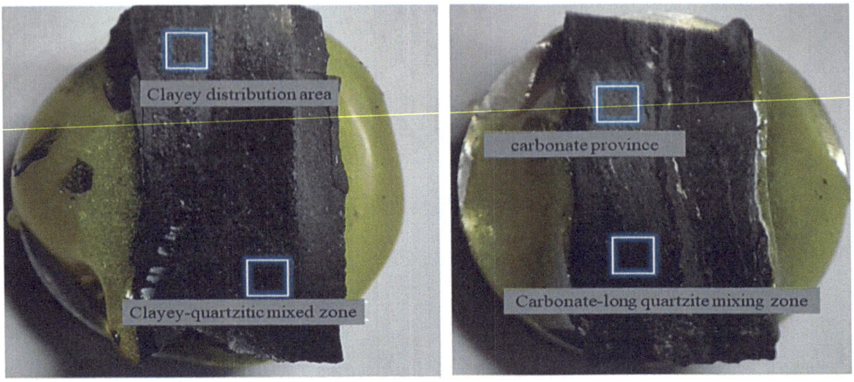

(a)Clay shale (b)Carbonate shale

Fig. 2.4 Nanoindentation shale samples. (**a**) Clay shale. (**b**) Carbonate shale

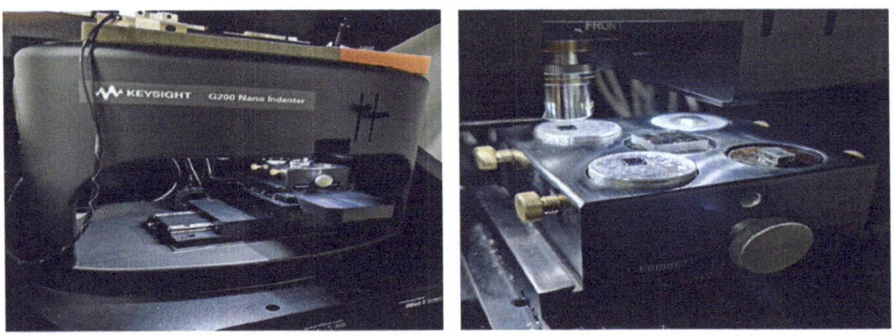

(a)Data output system (b) Scratch test platform

Fig. 2.5 Nanoindentation shale samples. (**a**) Data output system. (**b**) Scratch test platform

2.1.3 Nanoindentation Equipment

The mechanical device used for nanoindentation experiment is Agilent Nano Indenter G200 (Fig. 2.5), whose load resolution is 50 nN, the maximum load of standard test is 500 mN, the displacement resolution in the z direction is less than 0.01 nm, and the maximum indentation depth is greater than 500 μm. The displacement resolution in x and y directions is 1 μm, and the travel range is 100 mm × 100 mm. Use Berkovich indenters with tip curvature radius less than 20 nm. During the test, the indenter gradually approaches the surface of the sample with a load release rate of 30 nm/s. When the test system indicates that the indenter has touched the specimen surface, the system starts loading the specimen and automatically records the load and corresponding indentation depth.

2.1.4 Experimental Scheme

In order to further study the influence of feldspar minerals such as quartz and albite on the mechanical properties of continental shales, it is necessary to select a landmark test area before indentation testing. As shown in Fig. 2.6, using backscatter scanning electron (BSE) microscopy to analyze the mineral composition of the rock, both SEM and QEMSCAN images can be obtained to find the locations of different mineral distributions, thus helping to confirm the indentation test area. After determining the mineral distribution area of feldspar minerals, based on the distribution area of quartz and sodium feldspar minerals, a representative mineral distribution area of about 300 μm × 210 μm was determined for indentation test. In this nanoindentation test, four test regions, namely the clay mineral distribution region, the clay-quartz mineral mixture region, the carbonate mineral distribution region, and the carbonate-quartz mineral mixture region, were named A1, A2, B1, and B2, and a test point array of 88 indentations was performed for each test region (a total of 352 groups and a total of 4 samples). The indenter load reaches a peak of 100 mN at a loading time of 100 s, holds it for 4 s to measure creep properties, and then releases it within 10 s.

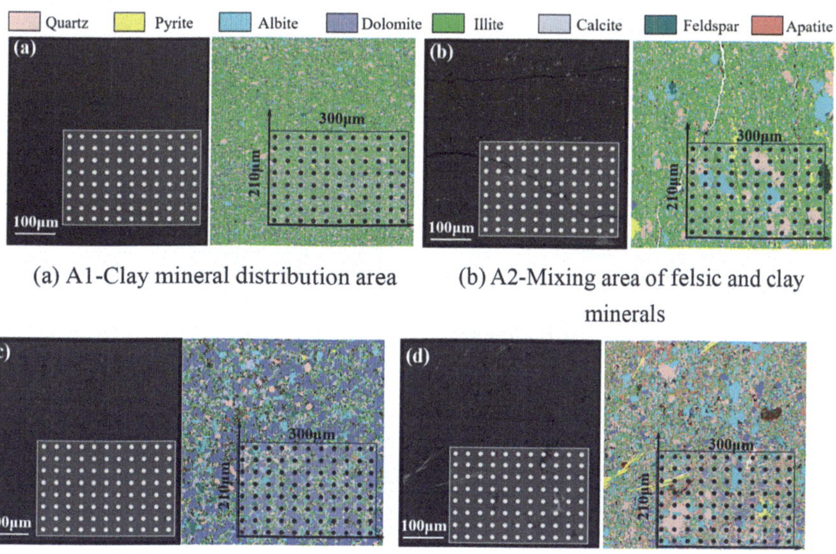

(a) A1-Clay mineral distribution area

(b) A2-Mixing area of felsic and clay minerals

(c) B1-Carbonate mineral distribution area

(d)B2-Mixed area of feldspar and carbonate minerals

Fig. 2.6 Nanoindentation test area. (**a**) A1: Clay mineral distribution area. (**b**) A2: Mixing area of felsic and clay minerals. (**c**) B1: Carbonate mineral distribution area. (**d**) B2: Mixed area of feldspar and carbonate minerals

2.2 Mechanical Properties of Microscopic Minerals

2.2.1 Distribution Characteristics of Minerals

The content and damage pattern of feldspar minerals in the indentation points were observed by BSE, as shown in Fig. 2.7. The microscopic structure and mineral distribution of the indentation points were plotted to illustrate the local morphology. The red ellipse surrounded the apparent bulk distribution of feldspar minerals. The indentation area of sample A1 is covered by a large amount of illite, quartz and albite minerals are distributed like fragments, and there are a few microcracks. The indentation of sample A2 is partially covered by large feldspar minerals, resulting in an increase in the microcrack area. Sample B1 is covered with dolomite and calcite minerals, mixed with brittle minerals such as quartz and albite, and has a few microcracks. Sample B2 contains a lot of feldspar minerals and is rich in organic matter. Samples A1 and B1 are tightly bound with almost no pores or microcracks. Due to the presence of feldspar minerals, samples A2 and B2 have complex pore structure, organic matter is well developed, and the number of microcracks increases significantly. Image Plus Pro software was used to conduct RGB recognition on QEMSCAN scan, and the content of feldspar minerals was calculated. As shown in Fig. 2.7, the content of feldspar minerals (quartz and albite) in samples A1, A2, B1, and B2 was 6.58%, 15.83%, 8.72%, and 22.69%, respectively. Samples A2 and B2 of the experimental group have significantly higher content of felsic minerals than samples A1 and B1 of the control group.

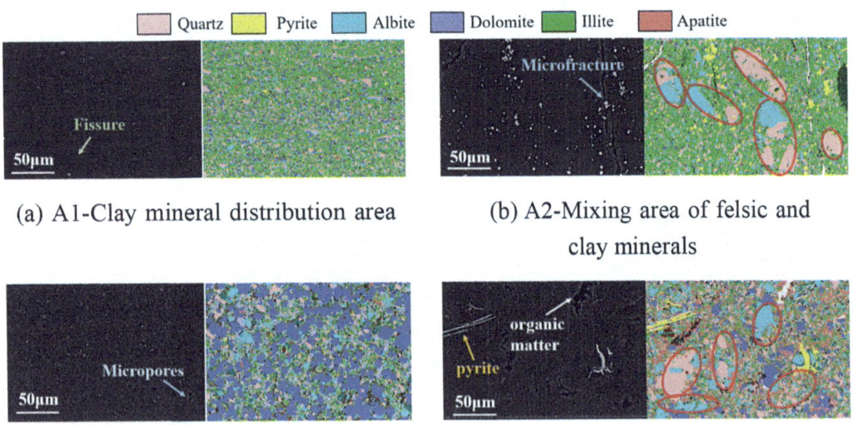

(a) A1-Clay mineral distribution area

(b) A2-Mixing area of felsic and clay minerals

(c) B1-Carbonate mineral distribution area

(d) B2-Mixed area of feldspar and carbonate minerals

Fig. 2.7 Distribution characteristics of Changying minerals. (**a**) A1: Clay mineral distribution area. (**b**) A2: Mixing area of felsic and clay minerals. (**c**) B1: Carbonate mineral distribution area. (**d**) B2: Mixed area of feldspar and carbonate minerals

2.2.2 Nanoindentation Failure Mode

The deformation of a material not only reflects the non-uniformity of the material but also provides information about the internal structure and distribution of mechanical properties of the material. Shale contains mineral particles of different hardness and toughness, so different deformation areas may appear in nanoindentation testing. The indentation damage pattern of shale is shown in Fig. 2.8. The indentation energy of viscous shale is better absorbed into the volume deformation in the indentation test, that is, there is material accumulation around the indenter. The failure mode of the clay-quartz mixed zone is shown in the figure. Brittle minerals do not break during the destruction process but fall into the clay texture, resulting in weak effects of brittle minerals on overall mechanical properties. The indentation damage pattern of carbonate shale is shown in the figure. Under the same stress state, dolomite and calcite are hard and the indentation depth is shallow, and organic materials such as asphalt and casein in shale are displaced during the indentation process, resulting in the appearance of holes. The damage pattern in the carbonate-quartz mixed zone is shown in the figure.

Due to the increase in the content of felsic minerals, the brittleness of the material increases, and the indentation energy is transformed into crack propagation and

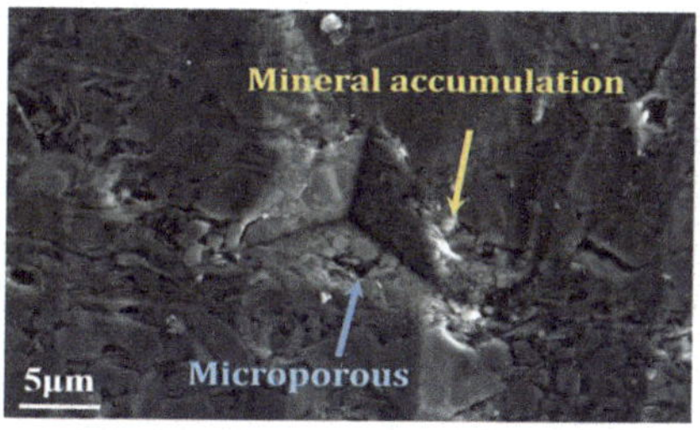

(a) A1-Clay mineral distribution area

(b) A2-Mixing area of felsic and clay minerals

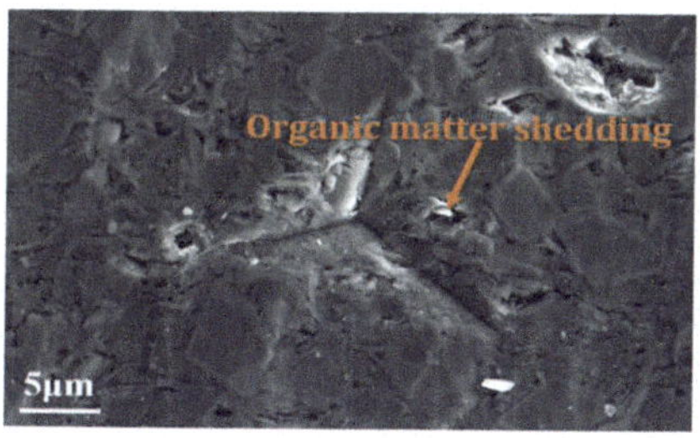

(c) B1-Carbonate mineral distribution area

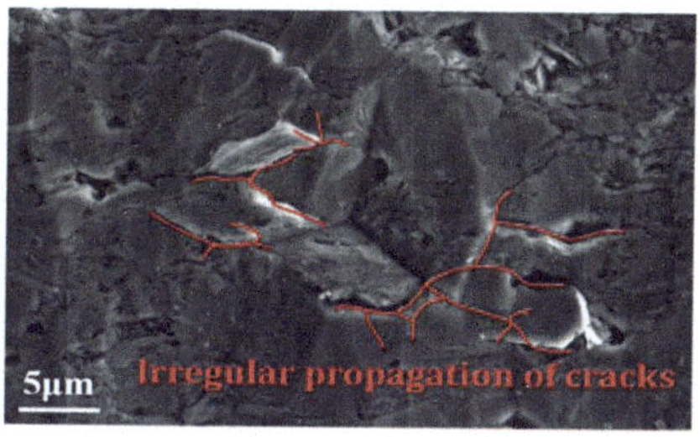

(d) B2-Mixed area of feldspar and carbonate minerals

Fig. 2.8 The indentation failure mode of continental shale under a load of 100 mN. (**a**) A1: Clay mineral distribution area. (**b**) A2: Mixing area of felsic and clay minerals. (**c**) B1: Carbonate mineral distribution area (**d**) B2: Mixed area of feldspar and carbonate minerals volume damage, and microcracks are formed on the specimen surface and extend irregularly in all directions. By observing the indentation morphology, important information about the fracture characteristics, plastic deformation, and microstructure of the material can be obtained. In nanoindentation tests, brittle minerals have higher elastic modulus and hardness, more dense structures, stronger connections between molecules, and are prone to fracture under external pressure, resulting in crack growth and volume failure. Plastic minerals can absorb part of the indentation energy and convert it into volume deformation, resulting in significant plastic flow and mineral accumulation when the indenter is applied.

2.2.3 Nanoindentation Curves

The load-displacement curve usually consists of elastic deformation, elastoplastic deformation, and plastic deformation stages. During loading, the slope of the curve increases rapidly in the elastic phase. When the yield strength of the rock itself is reached, elastoplastic deformation occurs and nanoindentation cracks begin to appear. When the indentation load exceeds the strength of the rock itself, plastic damage occurs and permanent cracks are formed. The typical indentation curve of laminated shale in Songliao Basin is shown in Fig. 2.9. The clay shale enters the elastoplastic stage from the elastic stage after 2000 nm. The load-displacement curve of the feldspar mineral-rich area is shown in the figure. When the displacement is 950 nm, the elastic phase begins to enter the elastoplastic phase. The load-displacement curve obtained in the feldspar mineral-rich area is not smooth, and there is a phenomenon of pop in the loading stage. The main reason for this phenomenon is that when the load is increased to the yield strength, microcracks

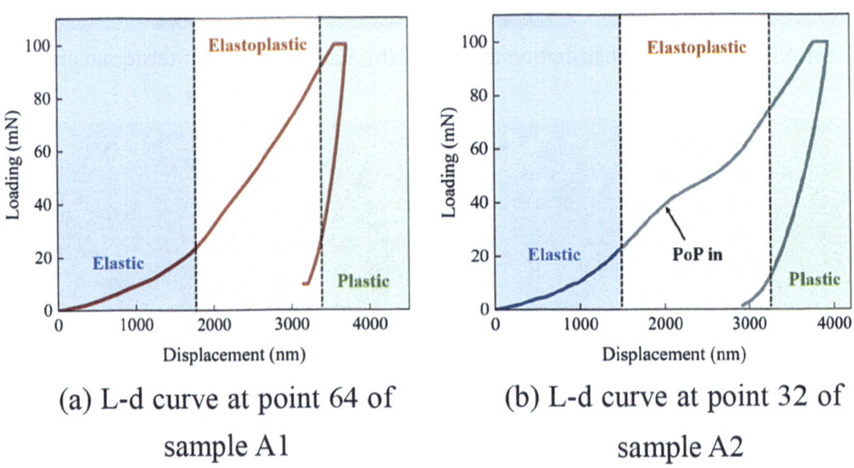

Fig. 2.9 Types of load-displacement (*L-d*) curves for continental shale. (**a**) *L-d* curve at point 64 of sample A1. (**b**) *L-d* curve at point 32 of sample A2

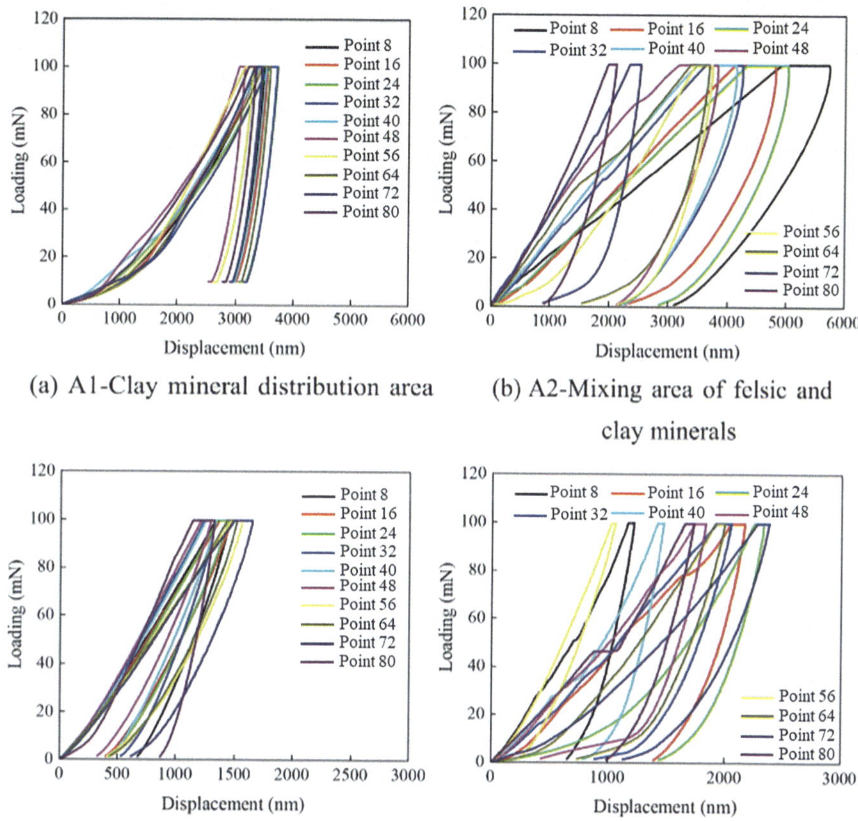

(a) A1-Clay mineral distribution area

(b) A2-Mixing area of felsic and clay minerals

(c) B1-Carbonate mineral distribution area

(d) B2-Mixed area of feldspar and carbonate

Fig. 2.10 Indentation curve of Gulong shale in Songliao Basin. (**a**) A1: Clay mineral distribution area. (**b**) A2: Mixing area of felsic and clay minerals. (**c**) B1: Carbonate mineral distribution area. (**d**) B2: Mixed area of feldspar and carbonate

appear in the rock, and the stress begins to spread inside the crack along with the crack tip. In addition, the burst phenomenon is likely due to the inhomogeneity of the shale itself. During the indentation experiment, the indenter passing through the matrix (such as cracks, micropores, and kerogen) can also cause the phenomenon of pop in.

The different internal microstructures of the rock lead to different characteristics of the loading part. To prevent the indentation curves from overlapping, 10 representative load-displacement curves were selected from each sample at equal intervals, as shown in Fig. 2.10. In the load increasing stage, the displacement gradually increases, resulting in plastic deformation and elastic deformation. Due to the creep characteristics of the rock, the constant load phase leads to a continuous increase in the depth of the indentation. At the unloading stage, with the load decreasing gradually, the elastic deformation part gradually recovers and maintains the residual displacement. Samples A1 and B1 have more uniform distribution of illite and dolomite minerals, less felsic minerals, more concentrated indentation spacing, and no burst phenomenon. Due to the different hardness of illite and dolomite minerals, the maximum indentation depth of A1 samples is distributed in the range of 3000–3700 nm, while the maximum indentation depth of B1 samples is distributed in the range of 1200–1700 nm. Due to the increase in felsic mineral content, the mineral cementation deteriorates, resulting in a significant increase in the dispersion of the indentation curve and a burst phenomenon. The maximum indentation depth of A2 sample is distributed in the range of 1950–5750 nm, and the maximum indentation depth of B2 sample is distributed in the range of 1000–2400 nm. The presence of felsic minerals increases the dispersion of load-displacement curves of clay shales and carbonate shales, but carbonate shales are less affected.

2.2.4 *Elastic Modulus and Hardness*

After the nanoindentation test was completed, the hardness data was obtained according to the formula calculation, and the planar thermal spectrum was generated, as shown in Fig. 2.11. Through careful comparative analysis, we can see the degree of influence of feldspar minerals on the hardness of clay shale and carbonate shale. Due to the wide distribution area of plastic minerals such as illite, the mechanical properties of brittle minerals such as quartz and albite have not been well characterized. In the A1 sample, only a small fraction of the measured indentation produced hardness greater than 1 GPa, ranging from 0.09 to 1.556 GPa, with an average hardness of 0.717 GPa. The hardness of clay shale increases slightly with the increase in feldspathic mineral content, ranging from 0.19 to 5.81 GPa, with an average hardness of 0.784 GPa. Sample B1 and B2 were further observed. The hardness distribution range of sample B1 was 0.19–5.81 GPa, and the average hardness was 1.447 GPa. At the same time, sample B2 shows a hardness distribution range of 0.36–9.72 GPa, showing a much higher average hardness of 4.481 GPa. It can be clearly seen that the increase in feldspar mineral content significantly increases the hardness of carbonate shales, while feldspar minerals have little effect on the hardness of clay shales.

Elastic modulus is an index of material's resistance to deformation. Figure 2.11 shows the distribution of elastic modulus of continental shale in Songliao Basin. Most of the elastic modulus of sample A1 is distributed in the range of 2.03–20.16 GPa, with an average value of 3.752 GPa. The elastic modulus of sample A2 in the distribution of brittle minerals will reach 4.12–23.49 GPa with an average of 6.414 GPa due to the increase in the content of felsic minerals. The influence of felsic minerals on the elastic modulus of carbonate shale is shown in Fig. 2.12. With the increase in mineral content, the elastic modulus of brittle mineral distribution in sample B2 increased to 4.91–61.68 GPa, with an average value of 31.714 GPa. The increase in the content of felsic minerals significantly increases the elastic modulus of carbonate shale, while the elastic modulus of clay shale only increases slightly.

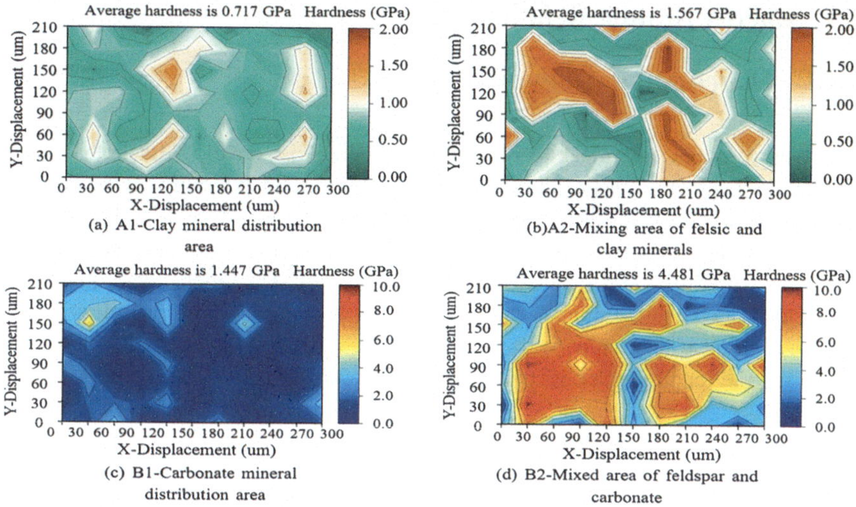

Fig. 2.11 Hardness distribution of nanoindentation test area. (**a**) A1: Clay mineral distribution area. (**b**) A2: Mixing area of felsic and clay minerals. (**c**) B1: Carbonate mineral distribution area. (**d**) B2: Mixed area of feldspar and carbonate

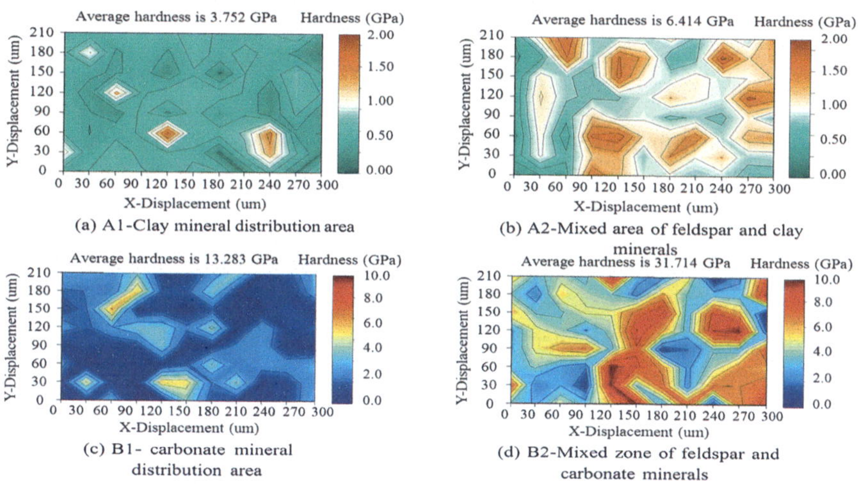

Fig. 2.12 Hardness distribution of nanoindentation test area. (**a**) A1: Clay mineral distribution area. (**b**) A2: Mixed area of feldspar and clay minerals. (**c**) B1: carbonate mineral distribution area. (**d**) B2: Mixed zone of feldspar and carbonate minerals

2.3 Energy Scaling Relationship

2.3.1 Dimensional Analysis

The energy scale relationship of laminated shale is greatly affected by the interbedding mineral elastic modulus, yield strength, and Poisson's ratio. For the dimensional analysis of indentation tests on multicomponent materials, the material is first simplified as a two-mineral system. The two minerals are characterized by their elastic moduli, yield stresses, and Poisson's ratios. The independent quantities are then selected as the fundamental variables, from which the dimensionless forms of F, H, U_e and U_t can be derived.

During the loading phase of the indenter, the load F can be expressed as:

$$F = f_F(E_1, E_2, \nu_1, \nu_2, Y_1, Y_2, \alpha, \theta, S, h) \tag{2.14}$$

There is an approximate linear relationship between the indentation depth and the indenter load F at the loading stage of the indentation curve. To ensure dimensional consistency, a new length parameter needs to be introduced to define the head load F. The loading force is influenced by the bedding width S and the mechanical parameters of various minerals in heterogeneous shales. Select elastic modulus E_1, E_2, indentation depth h, and cushion width S as basic quantities:

$$F = (E_1 + E_2)hS\Pi\left(\frac{Y_1}{E_1}, \frac{Y_2}{E_2}, \frac{S_1}{S_2}, \nu_1, \nu_2, \theta, \alpha\right) \tag{2.15}$$

Based on the definition of the total work of indentation based on the influence of the layer width, dimensionless parameter η is introduced:

$$U_t = \int_0^S F\,ds = \frac{\eta(E1 + E2)h_m S^2}{2}\Pi\left(\frac{Y_1}{E_1}, \frac{Y_2}{E_2}, \frac{S_1}{S_2}, \nu_1, \nu_2, \theta, \alpha\right) \tag{2.16}$$

At the end of unloading, there is no load on the compressor:

$$F = (E1 + E2)Sh_m\prod_\gamma\left(\frac{Y_1}{E_1}, \frac{Y_2}{E_2}, \frac{S_1}{S_2}, \nu_1, \nu_2, \theta, \alpha\right) = 0 \tag{2.17}$$

Plastic energy during unloading:

$$U_p = \int_0^S F\,ds = \frac{\eta(E1 + E2)h_m S^2}{2}\prod_p\left(\frac{Y_1}{E_1}, \frac{Y_2}{E_2}, \frac{S_1}{S_2}, \nu_1, \nu_2, \theta, \alpha\right) \tag{2.18}$$

From this, the following relationship between elastic energy and total energy can be obtained:

$$\frac{U_e}{U_t} = 1 - \frac{\prod_p\left(\frac{Y_1}{E_1}, \frac{Y_2}{E_2}, \frac{S_1}{S_2}, v_1, v_2, \theta, \alpha\right)}{\prod_T\left(\frac{Y_1}{E_1}, \frac{Y_2}{E_2}, \frac{S_1}{S_2}, v_1, v_2, \theta, \alpha\right)} \equiv \prod_e\left(\frac{Y_1}{E_1}, \frac{Y_2}{E_2}, \frac{S_1}{S_2}, v_1, v_2, \theta, \alpha\right) \tag{2.19}$$

2.3.2 Indentation Cavity Model

Johnson expansion cavity model (ECM) developed by Johnson was used to obtain analytical solutions for hardness, elastic work, and total work, and the Lame solution was used to obtain unloading work. In a multi-component spherically symmetric shale model, the number of variables to be solved depends on the mineral type. To simplify the calculations and take into account the heterogeneity of shale, the model assumes two materials: quartz and clay, representing brittle and plastic minerals, respectively. The elastic modulus, yield strength, Poisson's ratio, and bedding thickness of quartz are expressed as E_1, Y_1, v_1, and S_1, respectively. Assume that the displacement of the material under the indenter is only radial displacement or variable, and that the shear stress and shear strain components are zero. In the constitutive equation, it is necessary to consider the different stress-strain expressions of elastic and plastic phases. In the elastic phase, generalized Hooke's law is assumed, and in the plastic phase, the total relation of simple loads is assumed. In addition, the boundary or continuity conditions must be specified. The stress beneath the indenter should remain within a defined range. At the elastic–plastic boundary, the initial yield condition of the plastic zone should be defined. Within the plastic zone, yielding is described using the von Mises yield criterion. The displacement continuity condition should also be imposed along the boundary of the plastic zone. The extended cavity model of the stratified shale is shown in Fig. 2.13. Assuming that the two minerals are compressed in the core area and deform uniformly at different angles, the basic equations and specific forms of boundary conditions in double-laminated shales are given below.

In the core region, it is assumed that the deformation of the two minerals along the pressure head is the same at all angles, and there is no difference due to heterogeneity. The elastic modulus of the pressing process of brittle minerals and plastic minerals is calculated as an arithmetic average.

The geometric equation is:

$$\varepsilon_r = \frac{du}{dr} \tag{2.20}$$

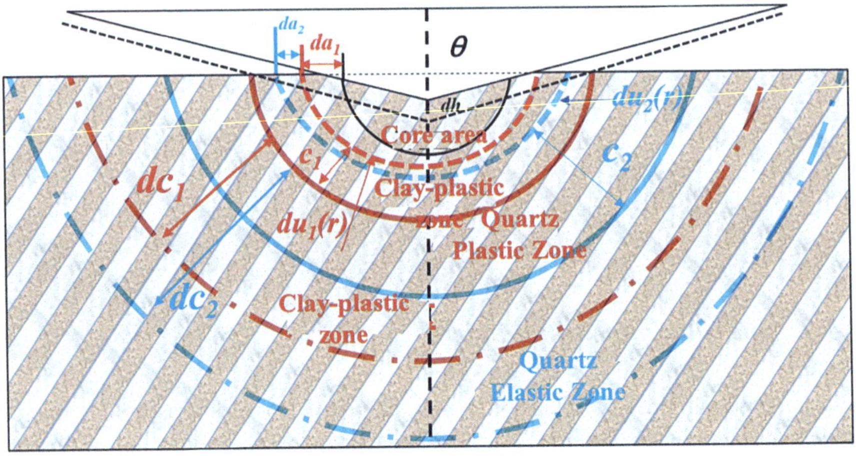

Fig. 2.13 Schematic diagram of the extended cavity model of rich layered shale

$$\varepsilon_\theta = \varepsilon_\varphi = \frac{u}{r} \tag{2.21}$$

The equilibrium equation is:

$$\frac{\mathrm{d}\sigma_r}{\mathrm{d}r} + 2\frac{\sigma_r - \sigma_\theta}{r} = 0 \tag{2.22}$$

According to the elastic modulus and Poisson's ratio of different materials, the generalized Hooke's law is comprehensively considered and modified by the constitutive equation for the elastic case, where v_1 is the value of the small Poisson's comparison:

$$\sigma_{ij} = 2G\varepsilon_{ij} + \lambda\varepsilon_{kk}\delta_{ij} = \frac{E1 + E2}{1 + \nu 1 + v2}\varepsilon_{ij} + \frac{(\nu 1 + v2)(E1 + E2)}{4(1 + \nu 1)(1 + 2\nu 2)}\varepsilon_{kk}\delta_{ij} \tag{2.23}$$

For the plastic region of each mineral in the assumption of the extended cavity model, the stress and strain are generally expressed as the spherical quantity and the deviant quantity in the corresponding three-dimensional coordinate system, and then the constitutive relations between the spherical quantity of stress and strain and the deviant quantity of stress and strain are established, respectively. The components of stress σ_{ij} and strain ε_{ij} are expressed as follows:

$$\sigma_{ij} = s_{ij} + \sigma_m\delta_{ij} = s_{ij} + \frac{1}{3}\sigma_{kk}\delta_{ij} \tag{2.24}$$

$$\varepsilon_{ij} = e_{ij} + \varepsilon_m \delta_{ij} = e_{ij} + \frac{1}{3}\varepsilon_{kk}\delta_{ij} \tag{2.25}$$

Taking the pressing problem as a simple loading problem, the relationship between strain and stress is redefined according to Poisson's ratio and elastic modulus of different minerals in multi-component rich laminated shale:

$$\varepsilon_{kk} = \frac{1-(\nu 1 + v2)}{(E1 + E2)}\sigma_{kk} \tag{2.26}$$

$$e_{ij} = \frac{3}{2}\frac{\widetilde{\varepsilon}}{\widetilde{\sigma}} s_{ij} \tag{2.27}$$

In the formula, $\tilde{\sigma}$ and $\tilde{\varepsilon}$ are equivalent stress and equivalent strain, respectively, which are defined as:

$$\widetilde{\sigma} = \sqrt{\frac{3}{2} s_{ij} s_{ij}} \tag{2.28}$$

$$\widetilde{\varepsilon} = \sqrt{\frac{2}{3} e_{ij} e_{ij}} \tag{2.29}$$

In the formula, σ_{ij} and ε_{ij} correspond to parameters σ_r, σ_θ, σ_φ, ε_r, ε_θ, ε_φ in spherical symmetric coordinates. s_{ij} and e_{ij} are the partial quantities of stress and strain, respectively. σ_{m} and ε_{m} are the spherical quantities of stress and strain, respectively, while σ_{kk} and ε_{kk} are the first invariants of stress and strain, respectively.

2.3.3 *Energy Density*

This section provides the basic equations required to solve this problem. The following are the boundary conditions and continuity conditions for determining the definite solution of the equation. It is assumed that the plastic deformation of the material does not cause a change in volume, so the volume of the plastic zone changes according to the elastic volume. The relationship between the equivalent stress and strain after the material enters the plastic state takes the form consistent with the uniaxial constitutive relationship and is denoted as:

There are seven boundary conditions and continuous conditions in this problem, including the stress at infinity is zero condition:

$$\sigma_r\big|_{r=\infty} = \sigma_\theta\big|_{r=\infty} = \sigma_\varphi\big|_{r=\infty} = 0 \tag{2.30}$$

Initial yield conditions of the plastic zone at the elastic-plastic boundary of minerals of different components:

$$\widetilde{\varepsilon}\big|_{r=c1}=\frac{Y_1}{E_1} \tag{2.31}$$

$$\widetilde{\varepsilon}\big|_{r=c2}=\frac{Y_2}{E_2} \tag{2.32}$$

$$\widetilde{\sigma}\big|_{r=c1}=Y_1 \tag{2.33}$$

$$\widetilde{\sigma}\big|_{r=c2}=Y_2 \tag{2.34}$$

According to the definition of equivalent stress, the stress at the yield point in the plastic zone should satisfy the corresponding constitutive relation. Therefore, the equivalent strain definition of Eq. (2.21) is equivalent to the Mises yield criterion:

$$(\sigma_r-\sigma_\theta)^2+\left(\sigma_r-\sigma_\varphi\right)^2+\left(\sigma_\theta-\sigma_\varphi\right)^2=(Y1+Y2)^2(\widetilde{\varepsilon})=2\widetilde{\sigma}^2 \tag{2.35}$$

For ideal elastic-plastic materials, the equivalent stress of the plastic zone element $\tilde{\sigma}=\sigma_r$ and the above equation degenerates to the commonly used Mises yield criterion. For hardened materials, the plastic zone element must meet the yield criterion of Eq. (2.29), and its yield stress is equal to the equivalent stress. In addition, the elastic-plastic boundary of brittle minerals and plastic minerals is in a critical yield state, see Eqs. (2.25)–(2.28), and apply the criterion of Eq. (2.29) to obtain:

$$\left[(\sigma_r-\sigma_\theta)^2+\left(\sigma_r-\sigma_\varphi\right)^2+\left(\sigma_\theta-\sigma_\varphi\right)^2\right]\Big|_{r=c1}=2Y_1^2 \tag{2.36}$$

$$\left[(\sigma_r-\sigma_\theta)^2+\left(\sigma_r-\sigma_\varphi\right)^2+\left(\sigma_\theta-\sigma_\varphi\right)^2\right]\Big|_{r=c2}=2Y_2^2 \tag{2.37}$$

On the elastic-plastic boundary (r = c), continuous conditions of displacement and equivalent stress can be adopted.

By solving the above basic equations, the deformation fields in the elastic region and the plastic region during the loading process can be, respectively, solved. For the elastic region, the displacement, stress, and strain components in the elastic region during the loading process are firstly solved, and then the elastic energy and total energy density in the elastic region are obtained by integral solution. The displacement method is used to solve this problem directly [4], and all the stress and strain components in the elastic region are obtained.

$$\sigma_r=-\frac{(Y1+Y2)}{3}\left(\frac{c1+c2}{2r}\right)^3 \tag{2.38}$$

$$\sigma_\theta=\sigma_\varphi=\frac{(Y1+Y2)}{3}\left(\frac{c1+c2}{2r}\right)^3 \tag{2.39}$$

$$\sigma_{\mathrm{m}} = 0 \tag{2.40}$$

$$\widetilde{\sigma} = \frac{(Y1 + Y2)}{2}\left(\frac{c1 + c2}{2r}\right)^3 \tag{2.41}$$

Considering that the energy transfer rate along the direction is greater than the φ and θ directions in the spherical symmetric coordinate approximation assumption, the three coordinate axis strain is as follows:

$$\varepsilon_r = -\frac{(1 + \nu 1 + v2)(Y1 + Y2)(c1 + c2)^3}{1.5\,(E1 + E2)r^3} \tag{2.42}$$

$$\varepsilon_\theta = \varepsilon_\varphi = \frac{(1 + \nu 1 + v2)(Y1 + Y2)(c1 + c2)^3}{1.5(E1 + E2)r^3} \tag{2.43}$$

$$\widetilde{\varepsilon} = \frac{(1 + \nu 1 + v2)(Y1 + Y2)(c1 + c2)^3}{1.5(E1 + E2)r^3} \tag{2.44}$$

Limiting the boundary condition strain sphere quantity to 0, Eqs. (2.35) and (2.36) are substituted into the energy densitometer formula to obtain:

$$w_{\mathrm{t}}^{\mathrm{e}} = w_{\mathrm{e}}^{\mathrm{e}} = \frac{1 + \nu 1 + v2}{3(E1 + E2)}\widetilde{\sigma}^2 + \frac{3(1 - \nu 1 - v2)}{2(E1 + E2)}\sigma_{\mathrm{m}}^2 \tag{2.45}$$

The energy density of the elastic region can be obtained by equivalent strain conversion:

$$\begin{aligned} w_{\mathrm{e}}^{e} &= \frac{1 + \nu 1 + v2}{6(E1 + E2)}\left[(Y1 + Y2)\left(\frac{c1 + c2}{2r}\right)^3\right]^2 \\ &= \frac{(1 + \nu 1 + v2)(Y1 + Y2)^2(c1 + c2)^6}{192(E1 + E2)r^6} \end{aligned} \tag{2.46}$$

2.3.4 *Analytical Solution*

In the plastic region, under the general constitutive relation $\widetilde{\sigma} = f(\widetilde{\varepsilon})$, the problem of the change of stress and strain with spatial position can be reduced to the spatial change expressed by the following two equations, so that the strain distribution in the plastic region can be expressed by the equivalent strain.

$$\widetilde{\varepsilon}+\frac{2(1-\nu 1-\nu 2)}{3(E1+E2)}\widetilde{\sigma}=\frac{C}{r^3} \tag{2.47}$$

where C is the undetermined constant and has no dimension. For the indenter, stiffness is much greater than the material unloading process, the material generally does not undergo plastic deformation or plastic deformation is weak. Assuming that the unloading is elastic recovery, the unloading phase can be described by the classical Lamé solution, and the unloading work can be obtained. According to the spherical symmetric coordinate approximation of the hole model, unloading is a process in which the boundary compressive stress in the sphere cavity gradually drops to zero. The unloading stage can be described by the elastoplastic field and the reverse superimposed elastic Lamé field at the end of loading. At the end of unloading, according to the boundary conditions of stress, the condition that the compressive stress is zero is satisfied. The stress superimposed in the unloading phase is elastic stress, so it is no longer necessary to distinguish between elastic and plastic regions.

On the sphere of $r = a$, the Lame elastic solution[10] under pressure P is of the form:

$$\sigma_r^* = -P\left(\frac{a1+a2}{2r}\right)^3 \tag{2.48}$$

$$\sigma_\theta^* = \sigma_\varphi^* = \frac{P}{2}\left(\frac{a1+a2}{2r}\right)^3 \tag{2.49}$$

By substituting the formula (2.42) and (2.43) into the stress sphere quantity and equivalent stress formula (2.21), respectively, we can get:

$$\sigma_{\mathrm{m}}^* = 0 \tag{2.50}$$

$$\widetilde{\sigma}^* = \frac{3}{2}\left(\frac{a1+a2}{2r}\right)^3 P \tag{2.51}$$

The residual stress field of the material in the loaded elastic region can be obtained by subtracting the equation of the loaded stress field (2.32), (2.33) from the equation of the loaded stress field (2.41):

$$\sigma_r^r = -\frac{(Y1+Y2)}{3}\left(\frac{a1+a2}{2r}\right)^3 + P\left(\frac{a1+a2}{2r}\right)^3 \tag{2.52}$$

$$\sigma_\theta^r = \sigma_\varphi^r = \frac{(Y1+Y2)}{3}\left(\frac{a1+a2}{2r}\right)^3 - \frac{P}{2}\left(\frac{a1+a2}{2r}\right)^3 \tag{2.53}$$

Since the unloading process is elastic, according to the calculation formula of elastic energy density (2.39) and the corresponding equivalent stress and stress sphere value formula (2.46) and (2.47), it can be obtained:

$$
\begin{aligned}
w_{\mathrm{u}} &= \frac{1+\nu 1+\nu 2}{3(E1+E2)}\left(\widetilde{\sigma}^{*}\right)^{2}+\frac{3(1-\nu 1-\nu 2)}{(E1+E2)}\left(\sigma_{\mathrm{m}}^{*}\right)^{2} \\
&= \frac{3}{4}\frac{1+\nu 1+\nu 2}{(E1+E2)}\left(\frac{a1+a2}{2r}\right)^{6}P^{2}
\end{aligned}
\tag{2.54}
$$

By integrating the elastic field of superposition, the work of pressing and unloading can be obtained:

$$
W_{\mathrm{u}}=\int_{a}^{\infty}2\pi r^{2}w_{\mathrm{u}}\mathrm{d}r=\frac{\pi(1+\nu 1+\nu 2)}{2(E1+E2)}P^{2}(a1+a2)^{3} \tag{2.55}
$$

The equation shows that the unloading energy is proportional to the Poisson's ratio, elastic modulus, and the square of the impingement load of the shale material.

2.3.5 Scaling Results

In order to avoid overlapping of a large number of indentation data points, 49 indentation points were selected from the indentation point array data of each sample as representatives. By analyzing the load-displacement curves of each indentation point, dimensionless hardness and dimensionless energy are obtained, and linear regression analysis is carried out. As shown in Fig. 2.14, the fitting correlation coefficient R^2 of all samples is greater than 0.97, indicating that the dimensionless relationship proposed by Cheng et al. [5] is valid for this experiment. The correlation coefficients of recovery for samples A1 and A2 were 0.225 and 0.242, respectively, while the dimensionless correlation coefficients for samples B1 and B2 were 0.157 and 0.184, respectively. The dimensionless energy recovery coefficient of clay shale is greater than that of carbonate shale, and the increase in dimensionless energy recovery coefficient is caused by feldspar minerals.

Cheng et al. deduced the relationship between indentation depth and total energy during indentation test of linear elastic power hardening materials, where U_t and h^3 showed an approximate linear relationship. As shown in Fig. 2.15, after sorting and summarizing the indentation energy and depth data obtained in the shale indentation test, the experimental results of U_t and h show an approximate linear relationship. In Fig. 2.16, the possible cause of this phenomenon was found by analyzing the indentation curve of the linear elastic energy hardening material and the shale. The slope of indentation curve of linear elastic power hardening material increases continuously during the loading process, and the indentation load F^2 is positively correlated with the indentation depth $h_{\max}$. The slope of indentation curve of

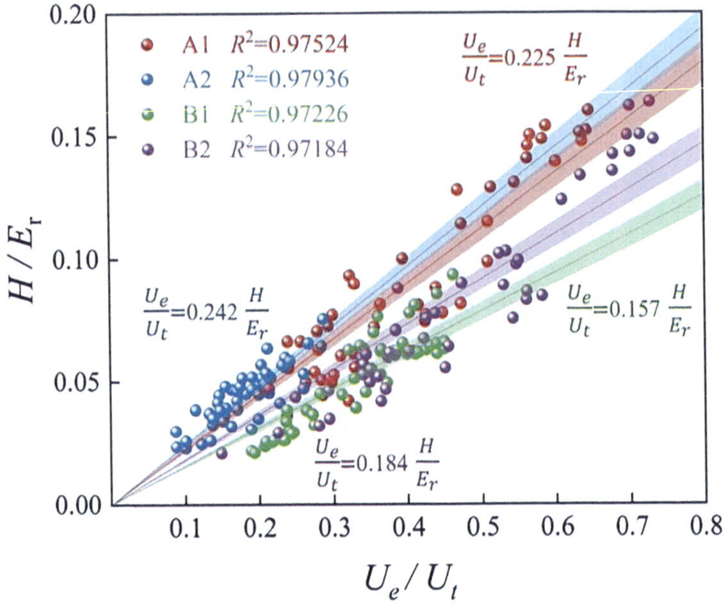

Fig. 2.14 Dimensionless relationship between hardness, elastic modulus, and indentation energy

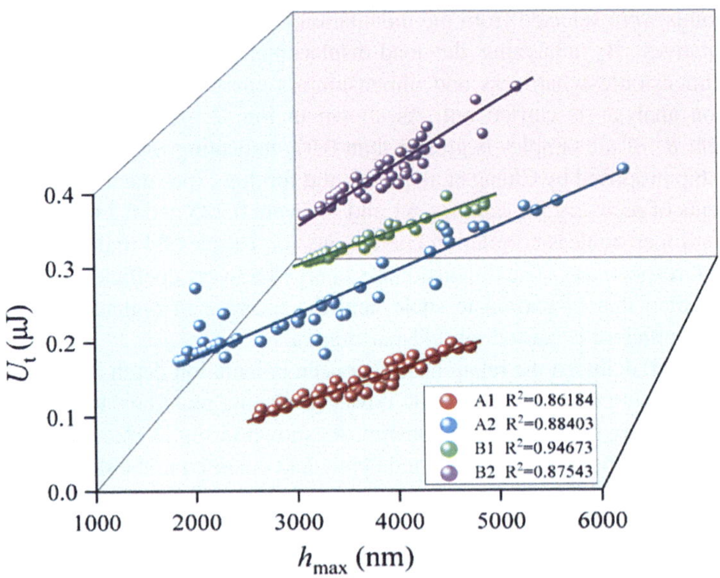

Fig. 2.15 Linear relationship between indentation energy and maximum indentation depth

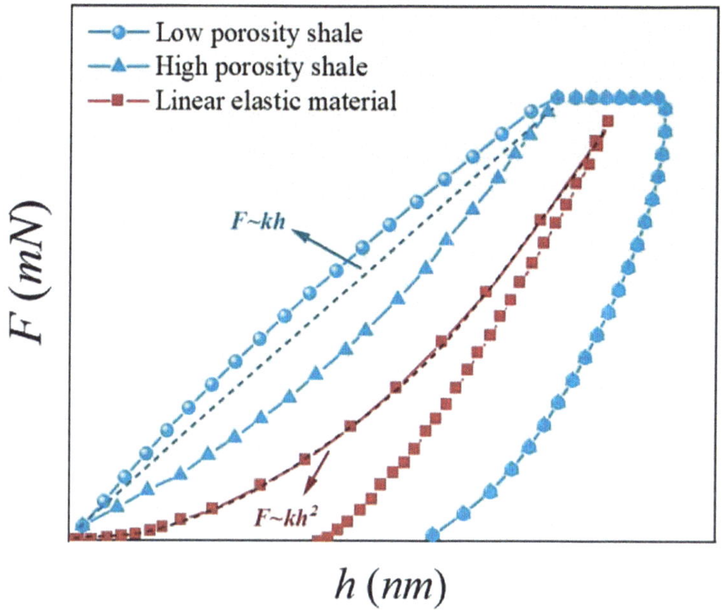

Fig. 2.16 Comparison of typical indentation curves between Gulong shale and linear elastic materials

continental shale is affected by pore space, mineral composition, and mechanical properties. The above differences may be caused by the heterogeneity of the shale minerals, the morphology of the bedding, or the properties of the minerals between the bedding. Therefore, the indentation process is theoretically derived based on the extended cavity model to clarify the influencing factors, and then the finite element method is used for further verification.

2.4 Indentation Brittleness Index

2.4.1 Calculation Method

As shown in Fig. 2.17, four regions (grid 1–4) were selected for point matrix indentation experiments after argon ion polishing of the shale samples. In order to facilitate the analysis of the change of brittleness index, SEM and mineral scanning analyses were carried out on the samples in different regions to clarify the correlation between brittleness index and mineral distribution.

The displacement-load curve corresponds to each indentation point, h_{max} is the maximum indentation depth, h_f is the indentation depth after absolute unloading, and h_s and h_c are the depths of irrecoverable deformation at the edge and center of the indentation, respectively.

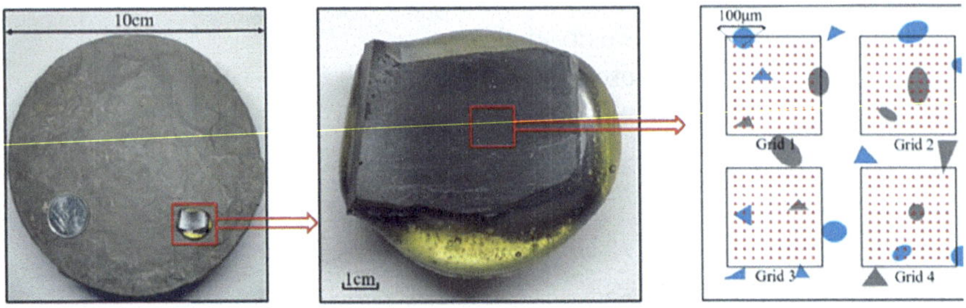

(a) (b) (c)

Fig. 2.17 Schematic diagram of sample acquisition: (**a**) full diameter core, (**b**) shale rock sample for nanoindentation, and (**c**) schematic diagram of grid indentation

The Oliver and Pharr method was introduced to calculate H and E for shale [6]. H can be calculated by

$$H = \frac{P_{\max}}{A_c} \tag{2.56}$$

where $P_{\max}$ is the maximum loading, N; and A_c is contacting area, m^2. For the ideal Bose indenter, it can be obtained by

$$A_c = 24.5h_c^2 \tag{2.57}$$

E can be calculated by

$$E_\gamma = \frac{\sqrt{\pi}S}{2\beta\sqrt{A_c}} \tag{2.58}$$

where S is the contact stiffness and N/m is the slope of the initial unloading stage. β is the parameter related to the shape of the indenter and the β of the ideal indenter is 1.034.

K_c is calculated by the energy method [7]:

$$K_c = \sqrt{G_c E_\gamma} \tag{2.59}$$

where E_r is the modified elastic modulus, Gpa; and G_c is the critical energy release rate.

In the calculation of G_c, the total energy in the indentation deformation is composed of elastic energy and plastic energy, and the plastic energy consists of pure plastic energy and the energy induced by fracture.

$$W = W_e + W_p = W_e + W_{pp} + W_f \tag{2.60}$$

$$\frac{W_{pp}}{W} = 1 - \left[\frac{1 - 3(h_f/h_{max})^2 + 2(h_f/h_{max})^2}{1 - (h_f/h_{max})^2} \right] \tag{2.61}$$

Based on above research, it can be obtained (Cheng et al. 2002):

$$G_c = \frac{\partial W_f}{\partial A_c} = \frac{W_f}{A_c} \tag{2.62}$$

In the fracture mechanic, the K_c can be determined by the length of the crack that resulted from residual indentation under set loading, which is shown as follows (Lee et al. 2012):

$$K_c = \frac{P_{max}}{l^{3/2}} \prod \left(\frac{E}{H}, v, \phi, \frac{l}{h} \right) \tag{2.63}$$

where l is the length of the crack, v is Poisson's ratio, and E is the shape parameter of the indenter:

$$\frac{K_c l^{3/2}}{P_{max}} = \prod \left(\frac{E}{H}, v, \phi, \frac{l}{h_c} \right) \tag{2.64}$$

Substituting Eqs. (2.56) and (2.57) to Eq. (2.64) yields:

$$\frac{K_c l^{3/2}}{H \cdot A_c} = \frac{K_c l^{3/2}}{H \cdot 24.5 h_c^2} = \prod \left(\frac{E}{H}, v, \phi, \frac{l}{h_c} \right) \tag{2.65}$$

$$\frac{K_c}{H \cdot l^{1/2}} = \prod \left(\frac{E}{H}, v, \phi, \frac{h_c}{l} \right) \tag{2.66}$$

$$\frac{K_c^2}{H^2 \cdot l} \cdot \frac{H}{E} = \frac{K_c^2}{HEL} = \prod \left(v, \phi, \frac{h_c}{l} \right) \tag{2.67}$$

Then, the B_1 can be regarded as:

$$B_1 = \frac{HEl}{K_c^2} \tag{2.68}$$

In the nanoindentation test, the length of the crack can be tested with a high precision electronic scanner, and the grid indentation results in a great deal of experimental work. Hence, the length of crack l is replaced by h_s, which is the test data directly from the indentation experiment:

$$h_s = \varepsilon \frac{P_{\max}}{S} \tag{2.69}$$

The horizontal projection of the deformation caused by the indentation at the edge is the crack length l. Due to the geometric similarity between h_s and l, it can be obtained:

$$h_s \propto l \tag{2.70}$$

So, the brittleness index B is defined by:

$$B = \frac{HEh_s}{K_c^2} \tag{2.71}$$

Substituting Eqs. (2.56), (2.59), and (2.62) to Eq. (2.71) yields

$$B = \frac{EHh_s}{K_c^2} = \frac{Eh_s P_{\max}/A_c}{EW_f/A_c} = \frac{P_{\max} h_s}{W_f} = 2\frac{W_s}{W_f} \tag{2.72}$$

where W_s is the elastic recoverable energy of the contacting area of the indenter and W_f represents the energy used to result in a fracture in the indentation deformation. The W_s/W_f represents the size of the friability of the indentation point, which further proves the applicability of B.

2.4.2 *Mineral Mechanics*

Mineralogy Information

In order to obtain the mineral component of the shale sample, the XRD tests were performed on the fragments within the experimental shale sample. As shown in Table 2.1, the mineral component of shale was composed of quartz, potash feldspar, plagioclase, calcite, dolomite, clay minerals, pyrite, and illite.

It should be noted that the proportion of carbonate minerals was 49.1% and accounted for the largest part. There are two traditional shale brittleness assessment methods based on mineral composition. The carbonate mineral was regarded as the brittleness mineral and the carbonate mineral was eliminated from the evaluation of brittleness. In this test, the micromechanical property was directly used to evaluate the brittleness of the shale.

As shown in Fig. 2.18, the surface of shale mostly consisted of dolomite particles, scattered quartz particles, and feldspar. The grains were filled with clay containing

Table 2.1 Mineralogical composition of shale sample

Mineral composition	Content/%
Quartz	12.5
K-Feldspar	3.0
Plagioclase	21.9
Calcite	23.2
Dolomite	25.9
Pyrite	1.4
Illite	7.34
Kaolinite	0.3
Chlorite	0.68
Kaolinite–montmorillonite	3.78

illite, chlorite, and a small amount of biotite, which corresponded to the results of the XRD tests. In grid 1, the quartz, feldspar, carbonate rock mineral, and clay mineral content were 13.79%, 14.09%, 48.16% and 17.43%, respectively. In grid 2, the quartz, feldspar, carbonate rock mineral, and clay mineral content were 9.52%, 14.38%, 50.72%, and 18.53%, respectively. In grid 3, the quartz, feldspar, carbonate rock mineral, and clay mineral content were 12.49%, 16.31%, 47.26%, and 17.43%, respectively. In grid 4, the quartz, feldspar, carbonate rock mineral, and clay mineral content were 8.43%, 17.3%, 51.52%, and 15.37%, respectively. Among the four grids, the quartz content in grid 1 was the largest, which was 5.36% higher than that in grid 4. Grid 4 had the highest feldspar content, which was 3.21% higher than grid 1 with the lowest feldspar content. The carbonate mineral content was the highest in grid 4, which was 4.26% higher than that in grid 3. The highest clay mineral content in grid 2 was 3.16% higher than that in grid 4. The difference in mineral content in the four grids was less than 6%, but the distribution of minerals was different, which resulted in the variations in the micro-properties of the shale.

Figure 2.19 shows the SEM images of the shale samples under different horizons, with magnifications of 1200×, 2000×, 5000×, and 8000× times, respectively. At 1200 magnification, the mineral particles were scattered on the surface of the shale, along with sporadic organic matter. In addition, the obvious micropores and pores formed after mineral exfoliation could be seen. When the magnification was 2000× times, the mineral contour was clear, and the clay minerals filled in the middle of the mineral particles could be observed clearly. The micropores in the clay minerals were more obvious, making the clay minerals honeycomb. When the magnification was 5000× times, the mineral shape was clearer. Combined with the mineral distribution obtained by QEMSCAN, the square dolomite particles could be clearly identified, but quartz, feldspar, and other particles could not be identified in this mode. In addition to the initial pores on the surface of the shale, there were also many pores caused by the whole or partial exfoliation of massive minerals. Beyond that, there were also microscopic fissures developed at the edges of the mineral grains and pores throughout the mineral grains. At 8000 magnifications, the bonds between minerals could be seen more clearly. In the middle of the lumpy mineral grains, clay minerals and organic matter filled them.

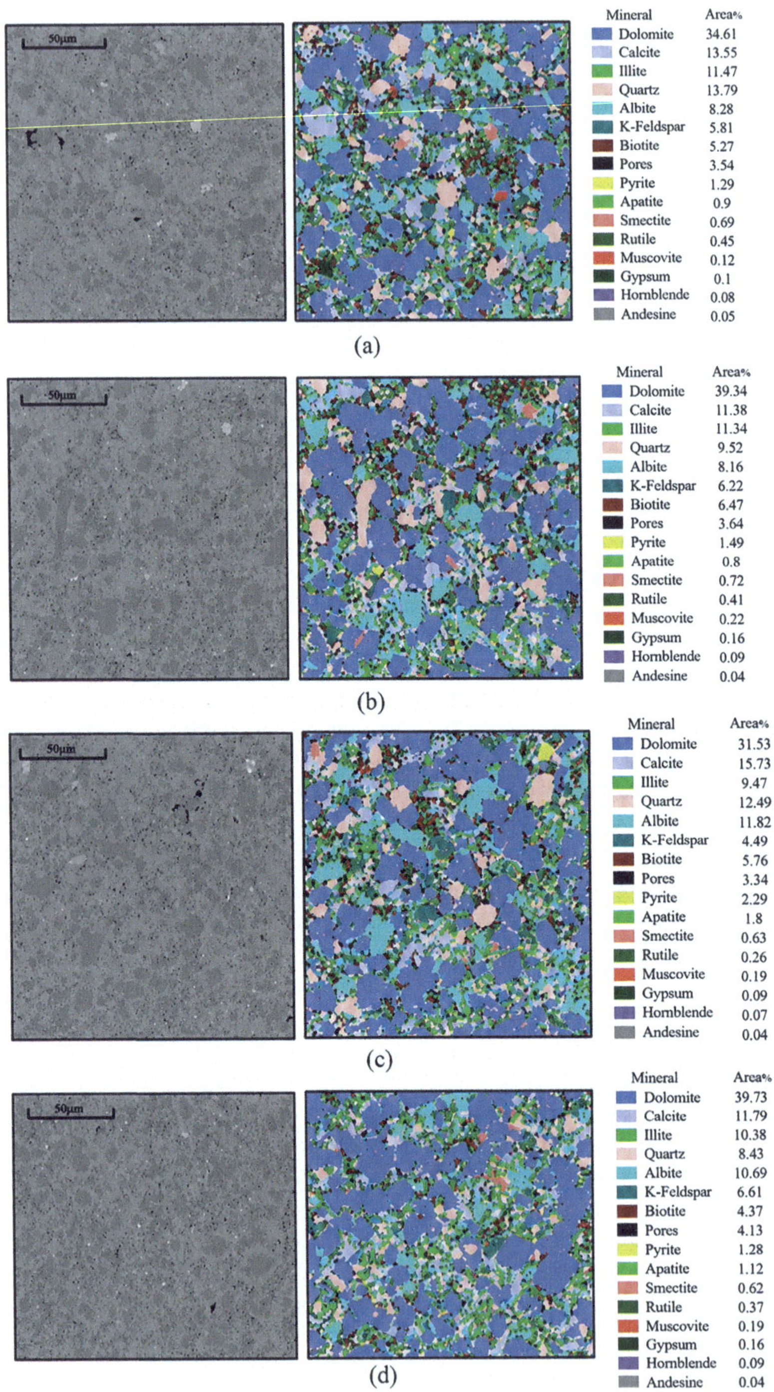

Fig. 2.18 Mineral distribution under QEMSCAN: (**a**) Nanoindentation grid 1, (**b**) nanoindentation grid 2, (**c**) nanoindentation grid 3, and (**d**) nanoindentation grid 4

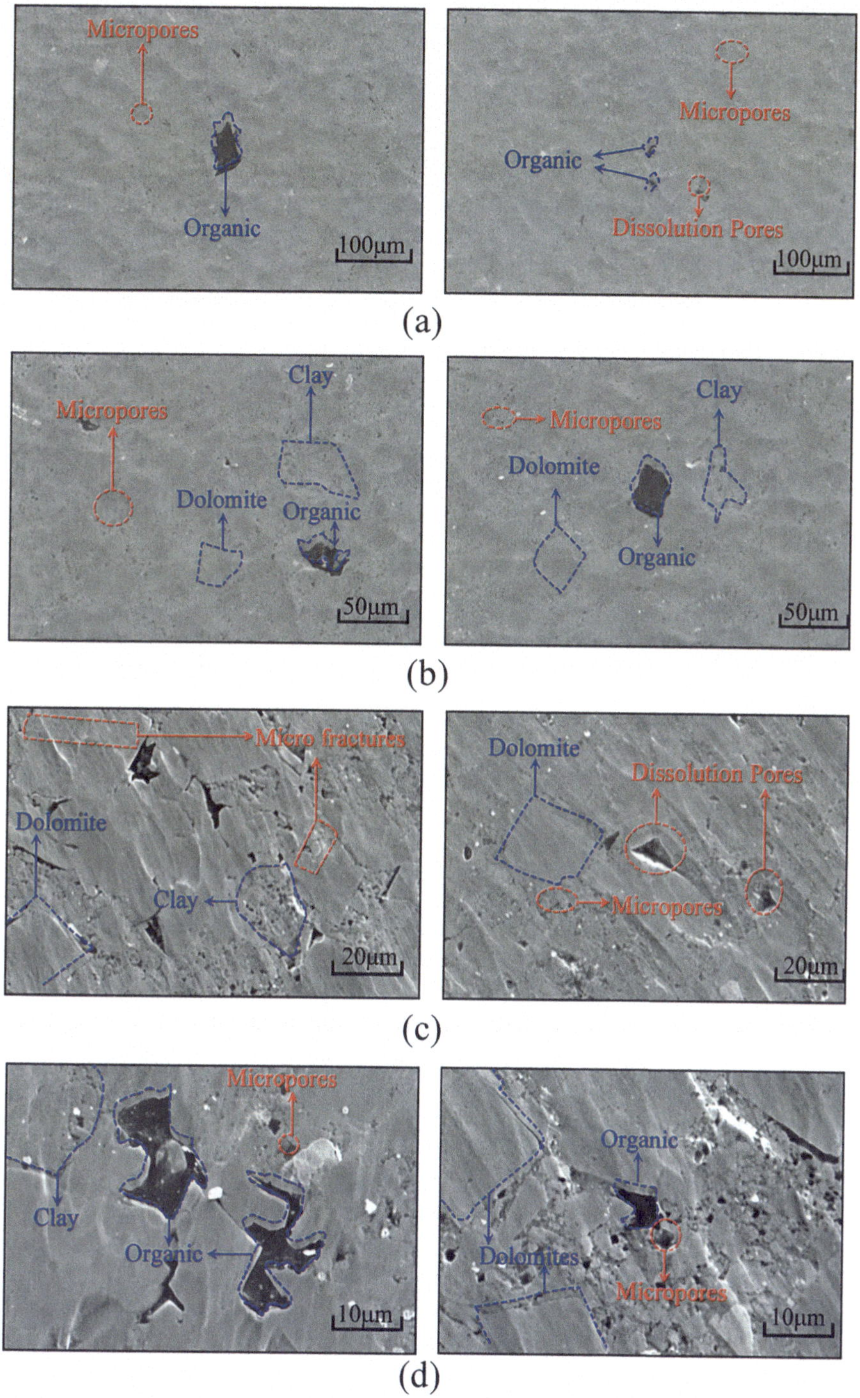

Fig. 2.19 SEM images in different fields of shale: (**a**) 1200× magnification, (**b**) 2000× magnification, (**c**) 5000× magnification, and (**d**) 8000× magnification

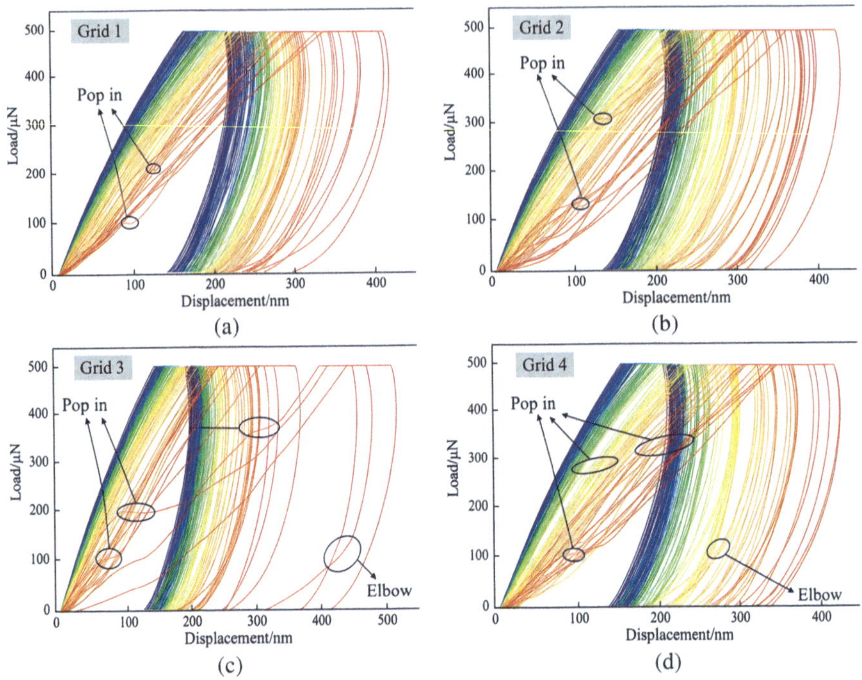

Fig. 2.20 Load–displacement curve: (**a**) nanoindentation grid 1, (**b**) nanoindentation grid 2, (**c**) nanoindentation grid 3, and (**d**) nanoindentation grid 4. Blue represents high elasticity and red represents low elasticity

Each indentation point corresponds to a load-displacement curve, which represents the mechanical behavior of the mineral composition at the indentation point. Under nanoindentation loading, the displacement loading curve presented a similar shape. As shown in Fig. 2.20, the displacement increased gradually with the increasing load in the loading stage. During the continuous loading stage, the displacement continued to increase. The elastic deformation recovered and left residual deformation h_f in the unloading stage. Under the same loading rate and peak load, the trends of the displacement load curves were roughly similar. However, the curve was slightly different due to diversity minerals, and the mechanical parameters were obtained through the analysis of the curve. Among the four grids, the maximum residual indentation depths were 260 nm, 328 nm, 325 nm, and 286 nm, respectively, and the minimum residual indentation depths were 50 nm, 52 nm, 45 nm, and 48 nm, respectively. It can be seen that h_{max} varied greatly, which owed to the different mineral compositions and shale micro-structures. The obvious "pop in" and "elbow" phenomena can be seen in curves, and the phenomenon of "pop in" was relatively common. Such a sudden increase in displacement may have resulted from the micro- and nanoscale pores. These may have been original or may have been microcracks produced by the indenter in pressing. "Pop in" mostly occurred at the indentation points with a large indentation depth, and the indentation points were mostly clay minerals. It can be seen from the SEM in Sect. 2.3.3 that there were many micropores developed in the clay minerals. When the indenter met these pores in the process of pressing, displacement mutation occurred. The elbow phenomenon occurred during the unloading process and was characterized by abrupt changes in the unloading slope. The occurrence of this phenomenon may have been the result of the phase transitions that occurred in the slow lifting of the indenter. This phenomenon can be explained by SEM and on the plane, and the clay minerals were used as substrates to fill between the large grains. However, in the longitudinal direction, when the indenters experienced a transition from matrix to brittle minerals in the unloading process, the "elbow" phenomenon occurred. It is worth noting that the indentation size was smaller than the grain size of the mineral, and the "elbow" phenomenon mostly appeared in grids 3 and 4.

The micromechanical properties of the shale were obtained by a series of nanoindentation experiments, i.e., E, H and K_c. To better analyze the relationships among E, H, and K_c, the analysis of the normal distribution on these was conducted by Eq. (2.73).

$$y_E = \frac{A}{\sqrt{2\pi}\sigma} \exp\left[-\frac{1}{2}\left(\frac{x-\mu}{\sigma}\right)^2\right] \tag{2.73}$$

where σ is the standard deviation and represents the amplitude of the distribution of data and μ is the mathematical expectation and characterizes the position of the normal distribution of the data.

Modulus of Elasticity

The frequency distribution of E is displayed in Fig. 2.21 and the fitting correlation coefficients of grid 1, 2, 3, and 4 were 0.6759, 0.5786, 0.6218, and 0.5106, respectively. The varying ranges of E of grid 1, 2, 3, and 4 were 3.58–12.07 GPa, 4.45–14.24 GPa, 3.82–16.48 GPa, and 2.74 ~ 17.16 GPa, respectively. The mean elastic modulus E^{-} of the four grids were 8.24, 8.54, 10.92, and 10.82 GPa, respectively, and the corresponding mathematical expectations were 8.24, 8.54, 11.14, and 10.95, respectively. It can be seen that the indenters varying from 7 to 10 GPa of E accounted for 52% in grid 1 and that E was the greatest in grid 3. However, the variance of grid 4 was much higher than that of grid 3, which can be accounted for with the discrete data in grid 4.

Hardness

The frequency distribution of H is displayed in Fig. 2.22 and the fitting correlation coefficients of grid 1, 2, 3, and 4 were 0.6087, 0.4961, 0.5266, and 0.5491, respectively. Except for grid 1, the other distributions in grid 2, 3, and 4 had an

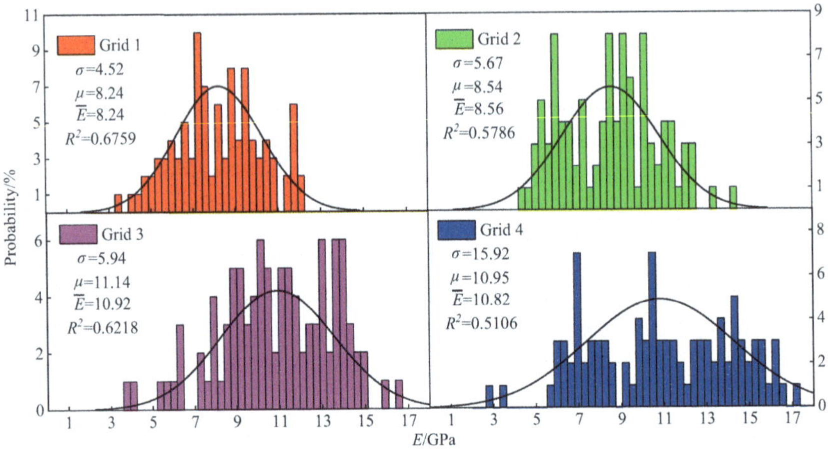

Fig. 2.21 Frequency distribution histogram of elastic modulus. The closer the value of R^2 is to 1, the better the fit of the regression line to the observed values; on the contrary, the smaller the value of R^2, the worse the fit of the regression line to the observed values

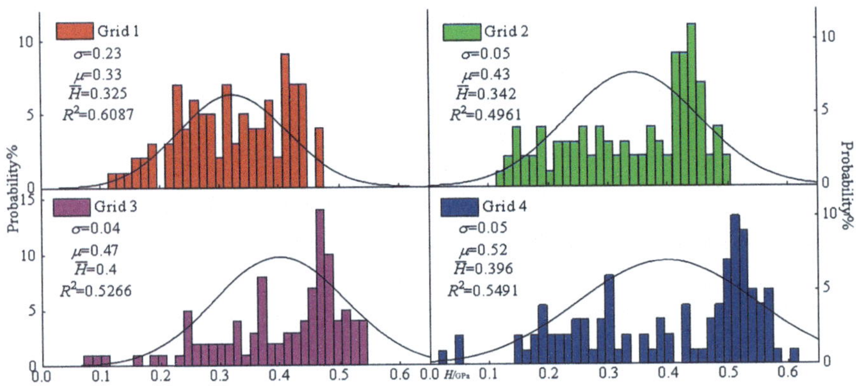

Fig. 2.22 Frequency distribution histogram of *H*. The closer the value of R^2 is to 1, the better the fit of the regression line to the observed values; on the contrary, the smaller the value of R^2, the worse the fit of the regression line to the observed values

obvious negative skew distribution, and the concentration areas of *H* were significant with range of 0.1 GPa. The varying areas of *H* of grid 1, 2, 3, and 4 were 0.122–0.476 GPa, 0.115–0.496 GPa, 0.081–0.546 GPa, and 0.146–0.604 GPa, respectively. The average *H* of grid 1, 2, 3, and 4 was 0.325, 0.342, 0.4, and 0.396 GPa, respectively. The concentration ranges of grid 1, 2, 3, and 4 were 0.4–0.5 GPa, 0.4–0.5 GPa, 0.45–0.55 GPa, and 0.45–0.55 GPa, respectively, and the proportions of the indenters were 27%, 48%, 46%, and 40%, respectively. Grid 4 had the largest hardness distribution interval and the largest average hardness. The

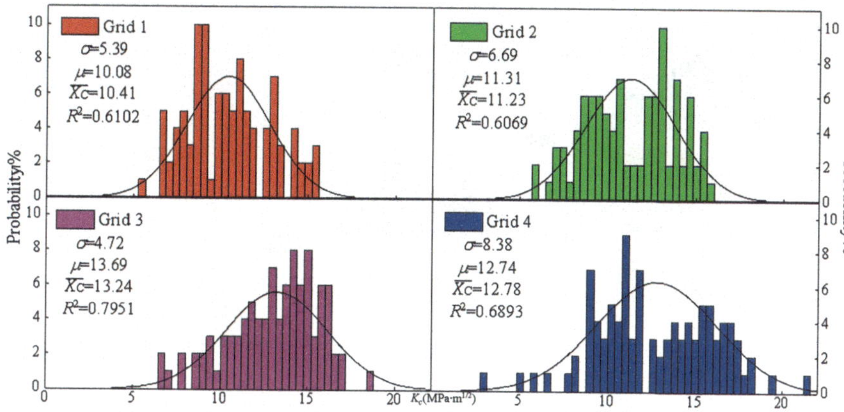

Fig. 2.23 Frequency distribution histogram of K_c. The closer the value of R^2 is to 1, the better the fit of the regression line to the observed values; on the contrary, the smaller the value of R^2, the worse the fit of the regression line to the observed values

mathematical expectations of grid 1, 2, 3, and 4 were 0.33, 0.43, 0.47, and 0.52, respectively, which were higher than H owing to the high-hardness minerals in the clay.

Fracture Toughness

Frequency distributions of K_c are displayed in Fig. 2.23 and the fitting correlation coefficients of grid 1, 2, 3, and 4 were 0.6102, 0.6069, 0.7951, and 0.6893, respectively. The varying ranges of K_c of grid 1, 2, 3, and 4 were 5.36–15.55 GPa, 5.6–15.71 GPa, 6.43–18.44 GPa, and 4.85–21.56 GPa, respectively. The mean fracture toughness $(K_c)^{-}$ of four grids was 10.41, 11.23, 13.24, and 12.78 MPa·m1/2, respectively, and that of grid 3 was the largest. The mathematical expectations of the four grids were 10.08, 11.31, 13.24, and 12.78, respectively. The mathematical expectation of grid 2 and grid 4 was same as the mean value. The mathematical expectation of grid 1 was less than its average value, and a positive skewness distribution was presented, while that of grid 3 was greater than its average value, which presented a negative skewness distribution.

Figure 2.24 gives the relationships among various mechanical properties of the four grids after statistics, and the mechanical property distributions mostly followed the normal distribution. The frequency distributions of E, H, and K_c are displayed on the diagonal. The correlation coefficients between E and K_c were higher than 0.8, which indicates a great fit. The correlation coefficient of H was a little worse with the value of 0.7348, but it was higher than that of the dispersed mesh. The mean values of E, H, and K_c were 9.64 GPa, 0.365 GPa, and 11.91 MPa·m1/2, respectively. The correlations among the mechanical properties are demonstrated in Fig. 2.25.

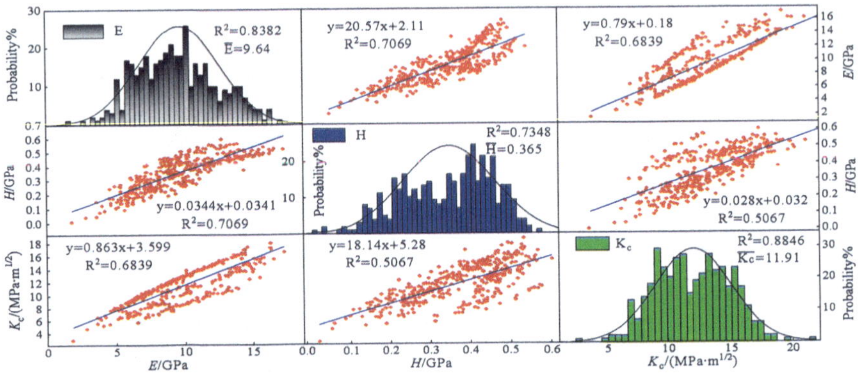

Fig. 2.24 The correlation between H and K_c was the lowest, and the correlation coefficient was 0.5067. In addition, the correlation coefficient between E and K_c was 0.6839. It can be observed that data points present discrete distribution, which was probably caused by mineral heterogeneity, micro-cracks, and micropores

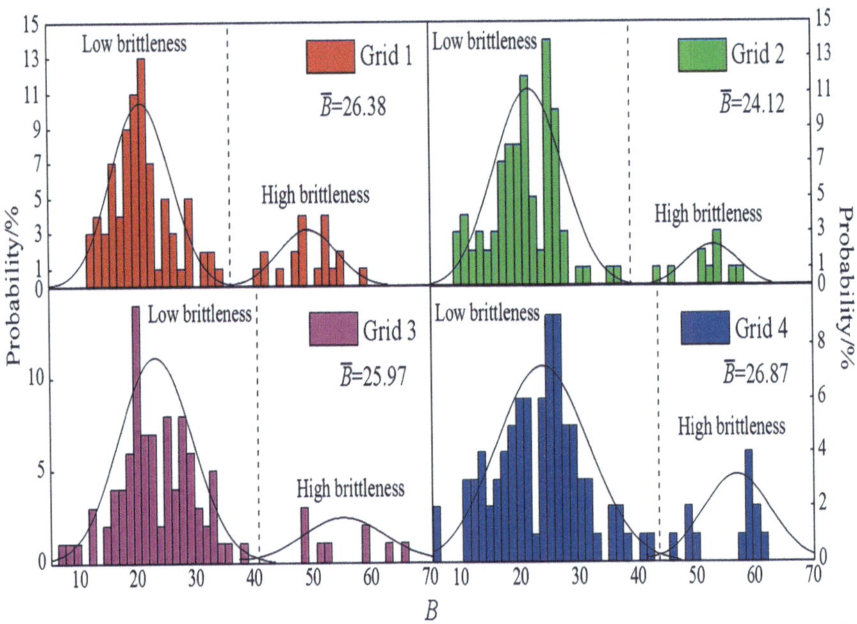

Fig. 2.25 Frequency index B

2.4.3 *Analysis of Brittleness Indices*

The micro-brittleness of the four grids can be obtained based on Eq. (2.16), and it reflects the friability of the indenter. By analyzing the brittleness index of numerous indenters, as displayed in Fig. 2.25, the varying ranges of the brittleness index of the four grids were 11.31–58.67, 9.16–57.07, 7.46–65.69, and 10.52–61.32, respectively. Additionally, the mean micro-brittleness index *B* of the four grids was 26.38, 24.12, 25.97, and 26.87, respectively. The friability of grid 4 was the best based on the above analysis on the brittleness index. However, the analysis was invalid due to it ignoring the integrity and heterogeneity of the shale. Hence, a deep analysis is necessary in the next section.

By investigating the distributions of the brittleness index in the four grids, it can be seen that the distribution of the brittleness index presented an obvious bimodal distribution, which could be divided into a low brittleness region and a high brittleness region. The boundary value of the low and high brittleness region in the four grids was 38, 40, 40, and 45, respectively. The mean brittleness indices of the low brittleness regions were 20.96, 21.05, 23.09, and 23.6, respectively. The mean brittleness indices of the high brittleness regions were 49.47, 51.77, 55.08, and 55.61, respectively. Additionally, the proportions of the high brittleness index were 19%, 10%, 9%, and 14%, respectively. It can be concluded that the mean value of the brittleness index was small in grid 1, while the proportion of the high brittleness index was the largest, indicating that the friability of grid 1 was positive and higher than the other grids.

The micro-brittleness index of a single point was obtained by pressing in and out of the nanoindentation head. In Sect. 2.4.1, the related statistical analysis was performed on the micro-brittleness index of a single point and the evaluation of the friability was finished. However, shale is a strong heterogeneity material, and it is difficult to represent the overall characteristics of shale only through the analysis of a single point. Even if the mesh statistics are the same, the micro-brittleness will be different due to the various mesh distribution. In this paper, the indenter was moved along a serpentine path of the array by a selected starting point. By giving the array length and spacing between adjacent indentations, the brittleness index was matched with the spatial position of the indentation. This spatial distribution provides a way to determine the micro-brittleness distribution on the shale surface.

Figure 2.26 shows the reference origin software and the method used was the difference method. As shown in the left of Fig. 2.26, the brittleness index was drawn as a hotspot map according to its position, and the strong heterogeneity of the shale was observed. The color of the grid represents the level of the brittleness index. The red color represents a high brittleness index, which is scattered in the indentation grid, just as quartz grains are scattered in shale. The existence of high brittleness indentation points improves the overall friability. However, it is difficult to transfer the crack to the next point when two high brittle indentation points are far away from each other. The cloud diagram of the brittleness index distribution was obtained with step fuzzy processing in the right of Fig. 2.26. When the crack extends to this point, it must extend to the high brittleness area. The high brittleness zones are connected to each other to form fracture propagation channels.

The content of the high brittleness indentation points was 19% in grid 1, which was higher than the other three indentation points. However, the concentration of the high brittleness indentation points led to a small number of crack expansion channels. When the crack expanded to grid 1, it was extremely easy to be absorbed by the

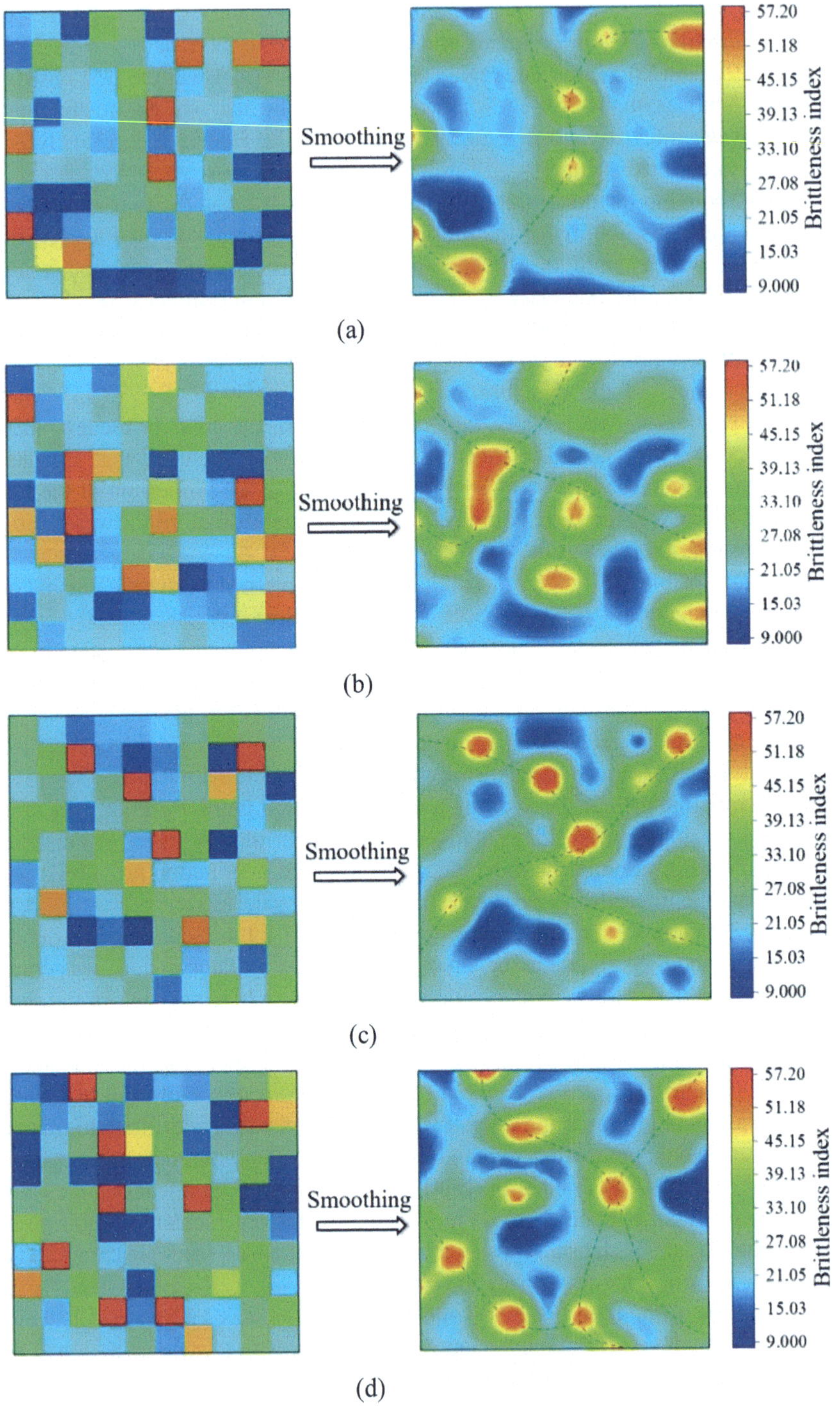

Fig. 2.26 Friability surface distribution: (**a**) nanoindentation grid 1; (**b**) nanoindentation grid 2; (**c**) nanoindentation grid 3; and (**d**) nanoindentation grid 4. The red dashed lines in Fig. 2.26 are connections between highly brittle regions that form crack propagation channels

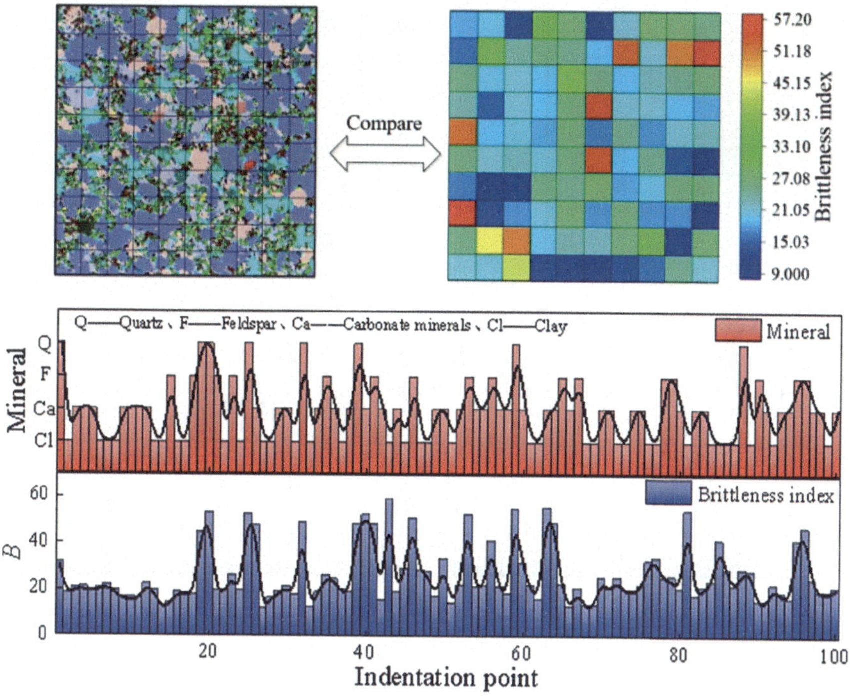

(a)

Fig. 2.27 The relationship between brittleness index and minerals: (**a**) nanoindentation grid 1; (**b**) nanoindentation grid 2; (**c**) nanoindentation grid 3; and (**d**) nanoindentation grid 4

low brittleness zone, resulting in the ability loss of the crack extension. Therefore, the high content of high brittleness in grid 1 was invaluable for friability. In grid 3, there were a few high brittleness indentation points (9%), but these high brittleness indentation points were scattered and connected with each other to form multiple channels, which were conducive to the propagation of cracks. The brittleness index distribution of the four grids was analyzed. There were two channels for grid 1, four channels for grid 2 and grid 3, and five channels for grid 4. Therefore, grid 4 had the best friability based on the distributions of the brittleness index.

In previous studies, the same mineral was considered to have specific mechanical properties. However, different mineral cementations have a strong influence on their mechanical properties. Nanoindentation tests were carried out by the grid indentation method, and an indentation point was placed in a grid with various mineral connections. The categories of the main minerals in the grid were compared with the brittleness index obtained by indentation tests. Figure 2.27 shows the corresponding relationship between the minerals and the brittleness index. In grid 1, the quartz

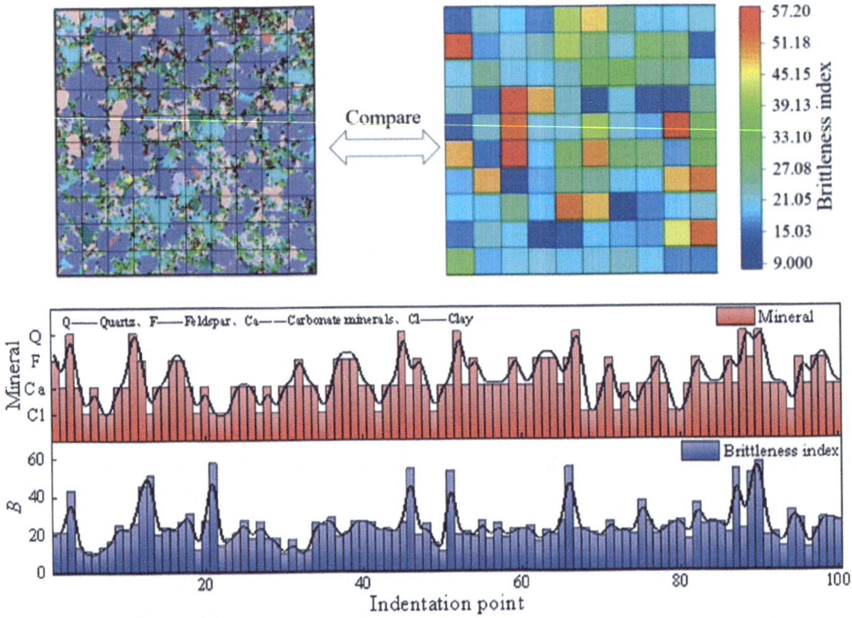

(b)

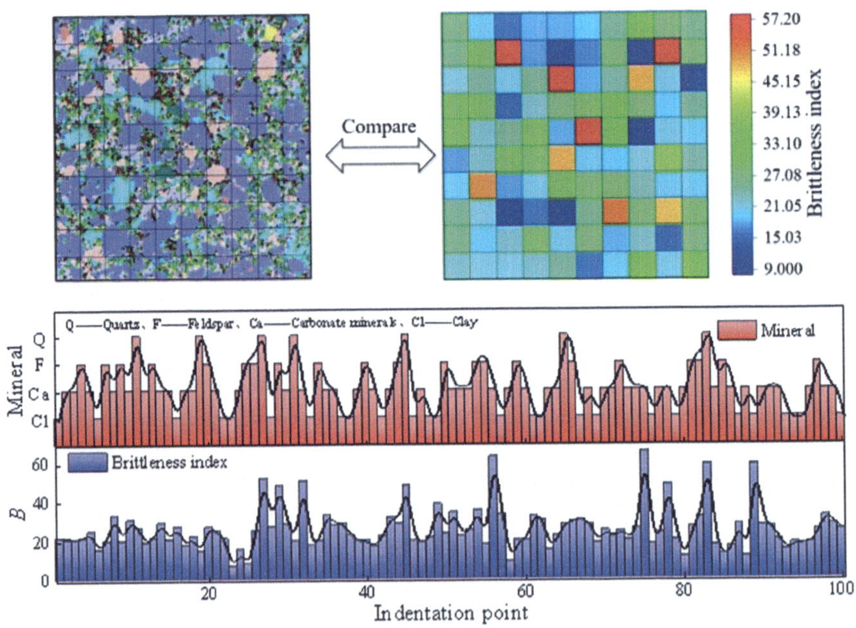

(c)

Fig. 2.27 (continued)

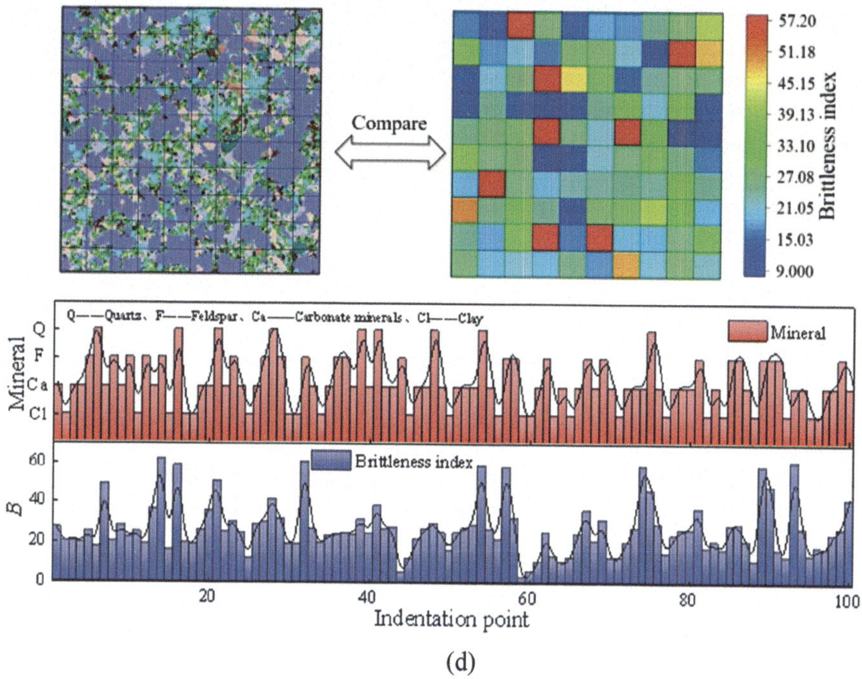

Fig. 2.27 (continued)

content, feldspar content, carbonate mineral content, and clay mineral content were 14.09%, 13.79%, 48.16%, and 17.43%, respectively. The indentation grid of the quartz-dominated grid accounted for 19%, basically corresponding to the high brittleness region. However, the brittleness index of the grid points by the feldspar-dominated and carbonate-dominated grids was in the middle. The grid points by the clay-dominated grid had a low brittleness index, but they also had a high brittleness index owing to the existence of brittle particles in the clay minerals. In grid 2, the quartz content was 9.52%, which corresponded to the high brittleness. The content of feldspar, carbonate minerals, and clay minerals was 14.38%, 18.53%, and 50.72%, respectively. In grid 3, the content of quartz, feldspar, carbonate minerals, and clay minerals was 12.49%, 16.31%, 47.26%, and 15.86%, respectively. The content of high brittleness was 9%, which was less than the quartz, and this was due to the low brittleness of some of the quartz surrounded by clay. In grid 4, the content of the quartz, feldspar, carbonate minerals, and clay minerals was 8.43%, 17.3%, 51.52%, and 15.37%, respectively. The proportion of high brittleness was 14%, which was much higher than that of quartz. This was because quartz particles are small and most of them were cemented with carbonate minerals, were part of the carbonate minerals, and showed high hardness.

It can be seen that quartz played a very important role in the evaluation of microscopic brittleness. Quartz is highly brittle, and when it is cemented with other minerals, the overall brittleness is improved. Quartz takes on different shapes, most of which are round shapes. These quartz particles are relatively large and are formed during the diagenesis process. They co-deposited with carbonate minerals and cemented with clay minerals to form shale. In addition, many clusters of quartz can be observed. These small particles of quartz are produced in clay minerals. They may not be formed by diagenesis, but by the precipitation of siliceous material during the transformation of clay to form quartz. This is why some clay minerals also exhibit a high brittleness.

References

1. Liu Y, Liu S, Kang Y. Probing nanomechanical properties of a shale with nanoindentation: heterogeneity and the effect of water-shale interactions. Energy Fuels. 2021;35(15):11930–46.
2. Bouzakis KD, Michailidis N, Hadjiyiannis S, et al. The effect of specimen roughness and indenter tip geometry on the determination accuracy of thin hard coatings stress–strain laws by nanoindentation. Mater Charact. 2002;49(2):149–56.
3. Akono AT. Energetic size effect law at the microscopic scale: application to progressive-load scratch testing. J Nanomech Micromech. 2016;6(2):04016001.
4. Yang L, Mao Y, Yang D, et al. The characteristic and distribution of shale micro-brittleness based on nanoindentation. Materials. 2022;15(20):7143.
5. Bandini A, Berry P, Bemporad E, et al. Role of grain boundaries and micro-defects on the mechanical response of a crystalline rock at multiscale. Int J Rock Mech Min Sci. 2014;71:429–41.
6. Broz ME, Cook RF, Whitney DL. Microhardness, toughness, and modulus of Mohs scale minerals. Am Mineral. 2006;91(1):135–42.
7. Liu Y, Deng J, Yue H, et al. Material removal behavior in processing green Al2O3 ceramics based on scratch and edge-indentation tests. Ceram Int. 2019;45(9):12495–508.

Chapter 3
Nanoscratch Experiments

Compared to nanoindentation experiment, the nanoscratch test method only needs to pass several scratch tests to obtain a large number of continuous data and reduce the measurement time to a large extent. At present, the identification of shale mineral composition and transition zone distribution length of laminar interface mostly relies on scanning electron microscopy, but this method can only obtain surface morphology and cannot further evaluate its mechanical properties. In this chapter, the failure mode and characteristic parameters of the nanoscratch are analyzed first, the continuity measurement of the nanoscratch is carried out, the mechanical parameters such as the depth of the scratch, the slope of the d^2-L curve, and the fracture toughness are determined, the mineral component identification method based on the nanoscratch curve is established, and the thickness change of the laminae transition zone is emphatically introduced.

3.1 Nanoscratch Method

3.1.1 Sample Preparation

The scratch test samples were taken from the Cretaceous Qingshankou Formation in Gulong Sag, Songliao Basin. The shale core was collected at a depth of about 1720 m and cut into cubes 20 mm long, 10 mm wide, and 10 mm high. Prior to the nanoscratch experiment, mechanical polishing was done twice with argon ions to obtain a smooth surface [1]. A Zeta-20 3D surface scanner was used to scan the rock surface to ensure that the roughness requirements of the experiment were met [2]. The sample is then ultrasonically cleaned and treated with acetone prior to the nanoscratch experiments to remove dirt such as rust [3]. Figure 3.1 shows the treated macroscopic surfaces of different groups of shales.

L. Yang, *Multiscale Scratch Mechanics of Laminated Shale*,
https://doi.org/10.1007/978-981-95-7041-6_3

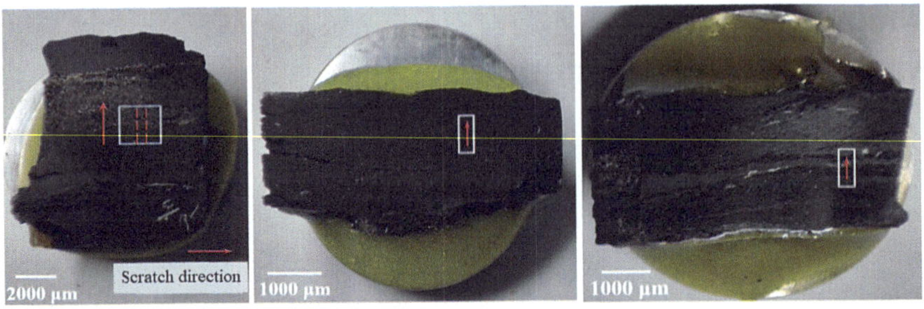

(a)Dolomitic nodule shale (b) Felsic laminated shale (c) Ostraco-laminated shale

Fig. 3.1 Nanoscratch experimental sample. (**a**) Dolomitic nodule shale. (**b**) Felsic laminated shale. (**c**) Ostraco-laminated shale

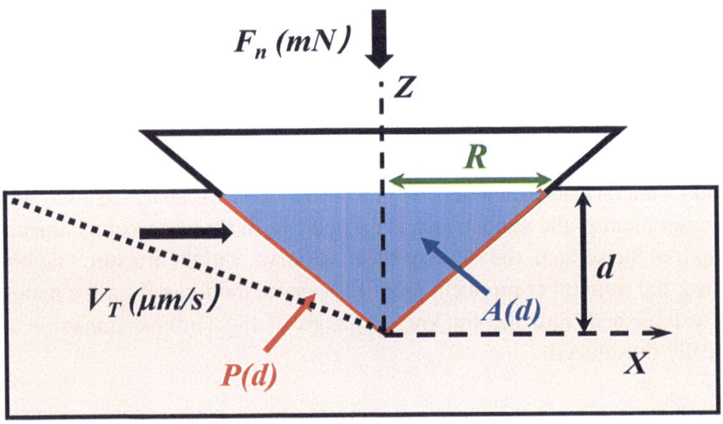

Fig. 3.2 Schematic diagram of nanoscratch process

3.1.2 Experimental Equipment

The mechanical equipment used in the nanoscratch experiment is still Agilent Keysight G200 nanoscratch test system, which sets a constant normal load of 50 mN and a constant tangential load of 30 μm/s for the scratch experiment process. Before conducting the nanoscratch experiment, JEOL JSM 6301F field-emission scanning electron microscope (see Fig. 3.2) was used to observe the laminar interface distribution on the surface of the shale sample to determine the experimental interval and analyze the mineral components. After the nanoscratch experiment, the scratch path is scanned by SEM to grasp the morphology of the shale surface after damage. In the SEM and QEMSCAN scanning results of shale, the distribution length of each mineral phase can be quantified by calculating the number of pixels in the SEM scanning image, which is calculated as follows:

$$\frac{N_{\mathrm{p}}}{N_{100}} = \frac{h}{100} \tag{3.1}$$

where N_p is the number of pixels in the range of length p, N100 is the number of pixels in the range of 100 μm, and h is the mineral phase distribution length. The pixel is the basic unit of the electron microscope scanning map, and the number of pixels per phase can be counted by Image-pro Plus software, which has distance measurement function and can automatically count the number of pixels.

Taking into account the impact of the test area on data collection, scratch tests were performed on three representative areas. The first section includes a typical illite-rich area, consisting mainly of large amounts of illite and small amounts of apatite and quartz. This area has the characteristics of low hardness, high ductility, stable mineral structure, few pores, and relatively uniform composition. The second section is a transitional zone between illite enrichment zone and dolomite nodule zone. The third part is the dolomite nodules area, mainly dolomite, quartz, and apatite, and all the minerals found in this area have a high level of brittleness and fracture toughness. Eight scratching experiments were carried out in different lithology areas. The length of each scratch is 500 μm, and the total scratch length is 4000 μm. The indenter collected mechanical parameter data at 1 μm intervals, which were sufficient to support the results of this paper.

3.1.3 Scratch Theory

The experimental process of nanoscratch is shown in Fig. 3.2, where X axis represents the direction of the transverse scratch, Z axis represents the direction of the vertical load, V_T represents the transverse scratch velocity, F_V represents the vertical load, F_T is the tangential force, d is the scratch depth, $A(d)$ is the projected area of the experimental volume perpendicular to the transverse scratch direction plane, and $p(d)$ is the length of the side of the projection area.

The friction coefficient during the scratch test can be calculated by the following formula:

$$\mu = \frac{F_{\mathrm{T}}}{F_{\mathrm{V}}} \tag{3.2}$$

Scratch hardness H reflects the material's ability to resist deformation:

$$H = \frac{F_n}{A(d)} \tag{3.3}$$

Fracture toughness is an important mechanical parameter, which describes the ability of a material to resist the instability and propagation of macroscopic cracks. In the nanoscratch test, fracture toughness K_c represents the energy absorption capacity per unit area of the material before fracture [4]. Previous studies have shown that the fracture of materials in scratch experiments is mainly caused by the damage caused by F_T. Hoover et al. proposed a fracture toughness measurement method based on scratch experiments, which can be used to characterize the fracture toughness of rock and concrete materials.

Fracture toughness K_c can be calculated based on Eq. (3.4), where d is the scratch depth and R is the radius of the scratch probe. When the value of d is large enough, K_c gradually approaches some constant indicating the fracture toughness of the material. It has been found that whether the value of K_c can represent the overall fracture toughness of shale minerals depends on the value of the indentation depth d. Only when the value of d is increased to a sufficiently large value can it represent the fracture toughness of the material as determined by scratch experiments. Under the load of 50 mN, the indentation depth of the two groups of samples in the scratch experiment in this paper is greater than 1000 nm. According to the theory of linear elastic fracture mechanics, the fracture toughness of shale samples can be expressed by parameter K_c, which is expressed as follows:

$$K_c = \frac{F_T}{\sqrt{2p(d)A(d)}} \tag{3.4}$$

3.1.4 Berkovich Indenter Parameter

For Berkovich indenters, the product of $p(d)$ and $A(d)$ largely depends on which side of the probe is aligned with the direction of the scratch, due to its positive trident nature. As can be seen from formula 3.4, parameters $p(d)$ and $A(d)$ of the indenter probe are required to calculate fracture toughness based on scratch data. Figure 3.3a shows the horizontal view of the Berkovich indenter, Fig. 3.3b shows the three-dimensional shape of the indenter at the depth of the indentation, and in Fig. 3.3c, the

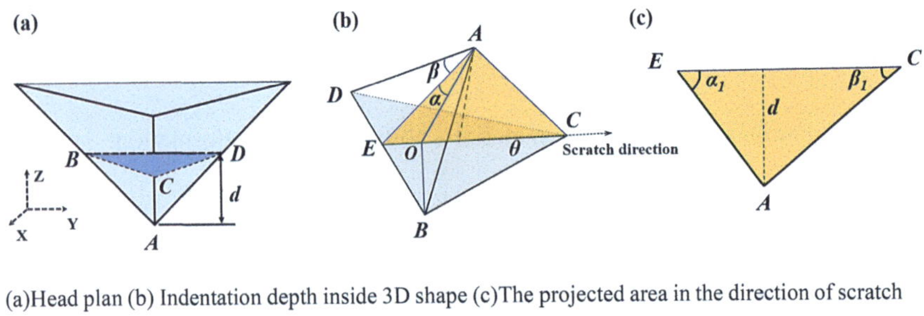

Fig. 3.3 Berkovich indenter geometry. (**a**) Head plan, (**b**) indentation depth inside 3D shape, and (**c**) the projected area in the direction of scratch

yellow plane shows the projection of the indenter in the direction of the scratch during the sliding process, where:

$$A(d) = S_{\Delta AEC} = \frac{\tan\beta}{\cos\alpha} \times [\sin\theta + \sin(60° - \theta)] \times d^2 \tag{3.5}$$

$$p(d) = AE + AC = d \times \left(\sqrt{\left(1 + \frac{4 \times \tan^2\beta \times \cos^2\theta}{3 \times \cos^2\alpha}\right)} + \sqrt{\left(1 + \frac{4 \times \tan^2\beta \times \sin^2(30° + \theta)}{3 \times \cos^2\alpha}\right)}\right) \tag{3.6}$$

The constants λ and ζ are defined as follows:

$$\lambda = \frac{\tan\beta}{\cos\alpha} \times [\sin\theta + \sin(60° - \theta)] \tag{3.7}$$

$$\zeta = \left(\sqrt{\left(1 + \frac{4 \times \tan^2\beta \times \cos^2\theta}{3 \times \cos^2\alpha}\right)} + \sqrt{\left(1 + \frac{4 \times \tan^2\beta \times \sin^2(30° + \theta)}{3 \times \cos^2\alpha}\right)}\right) \tag{3.8}$$

According to formulas 3.4 and 3.5, $A(d)$ is related to d^2 and $p(d)$ is related to the depth of indentation d. In addition, based on the projected surface of the triangular pyramid indenter in the direction of the scratch, the following conclusions can be drawn:

$$p(d) = d/\sin\alpha_1 + d/\sin\beta_1 \tag{3.9}$$

$$A(d) = (d/\tan\alpha_1 + d/\tan\beta_1) \times d/2 \tag{3.10}$$

$$p(d) \in \lambda d \tag{3.11}$$

$$A(d) \in \zeta d^2 \tag{3.12}$$

In Eqs. 3.10 and 3.11, λ and ζ are dimensionless linear correlation coefficients of $p(d)$ and $A(d)$ with respect to d. Based on the above analysis, the value of $A(d)$ is related to d^2, and the value of $p(d)A(d)$ depends on d^3.

During the scratch test, it was not possible to determine whether the Berkovich indenter had come into contact with the surface or side edge of the specimen. Therefore, it is necessary to calibrate the indenter parameters by conducting scratch tests on materials with known fracture toughness before conducting scratch tests. For the same Berkovich indenters, the relationship between the indentation depth d and the parameter $p(d)A(d)$ remains unchanged, independent of the scratch material. Under the conditions of Berkovich indenter, molten silica can be selected as the reference material for scratch experiments, and its known fracture toughness at room temperature is 0.6 MPa $m^{0.5}$, $p(d)\,A(d) = 2.2808d^3$. The same relationship between

$p(d)A(d)$ and d can be used to calculate the fracture toughness of different materials, which in turn can be used to obtain the relationship between $p(d)A(d)$ and K_c from formula 3.13.

$$p(d)A(d) = \frac{F_T}{2K_c^2} \tag{3.13}$$

3.2 Mechanical Characteristics

In the previous chapter, the principle of nanoscratch experiment and the method of indenter calibration are described, and the influence of triangular pyramid indenter on the collected mechanical parameters is analyzed. In this section, the results of nanoscratch experiments are analyzed, and the failure modes and mechanical parameters of different mineral distribution regions are mainly studied. The elastoplastic mechanical characteristics of minerals can be further discussed by analyzing the failure modes with scanning electron microscopy.

3.2.1 Damage Mode

In the scratch test, the failure modes of minerals with different lithologies are shown in Fig. 3.4. In the case of poor cement of plastic minerals, the scratch test will lead to the destruction of the mucous membrane of the matrix grains, and the broken rock chips will accumulate on the side of the residual scratches. Brittle minerals do not show obvious debris accumulation during the scratch process. When the indenter passes through a microcrack, the accumulated strain energy is released, causing rock debris to collapse and leave a crater. As the penetration depth increases, brittle failure causes microscale rock fracture, cracks propagate from the indenter tip, and rock fragments break apart, leaving a set of sharp jagged scratch paths.

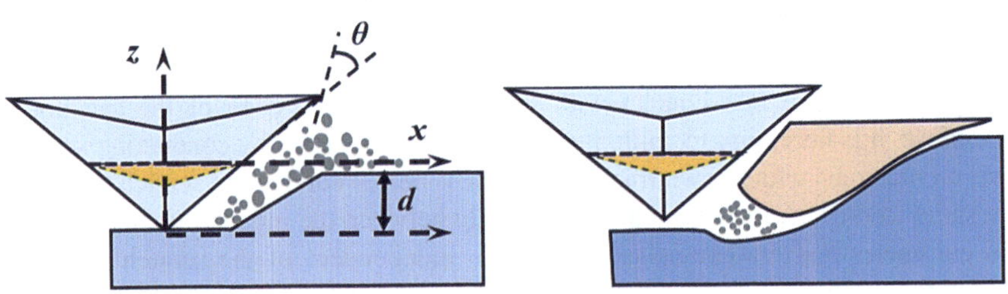

(a) Failure mode of plastic minerals (b) Failure modes of brittle minerals

Fig. 3.4 The failure modes of minerals from different rock types. (**a**) Failure mode of plastic minerals. (**b**) Failure modes of brittle minerals

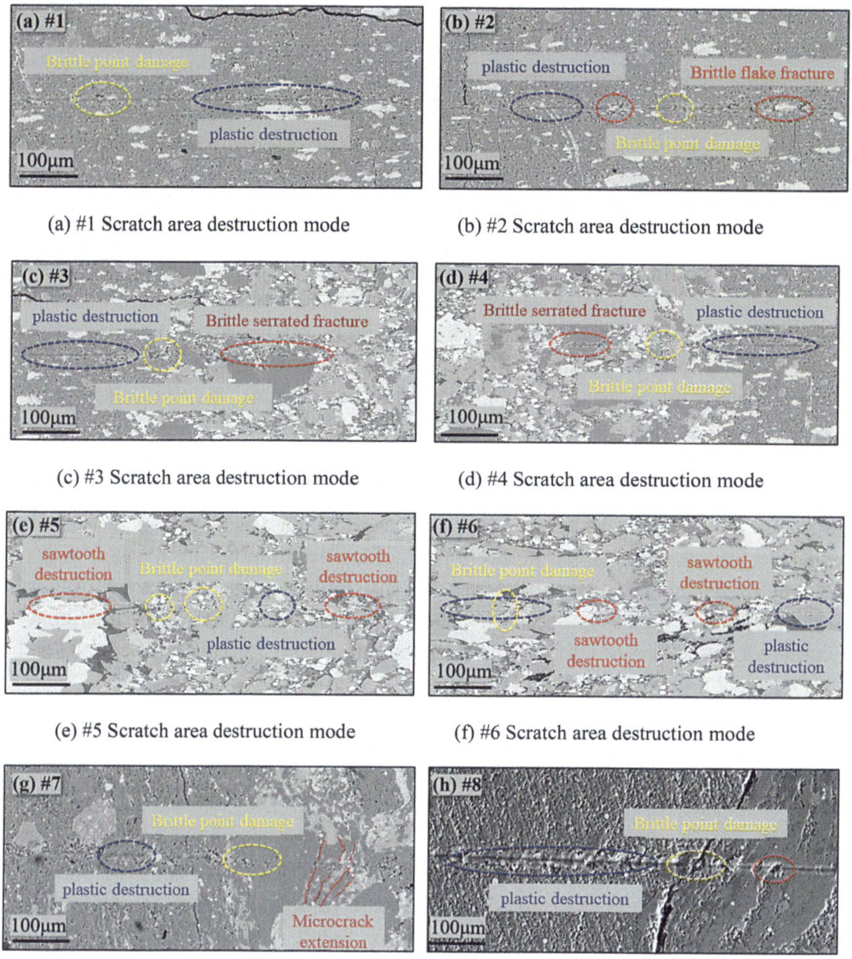

(a) #1 Scratch area destruction mode
(b) #2 Scratch area destruction mode
(c) #3 Scratch area destruction mode
(d) #4 Scratch area destruction mode
(e) #5 Scratch area destruction mode
(f) #6 Scratch area destruction mode
(g) #7 Scratch area destruction mode
(h) #8 Scratch area destruction mode

Fig. 3.5 Scratched surface morphology of Gulong shale. (**a**) #1 Scratch area destruction mode. (**b**) #2 Scratch area destruction mode. (**c**) #3 Scratch area destruction mode. (**d**) #4 Scratch area destruction mode. (**e**) #5 Scratch area destruction mode. (**f**) #6 Scratch area destruction mode. (**g**) #7 Scratch area destruction mode. (**h**) #8 Scratch area destruction mode

Figure 3.5 shows the SEM scanning results of 8 groups of laminated shale samples. Samples #1 and #2 mainly contain plastic minerals, such as illite and chlorite, while scattered quartz and pyrite further improve the heterogeneity. In the #1 and #2 regions, the damage pattern of scratches is mainly plastic, with occasional punctate brittle damage. After the indenter passes through the plastic mineral, debris accumulates on the side of the residual scratch, while the brittle mineral will show uneven point cracking and jagged crushing. Samples #3 and #4 are transition areas from illite-rich area to dolomite nodules. The first half of the scratches path is mainly filled with illite minerals, while the second half is mixed with brittle minerals, such as dolomite, quartz, and albitite. The plastic failure of the scratch path from illite-rich area gradually slides along with the indenter to dolomite nodule area resulting in brittle failure. The large distribution of dolomite in shale samples #5 and #6 has an important effect on the mechanical parameters collected during the scratch experiment. Because dolomite, quartz, and apatite are all hard minerals, they may contribute to the overall hardness of the shale. Sporadic inclusion of pyrite may lead to increased hardness in local areas, resulting in more complex scratch curves in scratch tests.

#7 SEM scanning results of the samples show that illite minerals are mainly distributed in the first half of the scratched path of the feldspar mineral-laminated shale, while the second half is divided into dolomite, quartz, albite, and other brittle mineral mixed areas. #8 Sample scanning results show that the first half of the striated shale path is an illite-rich area with fragmentary distribution of brittle minerals, such as albite and quartz, while the second half is a mixed area of brittle minerals, such as calcite, quartz, and albite. The grain cementation of laminated shale minerals is poor, and microfractures and pores are relatively developed. The failure modes of plastic minerals such as illite are marked with yellow ovals, while the dotted fracture and jagged failure modes of brittle minerals are marked with red and blue ovals. In scratch experiments, the distribution of heterogeneous minerals may lead to the change of friction and failure modes in scratch experiments. The accurate identification of surface mineral components is of great significance for the study of mechanical properties of shale.

3.2.2 Mineral Distribution

QEMSCAN mineral composition scanning results of laminated shale are shown in Fig. 3.6. Illite is widely distributed in two shale samples #1 and #2, while pyrite, dolomite, apatite, and quartz are sporadically included, which may affect the microstructure and pore characteristics of shale. Illite may form a continuous matrix with a dense matrix structure, while the fragmentary inclusion of brittle minerals may form microgranular gaps and fissures, affecting the pore structure and permeability. Different types of pore structure may affect the permeability and reservoir properties of shale. The distribution of shale minerals may result in complex microstructure and pore characteristics, which have an impact on pore structure, permeability, and reservoir properties and require further research and experimental analysis.

According to the results observed by electron scanning microscopy, shale samples #3, #4, #7, and #8 are the binding regions of the transition from illite-rich area to brittle mineral distribution. This means that there is an interfacial transition zone between illite minerals and brittle minerals such as dolomite and quartz on the surface of the two shale samples. In terms of microstructure, there may be microporous matrix in these two shale samples. The dense matrix structure may be formed in the illite rich area, while the microgranular gaps and cracks may be formed in the brittle mineral binding area, such as dolomite and quartz. In terms of the mechanical properties of the interfacial transition zone, due to the different physical properties of

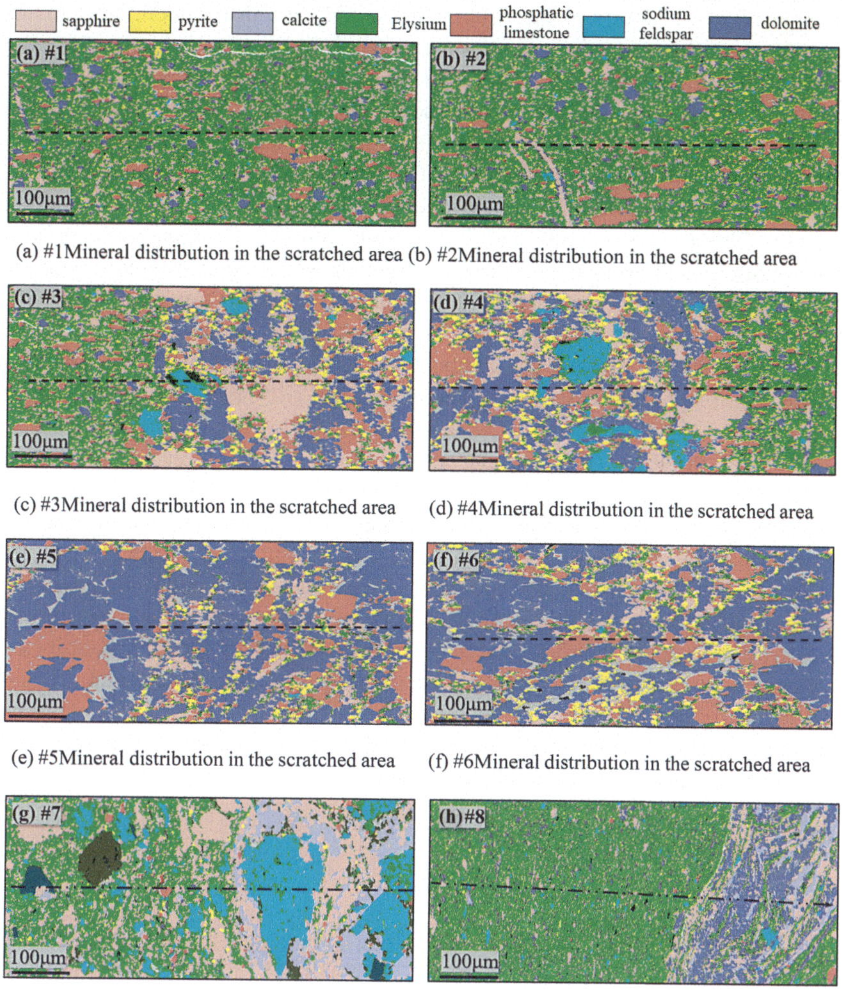

(a) #1Mineral distribution in the scratched area (b) #2Mineral distribution in the scratched area

(c) #3Mineral distribution in the scratched area (d) #4Mineral distribution in the scratched area

(e) #5Mineral distribution in the scratched area (f) #6Mineral distribution in the scratched area

(g) #7Mineral distribution in the scratched area (h) #8Mineral distribution in the scratched area

Fig. 3.6 QEMSCAN mineral composition scanning results. (**a**) #1 Mineral distribution in the scratched area. (**b**) #2 Mineral distribution in the scratched area. (**c**) #3 Mineral distribution in the scratched area. (**d**) #4 Mineral distribution in the scratched area. (**e**) #5 Mineral distribution in the scratched area. (**f**) #6 Mineral distribution in the scratched area. (**g**) #7 Mineral distribution in the scratched area. (**h**) #8 Mineral distribution in the scratched area

brittle minerals such as illite, dolomite, and quartz, the contact interface between them may have certain lithological effects, which further affect the mechanical properties of shale, such as the formation of pores and the expansion of microfractures in the range of interfacial cementation.

The surface of shale samples #5 and #6 has a large distribution of dolomite minerals, mixed with a small distribution of quartz and apatite. In addition, sporadic inclusion of pyrite improves the heterogeneity and pore distribution. In terms of microstructure, there may be microporous matrix in these two shale samples. The distribution of dolomite, quartz, and apatite forms complex microstructure, while the scattered inclusion of pyrite may lead to more microscopic pores and cracks, which are more likely to affect the mechanical properties and fracture resistance of shale.

The content of each mineral component of laminated shale is shown in Fig. 3.7. The shale samples #1, #2, and #8 are mainly composed of clay minerals, and the content of illite reaches 75.14% and 79.13%, which makes the shale samples have strong plasticity and viscosity. At the same time, the scattered distribution of quartz and apatite may help increase the hardness and compressive strength of the shale. Clay shale has high plasticity and weak fracture resistance. Shale samples #3 and #7 are mainly composed of clay and carbonate minerals, and illite mineral content is 37.13% and 36.68%, respectively. The quartz content is 17.49%, the apatite content is 12.06%, and the illite content is 37.13%. Due to the high content of illite, the #3 and #7 shale samples have strong plasticity, and the increase in quartz and apatite content helps to increase the hardness and compressive strength of the shale.

The #4, #5, and #6 shale samples are carbonate laminae, mainly composed of dolomite and brittle minerals. The high content of dolomite may make the samples have certain toughness. At the same time, the presence of apatite increases the hardness and fracture resistance of shale. The presence of quartz and pyrite will increase the strength of the shale, but their content is relatively low. Considering the composition content of various minerals, laminated shale may have certain toughness and hardness, but the mechanical performance of shale cannot be judged by scanning electron microscopy alone, and the specific mechanical properties need further experiments and tests to fully evaluate.

3.2.3 Micromechanical Properties

Nanoscratches can be used to test mechanical parameters such as surface hardness, scratch depth and fracture toughness of materials to further evaluate the deformation and fracture resistance of materials. By characterizing the mechanical properties of nanoscratches, the mechanical properties of shale materials can be understood more comprehensively, which provides an important reference for the design and process application of hydraulic fracturing schemes.

Load-Displacement Curves

In the scratch test of constant normal load, a larger scratch depth usually indicates that the material is soft or easy to break, while a smaller scratch depth indicates that the material is hard or has better deformation resistance. In this chapter, depth curves for eight groups of samples from #1 to #8 from various regions are plotted in Fig. 3.8. In group #1 and #2, softer illite accounts for more than 75% of the mineral

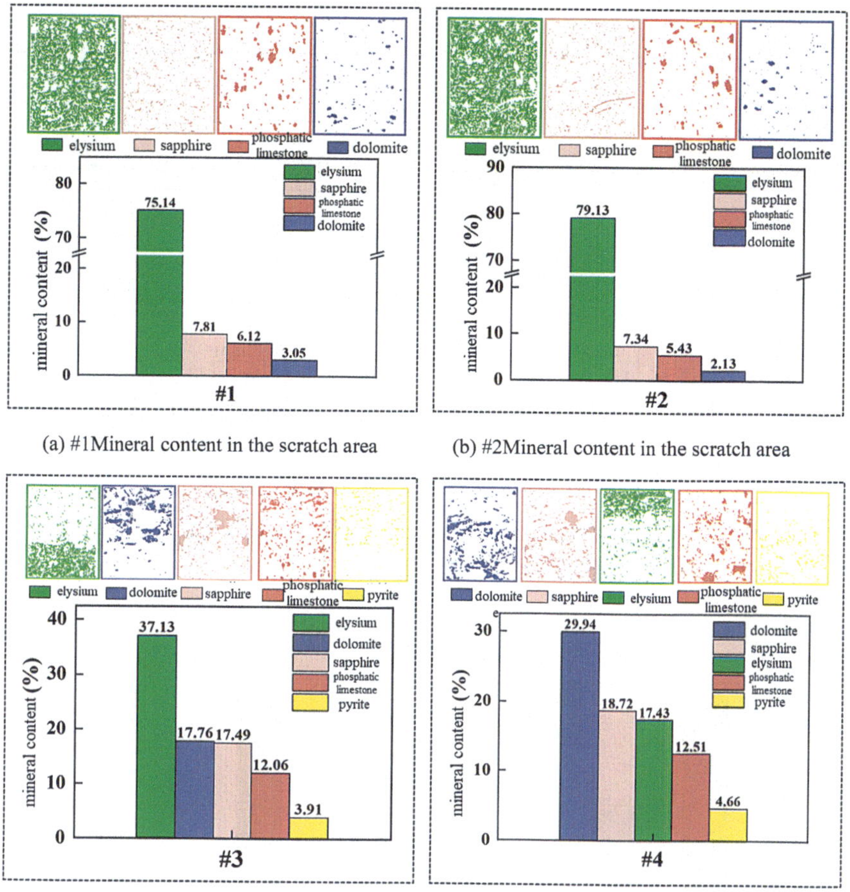

(c) #3Mineral content in the scratch area

(d) #4Mineral content in the scratch area

Fig. 3.7 Distribution and content of single point minerals in the scratch area. (**a**) #1 Mineral content in the scratch area. (**b**) #2 Mineral content in the scratch area. (**c**) # 3Mineral content in the scratch area. (**d**) #4 Mineral content in the scratch area. (**e**) #5 Mineral content in the scratch area. (**f**) #6 Mineral content in the scratch area. (**g**) #7 Mineral content in the scratch area. (**h**) #8 Mineral content in the scratch area

distribution in the scratch area, resulting in scratch depths of more than 2500 nm in both samples.

In samples #3 to #8, illite laminae and brittle mineral laminae are interleaved, dolomite, apatite, quartz, and calcite are fused with illite, and the content of brittle minerals is high. In addition, the presence of harder minerals such as pyrite and quartz along the scratch path gives the depth curve a very representative peak rise and fall. Compared with the illite grain layer, the mineral hardness of #3 to #8 area increases significantly. Under the same normal load, the subdivision depth decreases

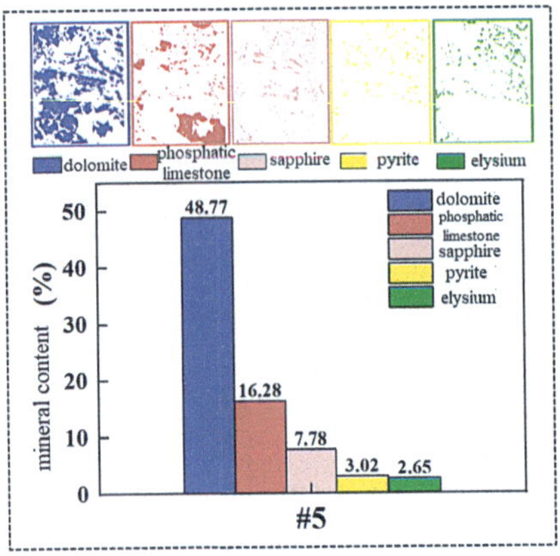

(e) #5Mineral content in the scratch area

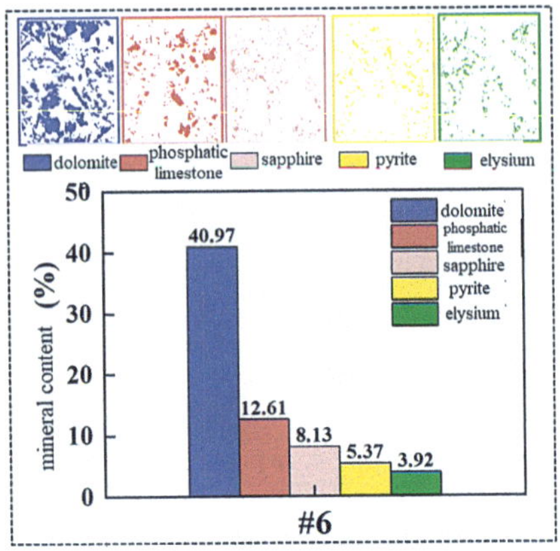

(f) #6 Mineral content in the scratch area

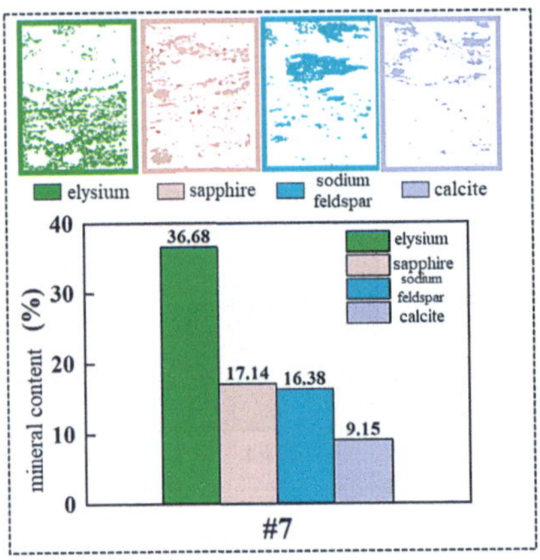

(g) #7Mineral content in the scratch area

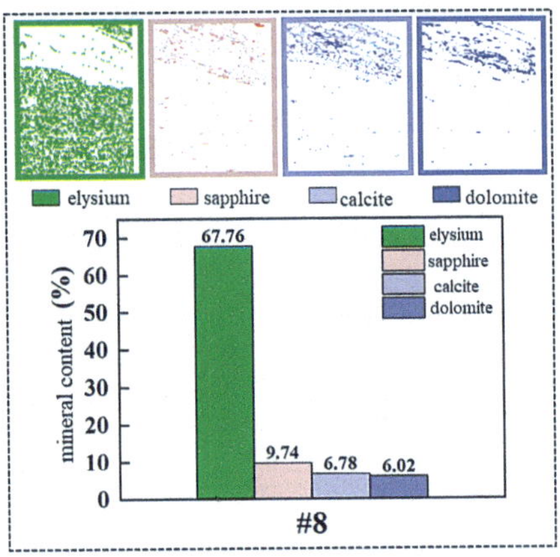

(h) #8Mineral content in the scratch area

Fig. 3.7 (continued)

greatly, and the subdivision depth ranges from 1000 nm to 1500 nm. Compared with the majority of illite distribution along the scratch path in the clay distribution area, #3 and #4 and #7 and #8 are the transition areas of mineral distribution in shale, which belong to the lithology change area and have unstable mechanical properties. The first half of the scratch path is divided into the distribution area of plastic minerals such as illite, and the second half is the mixed area of brittle minerals in shale. The amplitude of the curve in this region changed sharply with the experiment, and the fluctuation of the scratch depth in the early stage of the scratch experiment was not obvious, but the depth of the scratch depth showed a sharp increase in the later stage.

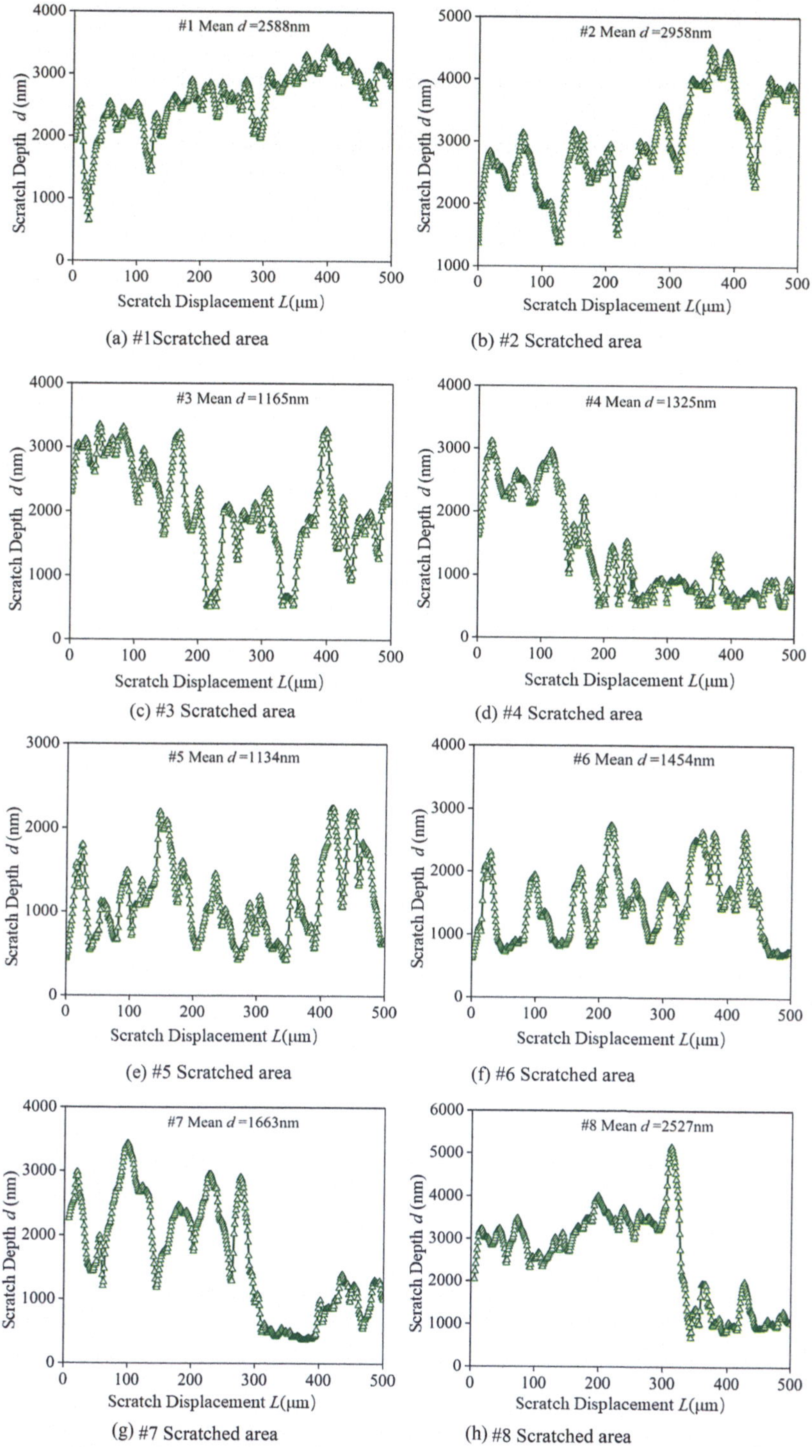

Fig. 3.8 Shale surface penetration depth curve. (**a**) #1 Scratched area. (**b**) #2 Scratched area. (**c**) #3 Scratched area. (**d**) #4 Scratched area. (**e**) #5 Scratched area. (**f**) #6 Scratched area. (**g**) #7 Scratched area. (**h**) #8 Scratched area

Fracture Toughness

Fracture toughness is an important parameter to describe the ability of materials to resist fracture. The fracture toughness curves of regions #1 to #8 are shown in Fig. 3.9, and the mineral composition in different regions has a significant influence on the fracture toughness of shale. In the #1 and #2 regions, the average fracture toughness is 0.24 GPa and 0.20 GPa, respectively, due to the high content of plastic weak minerals, such as illite. This indicates that the shales in these two areas are relatively prone to fracture when subjected to external stress, and their ability to resist fracture growth is low.

In region #3 and #4, there are lithologic transition regions of illite and dolomite nodules, with average fracture toughness of 0.41 GPa and 1.05 GPa, respectively. Compared with illite-rich areas, the average fracture toughness of these two areas is improved, which may be due to the presence of dolomite to enhance the ability of shale to resist fracture growth. Regions #5 and #6 are dolomite concretion regions with average fracture toughness of 1.03 GPa and 1.79 GPa, respectively. Although there are a large number of micropores and fractures in these two regions, the average fracture toughness of the shale is relatively high due to the wide distribution of quartz and dolomite. Regions #7 and #8 correspond to the interfacial transition zones of the long laminae and ostracod-rich laminae, respectively. Their average fracture toughness values are 0.78 GPa and 0.38 GPa, respectively. Due to the high content of brittle minerals, the average fracture toughness of felsic mineral-laminated shale is much higher than that of ostraco-laminated shale, which indicates that brittle minerals have an important effect on fracture toughness of shale.

Hardness

Hardness is one of the key physical characteristics of minerals and rocks, which indicates the rock's resistance to penetration, scratches, or permanent deformation. In this chapter, the hardness of mineral distribution along the scratch path in shale region is discussed and analyzed based on the data collected from scratch experiment. The hardness curve in the scratch area is shown in Fig. 3.10. Compared with the fracture toughness curve in Fig. 3.9, the region where the characteristic peak of the curve is located has the same place, and the scratch displacement corresponding to the peak value is basically the same. Samples #1 and #2 have a wide range of illite with an average hardness of 0.75 GPa along the path. At the effective scratch length of 125 μm, the drawing distance and hardness distribution curve reach the first maximum peak. Through the comparison of QEMSCAN electron microscope scanning results, it was found that the location of the scratch was just a small pyrite distribution area. Pyrite is a high-hardness mineral, and its hardness distribution range is generally about 6 GPa. The highest peak of the hardness curve in the pyrite distribution area here is only 3.5, which may be because the pyrite distribution area is

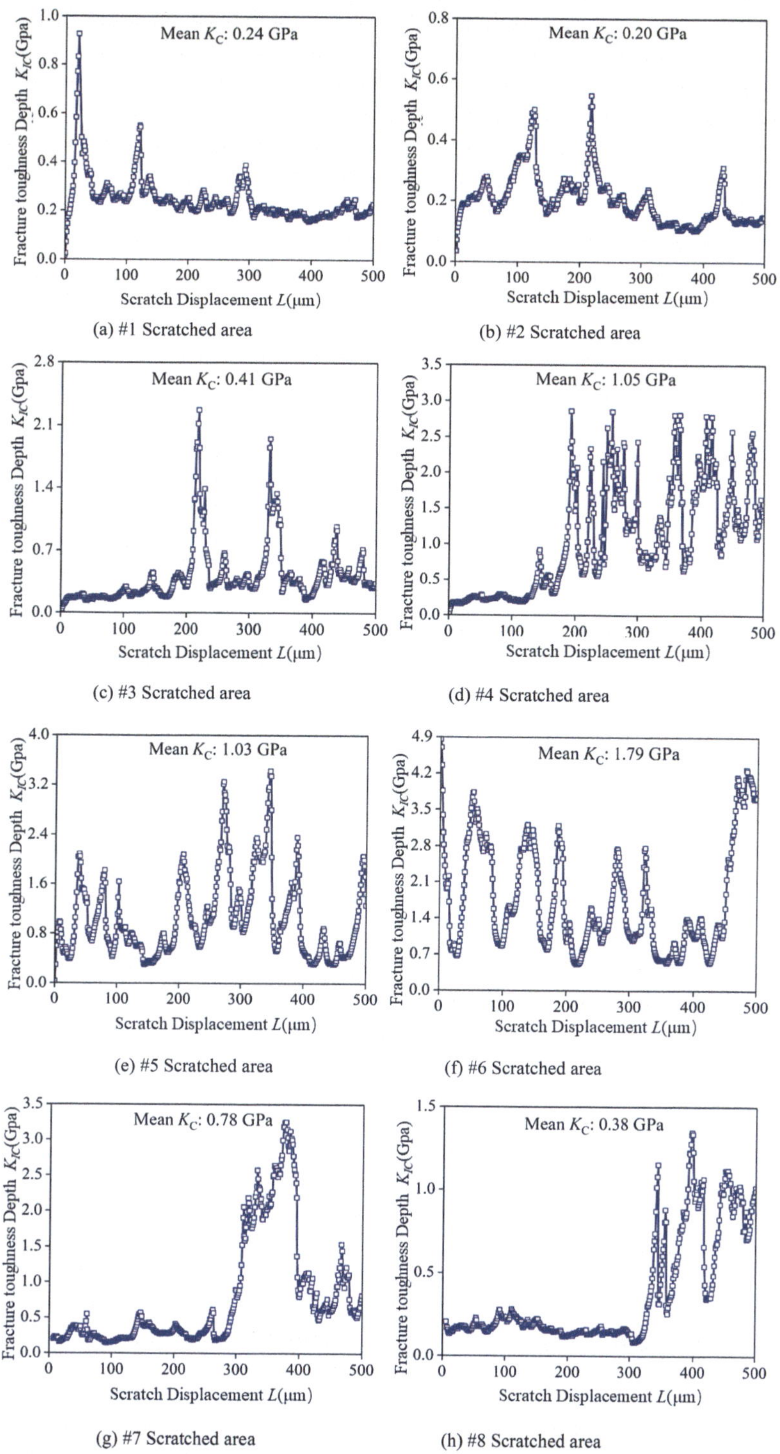

Fig. 3.9 Scratch hardness curve in #1 to #8 area. (**a**) #1 Scratched area. (**b**) #2 Scratched area. (**c**) #3 Scratched area. (**d**) #4 Scratched area. (**e**) #5 Scratched area. (**f**) #6 Scratched area. (**g**) #7 Scratched area. (**h**) #8 Scratched area

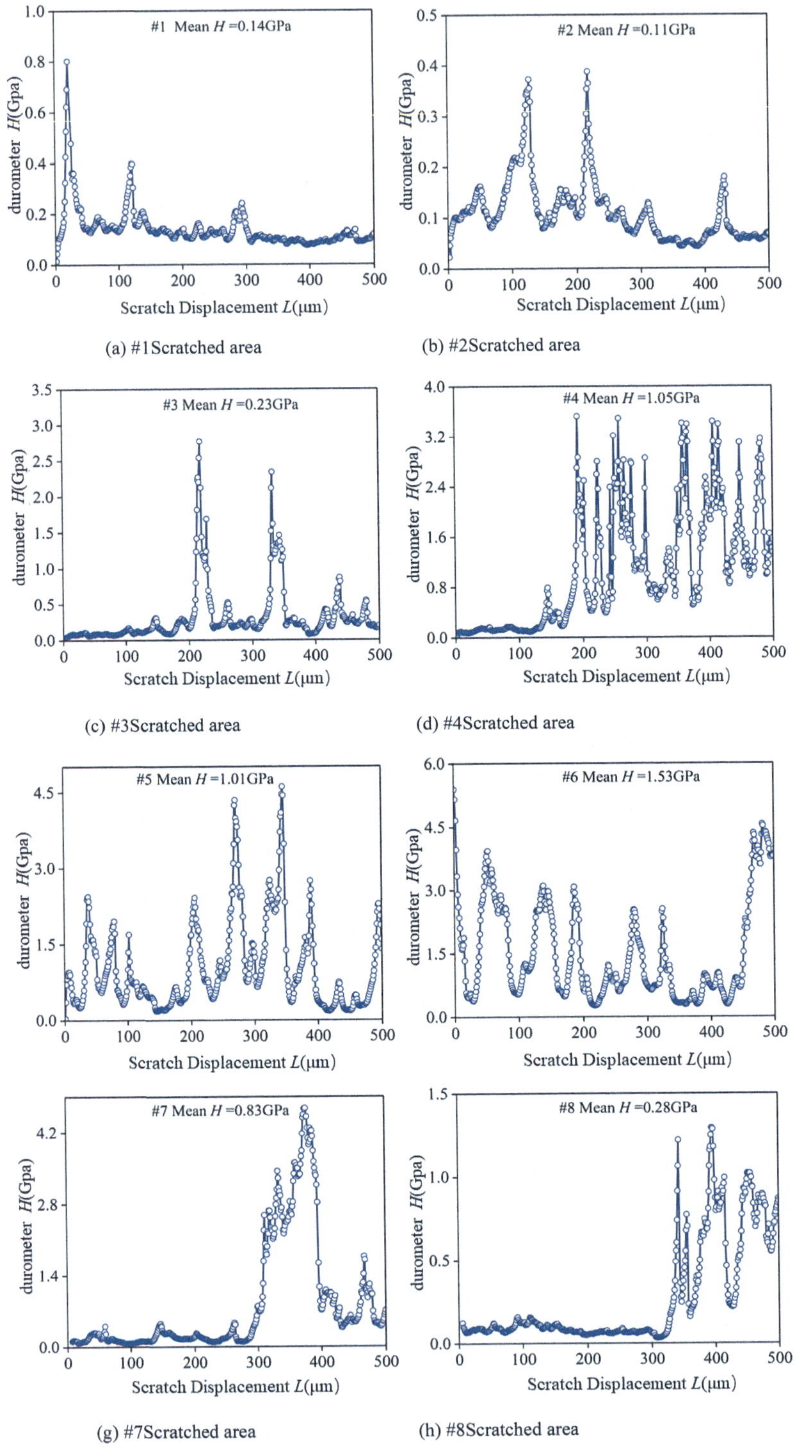

Fig. 3.10 Scratch hardness curve in #1to #8 area. (**a**) #1 Scratched area. (**b**) #2 Scratched area. (**c**) #3 Scratched area. (**d**) #4 Scratched area. (**e**) #5 Scratched area. (**f**) #6 Scratched area. (**g**) #7 Scratched area. (**h**) #8 Scratched area

small, and the pyrite and illite blend with each other, so that the data collected by the scratch indenter is the property of the fusion distribution of the two minerals.

For samples #3, #4, #7, and #8, the mineral distribution is basically pure illite along the scratch path of 125–220 μm, making the hardness of this part of the region about 0.12 GPa. At the scratch path of 220 μm, the hardness curve showed a second peak value of 0.4 GPa, and scanning electron microscopy showed that a small amount of pyrite was distributed in the illite range in this area. In the scratch path of 220–320 μm, the mineral hardness showed a trend of decreasing first and then increasing. By comparing the distribution of minerals, it is found that the distribution of illite decreases and the distribution area of brittle minerals increases gradually. For #5 and #6 samples, dolomite, illite, and quartz are interlaced along the scratch path of 320–600 μm, making the hardness of the path jump back and forth between 1.5 GPa and 4.5 GPa. The comprehensive discussion and analysis of the hardness of different mineral distribution areas are of great significance for the analysis of mechanical properties of laminar heterogeneous shales at microscale.

Relationship Between Hardness and Fracture Toughness

Hardness and fracture toughness are important indicators of shale fracture expansion ability, and there is a specific dimensionless relationship between them, which can be described by a dimensionless coefficient, the size of which can reflect the shale fracture expansion ability. Figure 3.11 shows the fitting results of the dimensionless relationship between hardness and fracture toughness in the #1 to #8 region. The fitting coefficients are all greater than 0.9, indicating that there is a stable correlation between scratch hardness and fracture toughness. In the #1 and #2 regions, the correlation coefficients are 0.741 and 0.569, respectively, due to the high content of plastic weak minerals, such as illite. The relationship between hardness and fracture toughness of shale is weak in the distribution range of illite laminae. In regions #3 and #4, which are lithologic transition regions of illite and dolomite nodules, the correlation coefficients are 0.935 and 1.075, respectively. Compared with illite-rich area, the correlation coefficient between these two areas is improved. This indicates that there is a close relationship between the hardness and fracture toughness of shale in these two regions.

The #5 and #6 regions are dolomite nodule regions, in which the mineral brittleness is relatively high, but accompanied by a large number of micropores and fractures, the correlation coefficients are 1.066 and 0.926, respectively, indicating that the correlation coefficients between the hardness and fracture toughness of shale are not only affected by mineral brittleness but also by micropores and fractures. Regions #7 and #8 correspond to the interfacial transition zones of the long laminae and ostracoid laminae, respectively. The corresponding correlation parameters are 1.243 and 0.841, respectively. Because of the high content of brittle minerals, the hardness and fracture toughness of felsic mineral-laminated shale are much higher than that of ostraco-laminated shale, and the correlation coefficient is also larger.

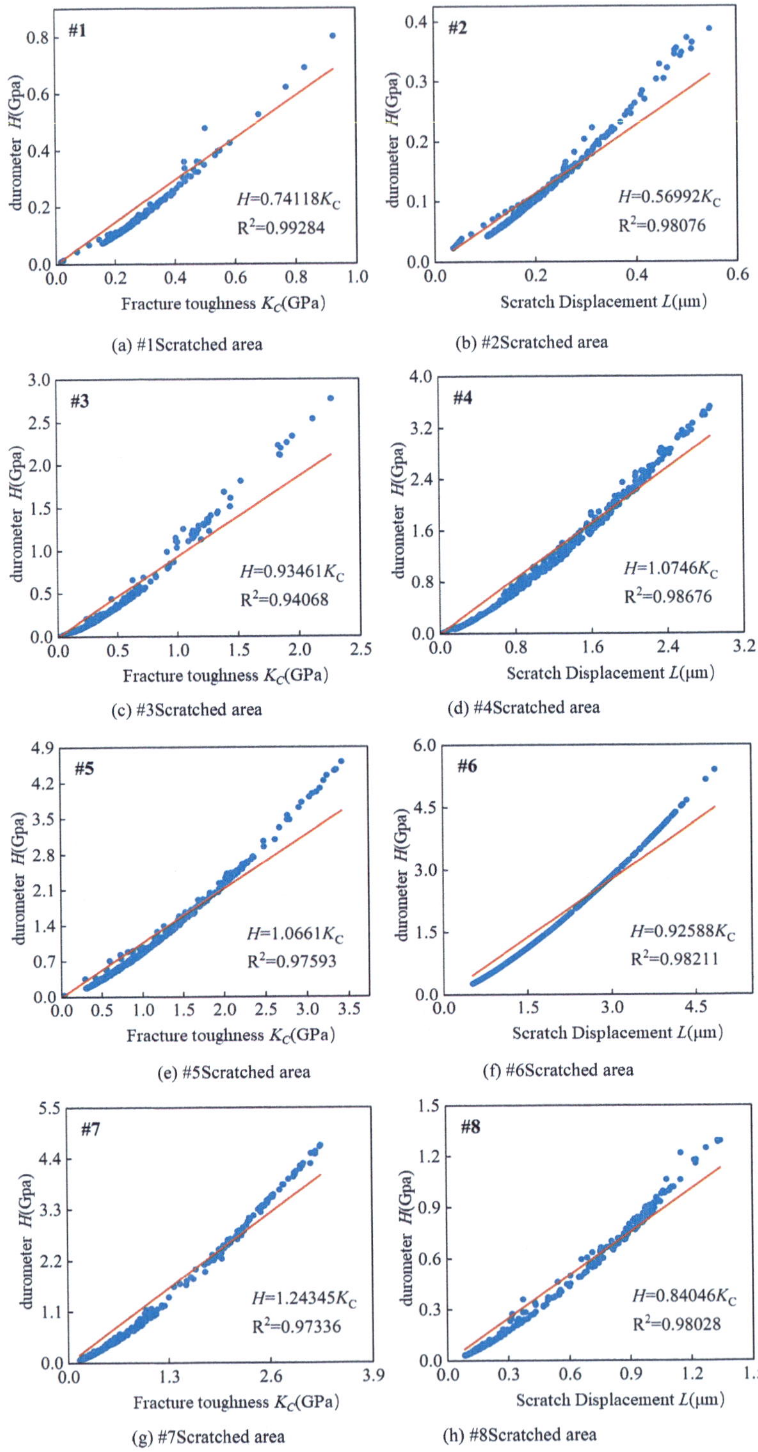

Fig. 3.11 Dimensionless relationship between hardness and fracture toughness in #1 to #8 region. (**a**) #1Scratched area. (**b**) #2Scratched area. (**c**) #3 Scratched area. (**d**) #4 Scratched area. (**e**) #5 Scratched area. (**f**) #6 Scratched area. (**g**) #7 Scratched area. (**h**) #8 Scratched area

3.3 Mineral Identification

3.3.1 Scratch Displacement Method

The feasibility of identifying mineral components according to scratch depth depends on the lithology of the mineral. The scratch depth of brittle mineral and plastic mineral is obviously different in scratch test. During scratch testing, brittle minerals generally exhibit higher hardness. Under the same normal load, they are less prone to deformation and therefore produce shallower scratches. Plastic minerals, due to their small hardness, flexibility, and ductility, will produce deeper scratches. The scratch depth can be used to initially identify the composition of the mineral, and the results of the SEM and QEMSCAN scan are combined for comprehensive analysis to ensure accuracy.

According to previous studies, the frequency distribution of scratch depth of a single mineral phase roughly follows a Gaussian distribution. Figure 3.12 shows the statistics and fitting of indentation depths for each group of samples, with the depth distributions of individual mineral phases grouped according to Gaussian trend lines. Under the vertical force of 50 mN, the scratch depth of pyrite is 0–575 nm, quartz mineral is 625–875 nm, dolomite is 925–1575 nm, apatite is 1625–2025 nm, and illite is 2075–3375 nm. The statistics of the scratch depth data showed that the illite content of samples #1 and #2 was 81.54% and 86.84%, respectively. Compared with the results of SEM, the illite content was higher according to the scratch depth. This may be due to the high frequency of large scratch depth values collected by the sensor due to micropores and cracks and the inability to accurately identify softer minerals and microcracks. According to the results of QEMSCAN, illite minerals are most widely distributed in the #3 scratch area, accounting for 37.13% of the mineral composition, while 53.13% of the mineral composition is identified as illite according to the scratch depth. The QEMSCAN result for area #4 contains 17.43% illite mineral composition compared to 35.24% illite ore composition based on delineation depth. The proportion of illite mineral composition determined by the value of scratch depth is larger than that of QEMSCAN electron microscope scanning results. This may be due to the interfacial transition zone between illite and dolomite nodules, which formed more complex microcracks and pores.

The stratified shale carbonate stratification is rich in dolomite, apatite, and quartz, while the presence of a small amount of pyrite greatly increases the inhomogeneity of the rock, and the brittle fracture dominates the mineral failure in zone #5 and #6. In QEMSCAN results, dolomite was most abundant in scratch area No. 5 and No. 6, 48.77% and 40.79%, respectively. In the scratch depth statistics of #5 and #6, the percentage of dolomite is 46.08% and 47.83%, respectively, which is close to the results of QEMSCAN. It is worth noting that the percentage of illite mineral distribution in the #5 and #6 scratch areas in the QEMSCAN scan results is 2.65% and 3.92%, respectively, while the content of illite in the #5 and #6 scratch depth statistics is 8.85% and 9.23%, respectively, which is inconsistent with the mineral distribution scanning results identified by SEM. In the #7 and #8 scratch areas, illite

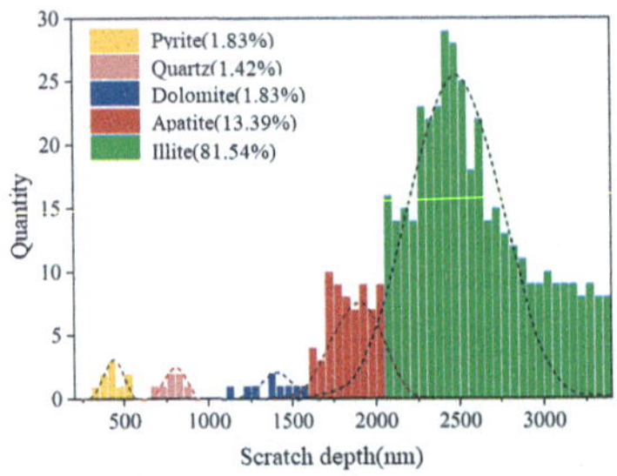

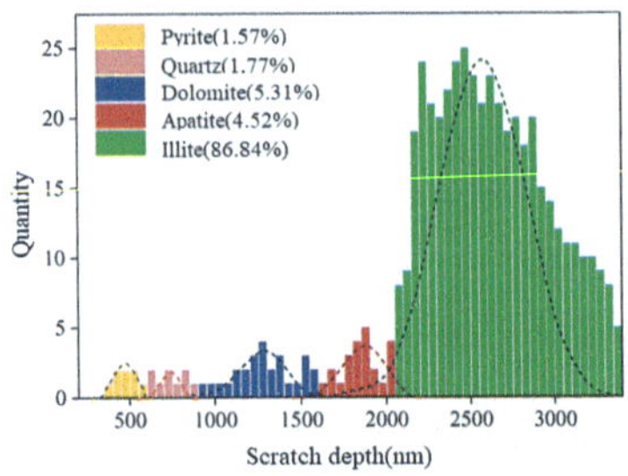

(a)#1Regions are divided into depth distributions (b) #2Regions are divided into depth distributions

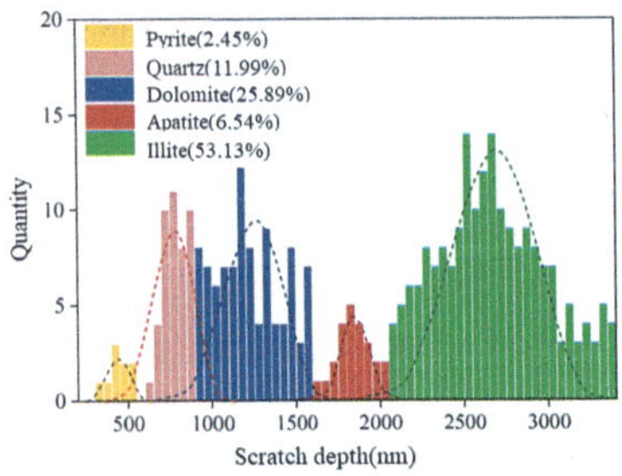

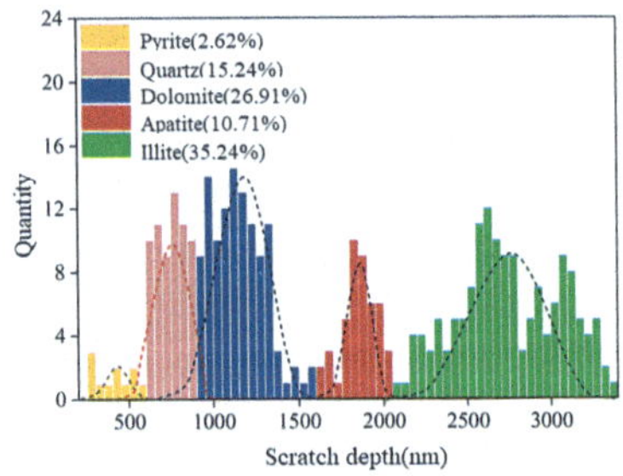

(c) #3Regions are divided into depth distributions (d) #4Regions are divided into depth distributions

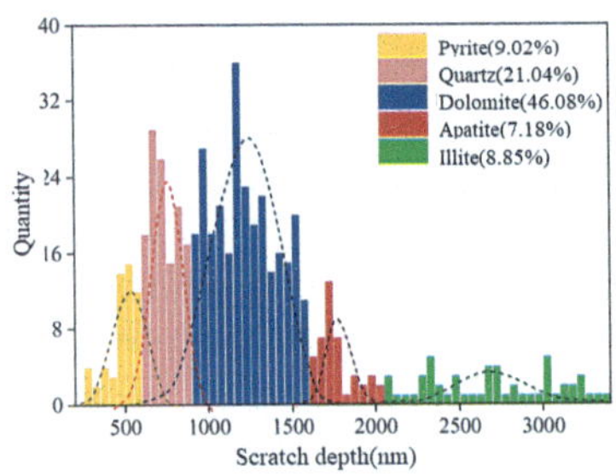

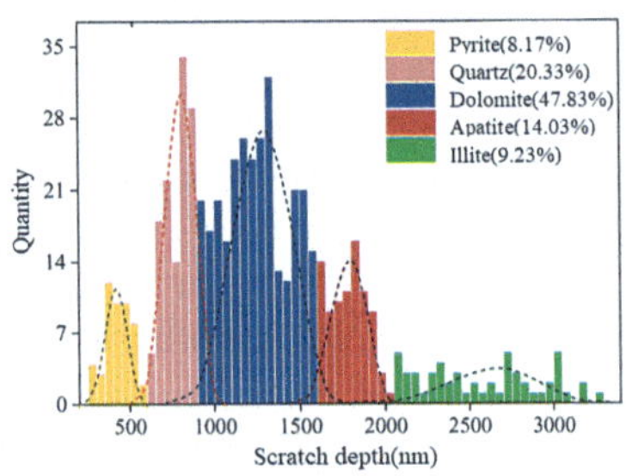

(e) #5Regions are divided into depth distributions (f) #6Regions are divided into depth distributions

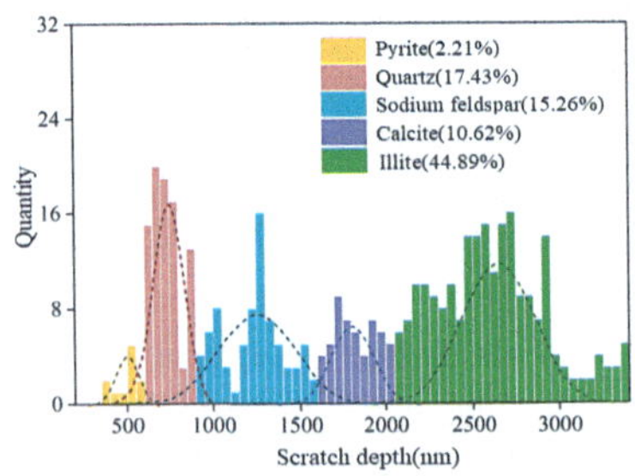

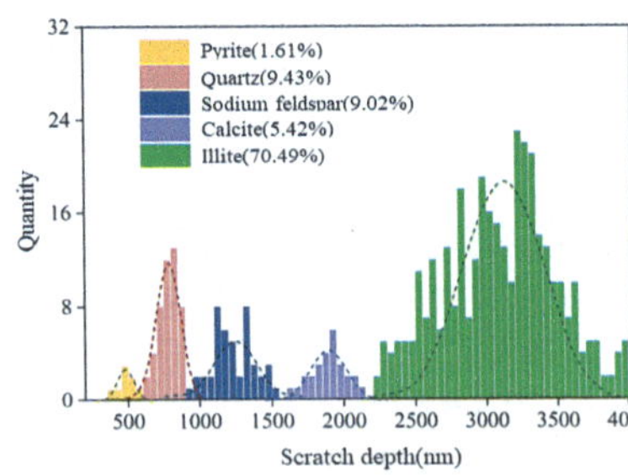

(g) #7Regions are divided into depth distributions (h) #8Regions are divided into depth distributions

Fig. 3.12 Mineral component identification results based on inclusion depth. (**a**) #1 Regions are divided into depth distributions. (**b**) #2 Regions are divided into depth distributions. (**c**) #3 Regions are divided into depth distributions. (**d**) #4 Regions are divided into depth distributions. (**e**) #5 mineral content identified by the depth of the scratch was also greater than the QEMSCAN scan. Shale is rich in micropores and fractures, and it is impossible to accurately distinguish whether large scratch depths are caused by clay minerals or microcracks and pores. In engineering applications, it is simple and convenient to judge mineral composition by scratch depth, but it is susceptible to microcracks when identifying softer minerals, such as clay, which causes some errors.

3.3.2 Slope Method

According to the tribological definition, the friction coefficient COF can be expressed as the ratio of tangential force F_T to vertical force F_n:

$$\mathrm{COF} = F_T / F_n \tag{3.14}$$

Through dimensional analysis, we can get:

$$[F_T] = MLT^{-2} \tag{3.15}$$

$$[H] = ML^{-1}T^{-2} \tag{3.16}$$

$$[dF_T / d_L] = MT^{-2} \tag{3.17}$$

$$[dF_T / dL / dH] = \Pi L = [dd^2 / dL] \tag{3.18}$$

During the nanoscratch experiment, F_n is a constant vertical loading force, and the change of friction coefficient is only related to the tangential force F_T obtained from the tip of the indenter. In the scratch test, MT^{-2} is a constant under the same indenter test conditions. Through dimensional analysis, the slope of d^2-L curve is a parameter related to friction coefficient, hardness, and scratch depth, which is closely related to the mechanical behavior of minerals. Using the slope of d^2-L curve to quantitatively identify mineral distribution is a theoretically feasible method.

Due to the large number of microcracks and pores in the shale, when the indenter passes through these microcracks, the scratch depth will also increase significantly. In order to reduce the mineral identification error caused by microcracks, the fluctuation amplitude of the friction coefficient curve can further determine the reason for the significant increase in scratch depth, thus eliminating abnormal data. Figure 3.13 corresponds to changes in *COF* and d^2 in the eight scratched areas. In the #1 and #2 scratch areas, the *COF* curve fluctuates slightly and remains within a

Fig. 3.12 (continued) Regions are divided into depth distributions. (**f**) #6 Regions are divided into depth distributions. (**g**) #7 Regions are divided into depth distributions. (**h**) #8 Regions are divided into depth distributions

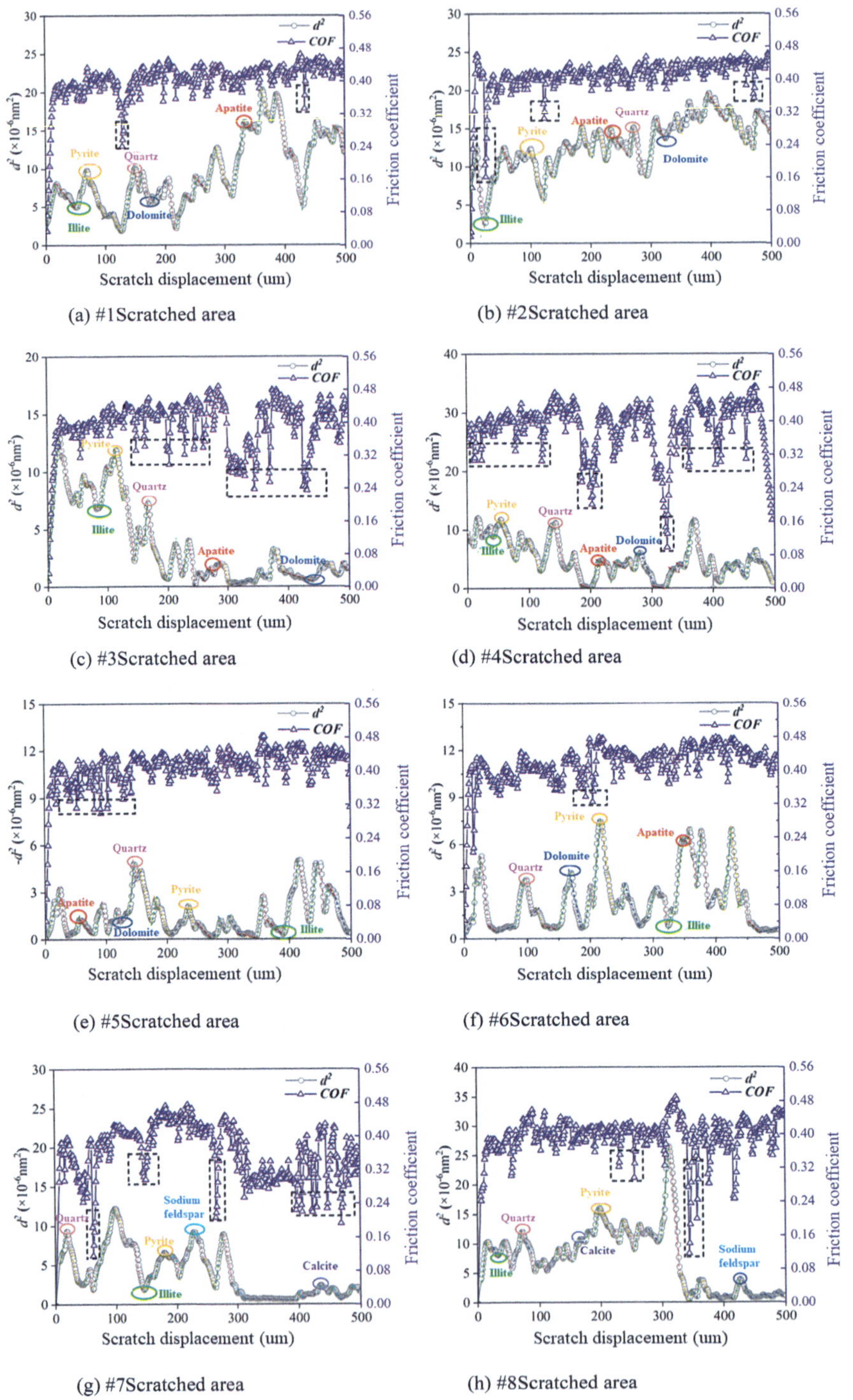

(a) #1Scratched area
(b) #2Scratched area
(c) #3Scratched area
(d) #4Scratched area
(e) #5Scratched area
(f) #6Scratched area
(g) #7Scratched area
(h) #8Scratched area

Fig. 3.13 Curves of d^2-L and COF-L. (**a**) #1 Scratched area. (**b**) #2 Scratched area. (**c**) #3 Scratched area. (**d**) #4 Scratched area. (**e**) #5 Scratched area. (**f**) #6 Scratched area. (**g**) #7 Scratched area. (**h**) #8 Scratched area stable overall distribution. This is due to the high plasticity of the illite rich area and the toughness of the overall mechanical properties. Due to the mixed distribution of plastic minerals and brittle minerals, the d^2-L curve shows a large slope transition and the scratch depth decreases. Due to the damage mode of the brittle mineral, tangential force tends to rise when the indenter passes through the brittle mineral. The friction coefficient curves in regions #3 and #4 show small fluctuations in the first half, but after passing through the transition region, the values show sharp fluctuations and a rapid increase in dispersion. The #7 and #8 scratch areas also have the transition stage of the rich areas of plastic minerals and brittle minerals, and the dispersion degree of the two parts of the curve is also differentiated. After passing through the interfacial transition zone, the indenter enters the dolomite nodule zone, and the discrete distribution of various brittle minerals leads to strong instability of the friction coefficient curve. After the indenter passes through the lithologic transition zone, the lithology in the brittle mineral nodule zone is basically stable. The *COF* curve fluctuation in the scratch area #5 and #6 decreased, and the fluctuation degree of the data was slightly greater than that of the illite mineral distribution.

For the scratch area #1 to #6, the minerals can be divided into illite, apatite, dolomite, quartz, and pyrite. Samples #7 and #8 contain little dolomite, but albite and calcite appear in the distribution area. After a statistical analysis of the d^2-L slope for each region, Figs. 3.1 and 3.2 shows the box plot distribution of mineral component percentages for the three scratched regions. Table 3.1 shows the distribution range of d^2-L slopes corresponding to different minerals. It is worth noting that the slope of dolomite minerals in the illite distribution area is always positive due to the high illite content. However, due to the increased presence of brittle minerals in the transition zone and dolomite nodule zone, the slope of dolomite or calcite minerals is negative when the indenter comes into contact with harder minerals such as quartz (Table 3.2). In addition, because the mineral distribution on the surface of continental shale is more discrete, and abundant microfractures and pores lead to continuous jumps in d^2-L curve, there is a big difference between the mineral distribution measured by this method and the scanning results of QEMSCAN (Fig. 3.14).

Table 3.1 The slope distribution of d^2-L for various minerals in the scratch range of #1 to #5

	#1	#2	#3	#4	#5
Quartz	−0.71 to −0.43	−0.76 to −0.34	−0.62 to −0.42	−0.59 to −0.41	−0.36 to −0.30
Pyrite	−0.27 to −0.18	−0.27 to −0.17	−0.39 to −0.25	−0.39 to −0.30	−0.29 to −0.20
Dolomite	0.17–0.30	0.16–0.27	−0.12 to 0.19	−0.16 to 0.08	−0.09 to 0.08
Apatite	−0.15 to 0.01	−0.15 to 0.01	−0.28 to 0.2	−0.29 to 0.21	−0.16 to 0.10
Illite	0.32–0.78	0.37–0.71	0.23–0.57	0.20–0.55	0.10–0.35

Table 3.2 The slope distribution of d^2-L for various minerals in the scratch range of #6 to #8

	#6		#7		#8
Quartz	0.37–0.30	Quartz	−0.49 to −0.38	Quartz	−0.89 to −0.35
Pyrite	0.29–0.21	Pyrite	−0.36 to −0.22	Pyrite	−0.33 to −0.25
Dolomite	0.03–0.08	Albite	−0.16 to 0.08	Dolomite	−0.22 to −0.14
Apatite	0.19–0.1	Calcite	−0.05 to 0.11	Calcite	−0.12–0.27
Illite	0.12–0.39	Illite	0.21–0.37	Illite	0.28–0.59

3.3.3 Fracture Toughness Method

Due to the heterogeneity of continental shales, sufficient data are needed for statistical analysis to obtain fracture toughness of various mineral phases. In this study, a total of 4000 scratch data were obtained and the corresponding fracture toughness distribution was statistically analyzed, as shown in Fig. 3.15. According to previous literature, the fracture toughness distribution of each phase is consistent with the Gaussian distribution. This is because the indenter passes through the mineral-containing region at a uniform rate, so the fracture toughness collected from the minerals distributed in a particular region will first reach the edge of the mineral-rich region. It will then reach the center of the mineral distribution, and finally, the indenter will move through the mineral to another area of the mineral distribution. Continental shale has strong heterogeneity and there are many chaotic peaks in the frequency distribution of different minerals. By fitting the trend line to the Gaussian peak curve of the data, the different mineral components can be identified based on the Gaussian peak line of the fracture toughness of each phase.

The criterion for the applicability of scratch technology in phase recognition is to ensure that the ratio of scratch depth to phase feature size should be less than 0.1 in order to avoid the impact of phase overlap on measurement. At a normal load of 50 mN, the maximum scratch depth obtained in this study is about 5000 nm. The scratch method proposed in this study is effective for mineral identification because it avoids the effect of overlapping of different relative measurements. The slope of the d^2-L curve can be used to identify different phases. In complex mineral compositions, the positive or negative slope depends on the mechanical properties of previous minerals and microcracks, and pores can also lead to errors in identification results. Unlike the scratch depth and slope methods, the fracture toughness of each phase and its fracture toughness frequency distribution are calculated independently of the complex mineral composition and micropores. The mineral composition identified by this method is closer to the result of QEMSCAN, and it is a reliable method for mineral composition identification.

The fracture toughness distribution of a single mineral phase is shown in Fig. 3.16, in which the fracture toughness of each phase in the illite distribution zone is weak. The indenter enters the transition area and dolomite nodule area and gradually enters the distribution of harder lithology, and the fracture toughness is generally improved. Tables 3.3 and 3.4 show the distribution range of fracture toughness of each phase in different mineral composition regions, and the fracture

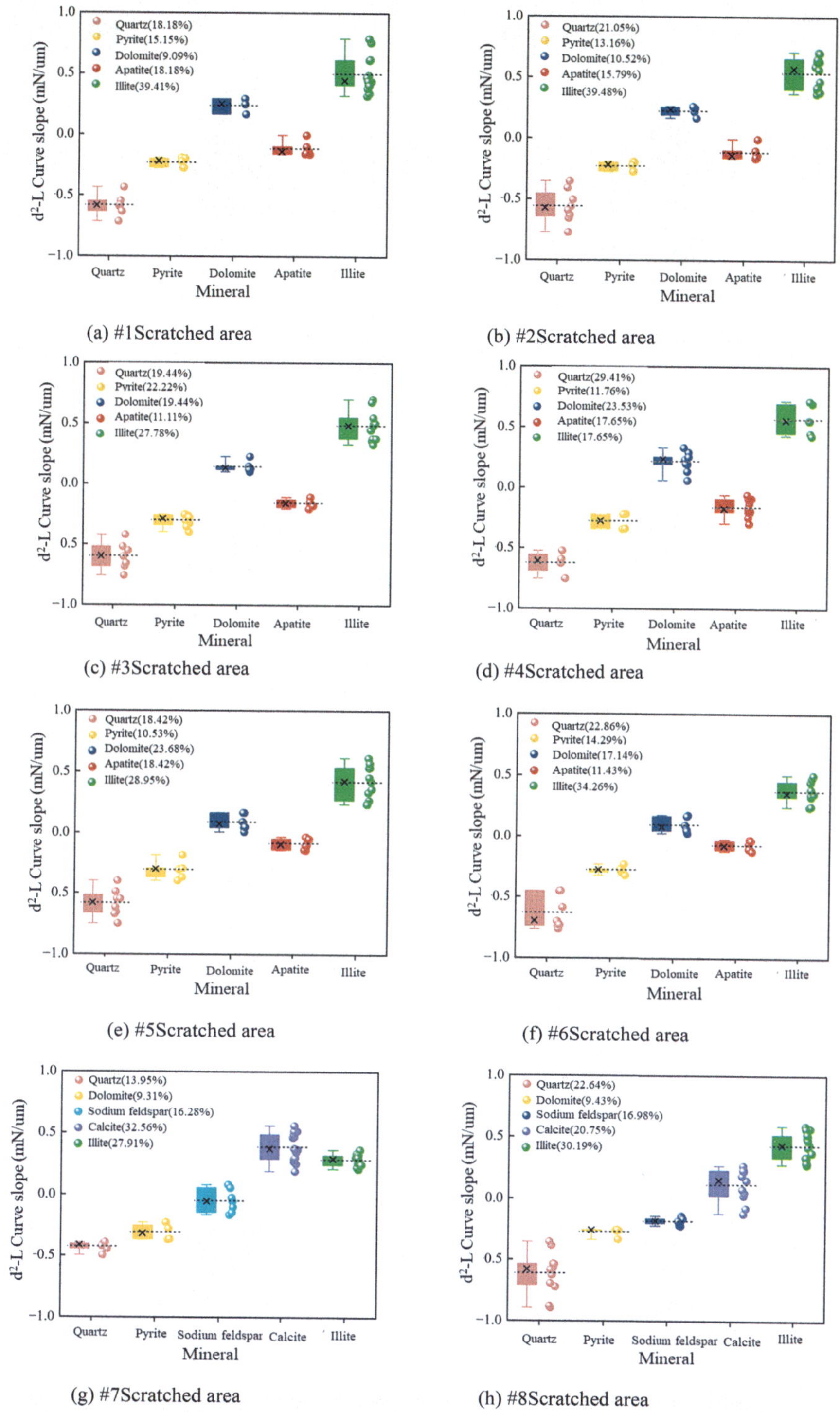

Fig. 3.14 Box plot of the slope of the d^2-L curve of minerals in different areas of the scratch. (**a**) #1 Scratched area. (**b**) #2 Scratched area. (**c**) #3 Scratched area. (**d**) #4 Scratched area. (**e**) #5 Scratched area. (**f**) #6 Scratched area. (**g**) #7 Scratched area. (**h**) #8 Scratched area

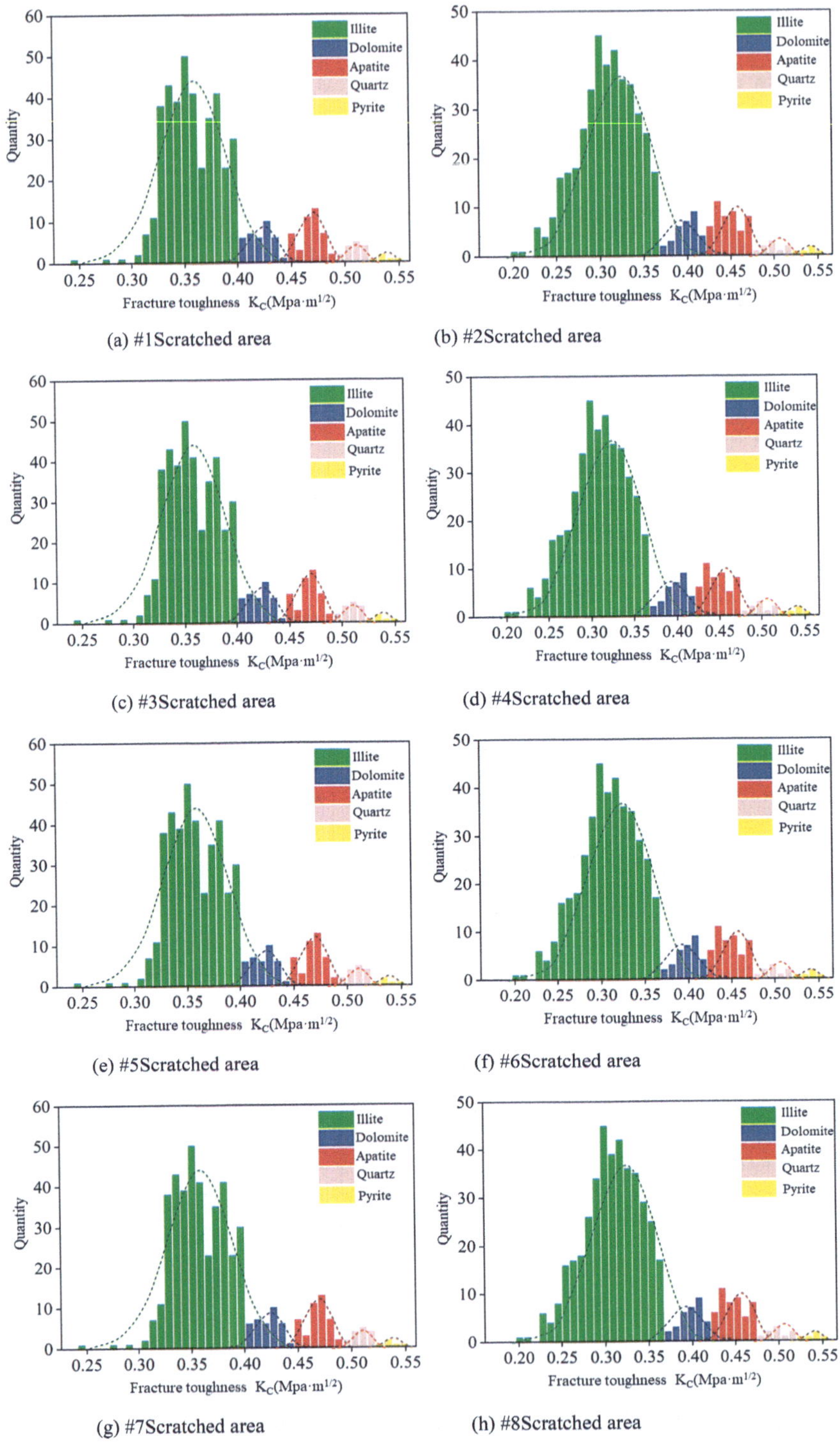

Fig. 3.15 #1 to #8 Distribution of fracture toughness in the scratch area. (**a**) #1 Scratched area. (**b**) #2 Scratched area. (**c**) #3 Scratched area. (**d**) #4 Scratched area. (**e**) #5 Scratched area. (**f**) #6 Scratched area. (**g**) #7 Scratched area. (**h**) #8 Scratched area

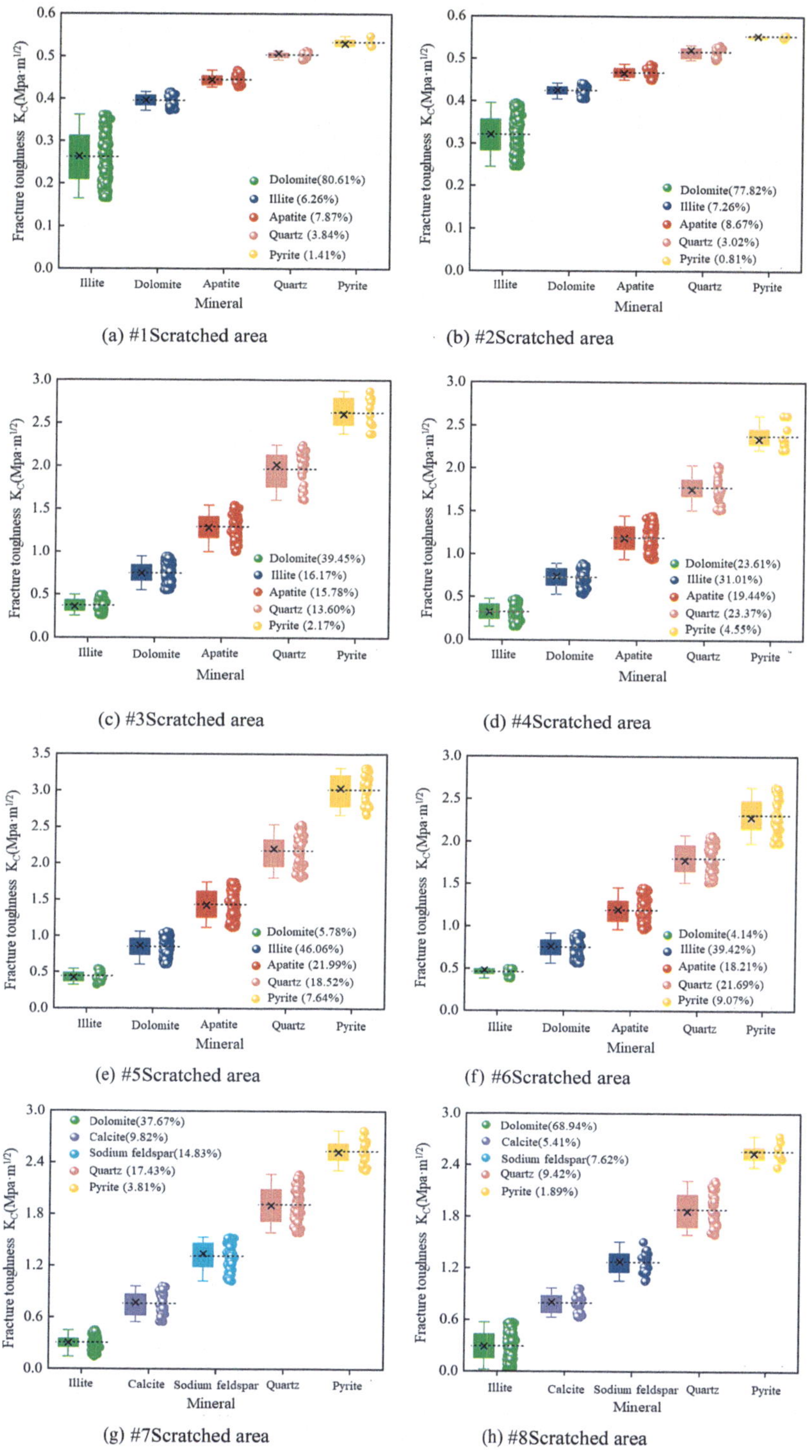

Fig. 3.16 Box plot of fracture toughness in the scratch area. (**a**) #1 Scratched area. (**b**) #2 Scratched area. (**c**) #3 Scratched area. (**d**) #4 Scratched area. (**e**) #5 Scratched area. (**f**) #6 Scratched area. (**g**) #7 Scratched area. (**h**) #8 Scratched area

Table 3.3 Distribution range of fracture toughness in scratch test #1 to #5 zone (MPa · $m^{0.5}$)

Mineral class	#1	#2	#3	#4	#5
Pyrite	0.52–0.56	0.53–0.56	2.37–2.88	2.22–2.63	2.67–3.31
Quartz	0.47–0.51	0.49–0.52	1.61–2.25	1.52–2.05	1.80–2.54
Dolomite	0.36–0.42	0.40–0.45	1.02–1.55	0.53–0.90	0.61–1.07
Apatite	0.42–0.47	0.45–0.49	0.55–0.96	0.94–1.49	1.12–1.74
Illite	0.15–0.36	0.23–0.40	0.25–0.50	0.15–0.48	0.32–0.56

Table 3.4 Distribution range of fracture toughness in scratch test #6 to #8 zone (MPa·m^{05})

Mineral class	#6	Mineral class	#7	Mineral class	#8
Pyrite	1.98–2.66	Pyrite	2.31–2.78	Pyrite	2.38–2.75
Quartz	1.51–2.08	Quartz	1.59–2.26	Quartz	1.59–2.23
Dolomite	0.96–1.46	Albite	1.02–1.53	Dolomite	1.05–1.51
Apatite	0.56–0.92	Calcite	0.55–0.96	Calcite	0.63–0.97
Illite	0.38–0.51	Illite	0.15–0.45	Illite	0.03–0.57

toughness of minerals in illite-rich regions is low. Although in QEMSCAN results, high-hardness materials such as quartz and pyrite are scattered in the illite distribution area, they do not show strong fracture toughness. Existing literature has shown that the metamorphism of plastic minerals during shale formation changes the lithology of these hard minerals, resulting in minerals such as quartz and apatite exhibiting weak mechanical properties similar to illite.

3.3.4 Comparison and Verification

Table 3.5 shows the summary of recognition results of illite minerals by different methods. The recognition results of illite minerals based on subdivision depth are obviously greater than those of the other three methods. This is because although the scratch depth method is more convenient and saves the time of data processing, it is difficult to distinguish cracks from soft minerals. Table 3.6 shows the summary of recognition results of quartz minerals by different methods. The recognition results of illite minerals based on subdivision depth are obviously greater than those of the other three methods, and the recognition results of quartz minerals by slope method of d^2-L curve are significantly greater than those of the other three methods, especially in samples #1 and #2. This is because in the illite distribution area, when the indenter passes over minerals with greater hardness than illite, the slope of the d^2-L curve will rise significantly. Moreover, the hardness of quartz and other hard minerals in the illite enrichment area will change due to the influence of the lithology of plastic minerals, resulting in the failure to accurately identify quartz minerals and other hard minerals. Although the slope of d^2-L curve can identify the mineral composition, its accuracy is greatly affected by the heterogeneity of shale.

Table 3.5 Summary of identifying illite minerals (%)

Sample number	Subdivision depth method	Curve slope method	Fracture toughness method	QEMSCAN
#1	81.54	39.41	80.61	75.14
#2	86.84	39.48	77.82	79.13
#3	53.13	27.78	39.45	37.13
#4	35.24	17.65	23.61	14.43
#5	8.85	28.95	5.78	2.65
#6	9.23	34.26	4.14	3.92
#7	44.89	27.91	37.67	36.68
#8	70.49	30.19	68.94	67.76

Table 3.6 Summary of identifying quartz minerals (%)

Sample number	Subdivision depth method	Curve slope method	Fracture toughness method	QEMSCAN
#1	1.42	18.18	3.84	7.81
#2	1.77	21.05	3.02	7.34
#3	11.99	19.44	13.60	17.49
#4	15.24	29.41	23.37	18.72
#5	21.04	18.42	18.52	7.78
#6	20.33	22.86	21.69	8.13
#7	17.43	13.95	17.43	17.14
#8	9.43	22.64	9.42	9.74

Table 3.7 Identification results of carbonate minerals (%)

Sample number	Subdivision depth method	Curve slope method	Fracture toughness method	QEMSCAN
#1	1.83	9.09	6.26	3.05
#2	5.31	10.52	7.26	2.13
#3	25.89	19.44	16.17	17.76
#4	26.91	23.53	31.01	29.94
#5	46.08	23.68	46.06	48.77
#6	47.83	17.14	39.42	40.97
#7	10.62	32.56	9.82	9.15
#8	9.02	20.75	7.62	6.02

Table 3.7 shows the summary of identification results of carbonate minerals by different methods. The slope method of d^2-L curve still identifies more carbonate minerals than the other three methods. This is because the slope of carbonate minerals in illite enrichment area is positive, while the slope of carbonate minerals in brittle mineral nodules area is negative. The composition content of carbonate minerals cannot be accurately identified. Figure 3.17 shows the summary of

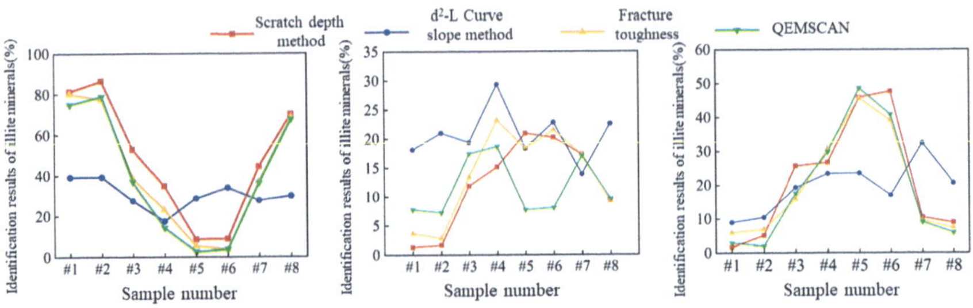

Fig. 3.17 Summary of four methods for mineral identification of Gulong Shale

identification results of laminar shale minerals by four methods. The curves of mineral content identification by fracture toughness method and QEMSCAN method are close to each other and accurately identify minerals by their ability to resist fracture. In the actual application process, scanning electron microscopy and nanoscratch methods should be combined to identify minerals while collecting mechanical parameters, which is conducive to more rapid and efficient evaluation of shale reservoir minerals and mechanical properties.

3.4 Thickness of Transition Zone

3.4.1 Mineral Distribution Length

The distribution of minerals in the transition zone was observed by SEM and QEMSCAN. It was found that the peak position of fracture toughness curve corresponded with the mineral composition in QEMSCAN image to some extent. The BITZ distribution lengths of #3 and #4 shale samples are identified by electron scanning microscopy and fracture toughness parameters, as shown in Fig. 3.18. The lithology distribution around the interfacial transition zone was determined by QEMSCAN, and pixel identification was performed by Image-pro Plus software according to SEM results. The distribution length of illite and dolomite zones was quantified. In sample #3, the distribution lengths of illite and dolomite minerals are 194.25 μm and 305.75 μm, respectively. In the BITZ distribution length based on fracture toughness parameters, the illite region is 144.31 μm, and the dolomite nodal region is 333.73 μm. In sample #4, the distribution lengths of illite and dolomite nodules measured by electron microscopy are 150.72 μm and 349.28 μm, respectively. Based on fracture toughness parameters, the BITZ distribution length of shale is 120.21 μm for illite and 325.76 μm for dolomite nodules.

From the above analysis, it can be clearly seen that the mineral distribution length of illite and dolomite quantified by SEM and QEMSCAN is significantly different from the results quantified by fracture toughness. This is due to the inability of SEM and QEMSCAN methods to accurately identify the BITZ region where fracture

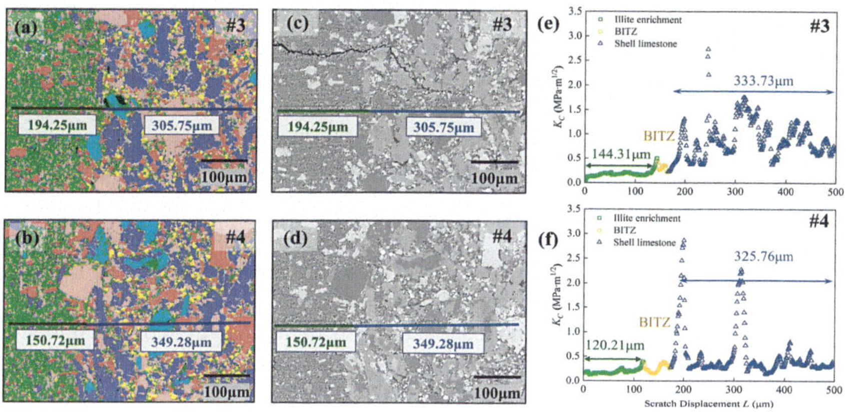

Fig. 3.18 Identification results of mineral distribution length

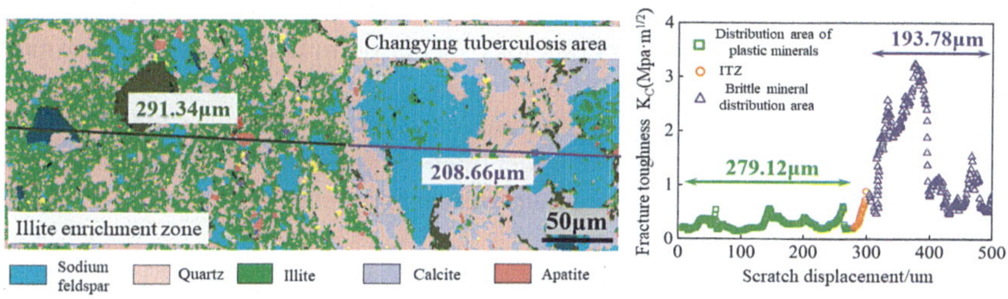

Fig. 3.19 Quantitative results of mineral distribution length in Changying Layered Shale

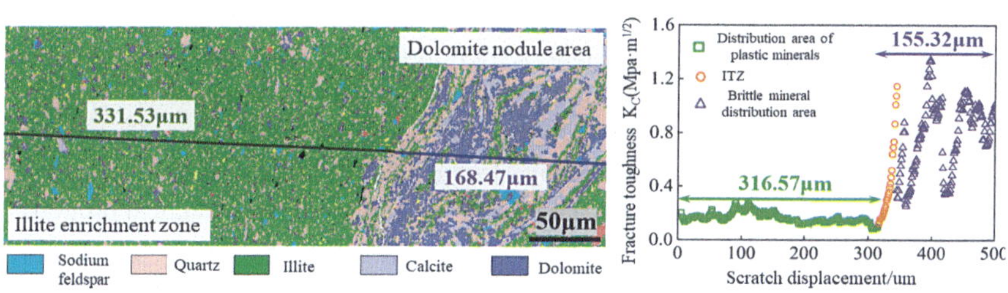

Fig. 3.20 Quantitative results of mineral distribution length in stratification of ostracods shale

toughness has changed, even though the mineral morphology in this region has not changed. It is a more accurate result to classify the distribution of different minerals based on the mechanical parameter data collected at the tip of the indenter.

Shale samples #7 and #8 are felsic and ostraco-laminated shale, respectively. As shown in Figs. 3.19 and 3.20, QEMSCAN results show that their mineral composition is still mainly composed of quartz minerals, carbonate minerals, phosphate minerals, and clay minerals. In addition, they also contain a small amount of white mica, pyrite, and amorphous minerals that are difficult to distinguish. Based on the pixel method, the mineral distribution lengths in the QEMSCAN results were quantitatively analyzed. The distribution lengths of plastic and brittle minerals in the felsic laminated shale are 291.34 μm and 208.66 μm, respectively, and the distribution lengths of plastic and brittle minerals in the amoebium laminated shale are 331.53 μm and 168.47 μm, respectively. At the same time, the mineral distribution length is quantified according to the fluctuation interval of fracture toughness curve. The plastic and brittle mineral distribution lengths of felsic laminate shale are 279.12 μm and 193.78 μm, respectively, and the plastic and brittle mineral distribution lengths of amoeba laminate are 316.57 μm and 155.32 μm, respectively.

It can be seen from Figs. 4.2 and 4.3 that the distribution length of plastic minerals measured by scanning electron microscopy is greater than the quantitative result of fracture toughness. This is because QEMSCAN collects information about the composition of a single element on the surface of a sample and then compares this information to an existing mineral database to determine the type of mineral at a single point. However, scanning electron microscopy cannot accurately identify the interface transition region in the transition region from plastic minerals to brittle minerals. Although the surface morphology of plastic minerals near the transition region has not changed significantly, its mechanical properties have been affected by brittle minerals. Obviously, the distribution length of minerals near the interfacial transition region cannot be accurately identified by scanning electron microscopy, and the quantitative basis of fracture toughness is more reliable.

3.4.2 Traditional Method

As shown in Fig. 3.21, for concrete materials, discontinuous data of ITZ boundary can be determined based on coefficient of friction (COF) curve and linear fitting can be carried out. ITZ thickness can be determined by the *X* axis projection length of intersection point of interval linear fitting line and mean friction coefficient line. Continental shales are characterized by strong heterogeneity and usually include many kinds of minerals in the interfacial transition zone. Their mechanical properties affect each other and the friction coefficient curves fluctuate greatly. A new method for quantifying BITZ distribution range of shale is proposed by extracting the characteristic points of friction coefficient curve.

The *COF* data of the lithologic transition region of the shale is shown in Fig. 3.22, and its fluctuation shows a steady-oscillation-steady trend. Statistical analysis is carried out on the positive and negative intervals of the slope between two points in the oscillating data. If the slope of adjacent data points in a certain region is the same, linear fitting is performed on the data segment and its slope is named k_i. *COF* curve shows a monotonically decreasing trend before entering the BITZ region. Origin 2023 Pro software is used to perform linear regression on the data, and the extreme point ($k_1 = 0$) at the end of the fitting interval is used as the starting point L_1 of the data in the BITZ interval. When the scratch displacement $L > L_1$, it is

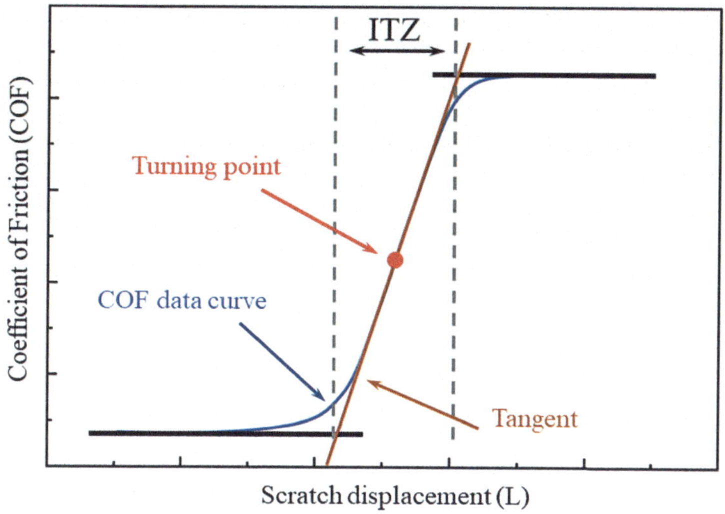

Fig. 3.21 Quantitative method for ITZ of concrete materials

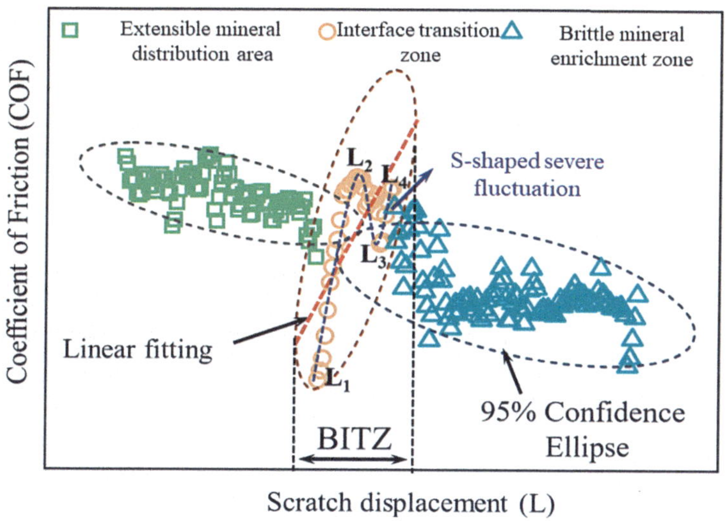

Fig. 3.22 Quantitative method for distribution length of BITZ in continental shale

considered that the scratch indenter has entered the interface transition zone. Within a short displacement after passing through the L_1 point, the *COF* curve rises rapidly and then falls rapidly again. Within this range, the extreme value point ($k_2 = 0$) corresponds to the displacement of L_2. In the displacement range from L_1 to L_2, it is considered that the indenter is classified into BITZ in the initial stage. When the scratch displacement $L > L_2$, *COF* curve shows a trend of decreasing first and then increasing, and the extreme point ($k_3 = 0$) in this range corresponds to L_3 displacement. After the indenter is displaced by L_3, the *COF* curve shows a trend of increase and then decrease and stops violent fluctuations. The displacement corresponding to the extreme value point ($k_4 = 0$) in this range is L_4. At this time, the scratch test indenter has entered the brittle mineral distribution area through *COF*. According to the fluctuation amplitude of *COF* curve and the slope change trend fitted by linear regression, *L-COF* curve can be divided into plastic mineral enrichment area, initial transition stage of BITZ, middle transition stage, late transition stage, and brittle mineral distribution area through BITZ. The corresponding vibration range of the friction coefficient curve ($0 \sim L_1$), ($L_1 \sim L_2$), ($L_2 \sim L_3$), ($L_3 \sim L_4$) and ($L_4 \sim \infty$).

3.4.3 Mechanical Method for Thickness Identification

As shown in Fig. 3.23, continental shales in Songliao Basin present multi-layer structure on the macro-scale. In the mesoscale fracture propagation, the fracture passes through the shale laminar interface BITZ, and the fracture propagation length is related to the mechanical parameters of BITZ. On the microscopic scale, the morphology of crack propagation at the interface changes significantly after crack propagation to the BITZ region. In this chapter, four groups of shale samples with obvious laminar interface transition zones are used, namely samples #3, #4, #7, and #8 in Sect. 3.6. Based on the results of nanoscratch experiments, the applicability and accuracy of various mechanical parameters used to quantify the distribution length of the transition zone at the rock laminar interface are discussed.

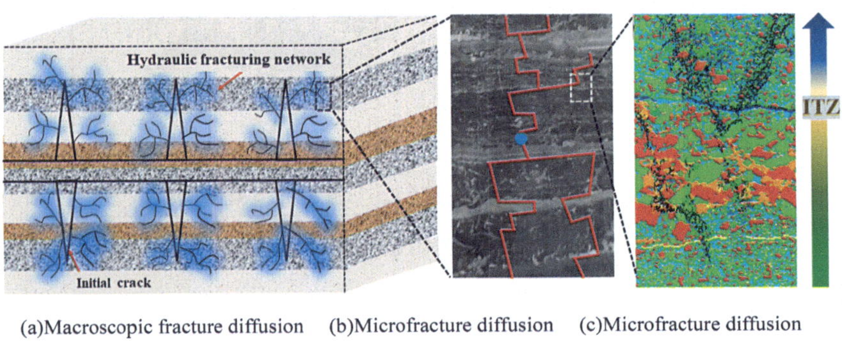

Fig. 3.23 Crack propagation of shale in the Songliao Basin during hydraulic fracturing. (**a**) Macroscopic fracture diffusion. (**b**) Microfracture diffusion. (**c**) Microfracture diffusion

BITZ Thickness Identification of Shale Based on Friction Coefficient Method

After the friction coefficient curves in the BITZ interval are initially divided, the friction coefficient data in the L_1–L_4 interval are linear regression to determine the BITZ thickness of laminae. In order to reduce the influence of the discreteness of the data in the interval on the BITZ thickness assessment, the discrete data are overwritten with a 95% confidence ellipse. The projected length of the intersection of the fitting line and the confidence ellipse in the horizontal direction are used as the final BITZ thickness of the shale layer. COF curve of shale samples is shown in Fig. 3.24, which fluctuates sharply. Due to the influence of surface roughness and porosity on the friction coefficient, there is no obvious distinction between the friction coefficient in the distribution area of plastic minerals and brittle minerals. The improved method is used to quantify the thickness of the interfacial transition zone of shale. The distribution lengths of the interfacial transition zone of shale #3 and #4 samples are 37.43 μm and 79.02 μm, respectively, and the linear fitting confidence levels are $R^2 = 0.20249$ and 0.17939, respectively. The BITZ thickness of #7 shale sample is 49.81 μm, and the correlation coefficient $R^2 = 0.45754$; the BITZ thickness of #8 shale sample is 47.06 μm, and the correlation coefficient $R^2 = 0.15038$. It can be seen from the fitting results that the fitting confidence of quantifying BITZ thickness of shale based on friction coefficient is low, and the results are likely to be inaccurate.

BITZ Thickness Identification of Shale Based on Fracture Toughness Method

The BITZ thickness of shale can be quantified by the friction coefficient, but the friction coefficient collected by the scratch test largely depends on the surface roughness of the sample. Due to the micropore development and strong heterogeneity of continental shale, although the sample has been polished to meet the experimental requirements, its *COF* data is relatively discrete, and the linear regression correlation coefficient of the data in the BITZ interval is low. It has been shown that properties such as elastic modulus or hardness can be used to quantify the thickness of the interfacial transition zone. As an important mechanical property, fracture toughness has not been used to quantify the BITZ thickness of shale striates. The fracture toughness of plastic and brittle minerals near shale striates is very different, which is the basis for identifying BITZ thickness. The fracture toughness curve of samples in the interfacial transition area of laminated shale is shown in Fig. 3.25. For the scratch area #3 and #4, the fracture toughness data curve within BITZ still shows an obvious "S"-shaped fluctuation trend. The quantification lengths of BITZ were 34.67 μm and 80.93 μm, respectively, and the linear regression confidence levels R^2 were 0.23472 and 0.19586, respectively. Compared with the friction coefficient curve, the continuity of fracture toughness data is improved in the "*S*" interval, and R^2 is slightly increased.

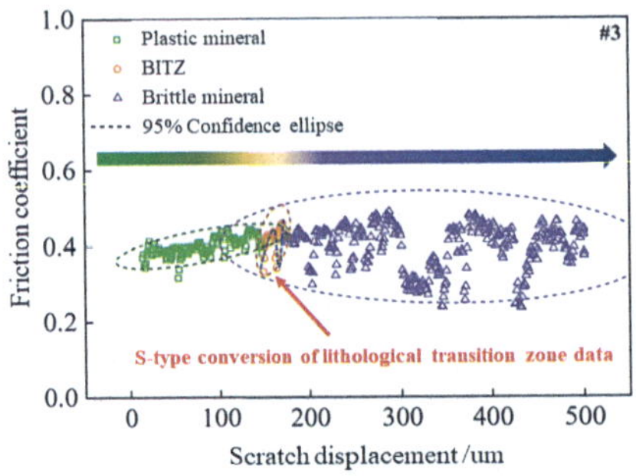

(a) #3Sample friction coefficient curve

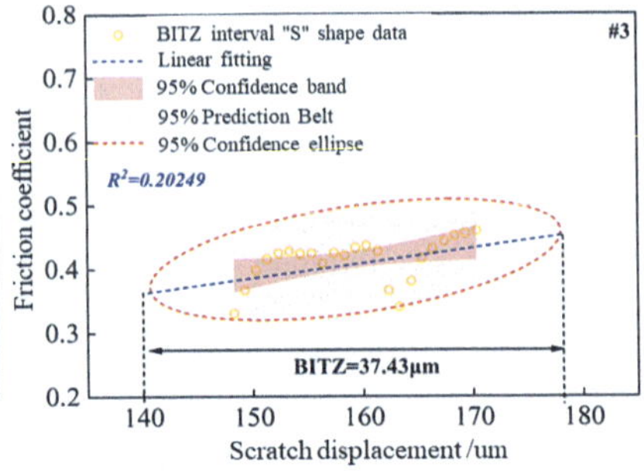

(b) #3BITZ quantitative results of samples

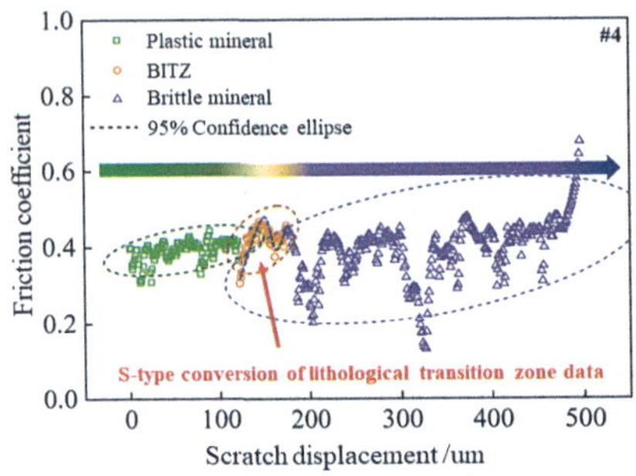

(c) #4Sample friction coefficient curve

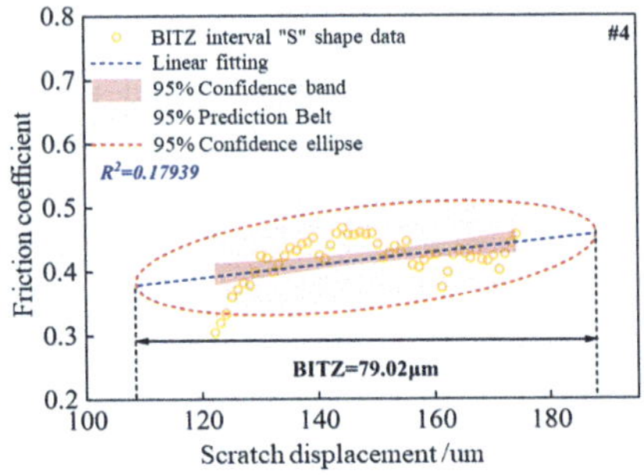

(d) #4BITZ quantitative results of samples

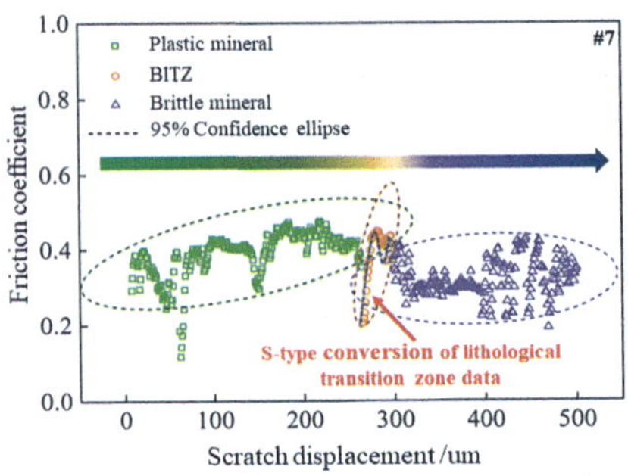

(e) #7Sample friction coefficient curve

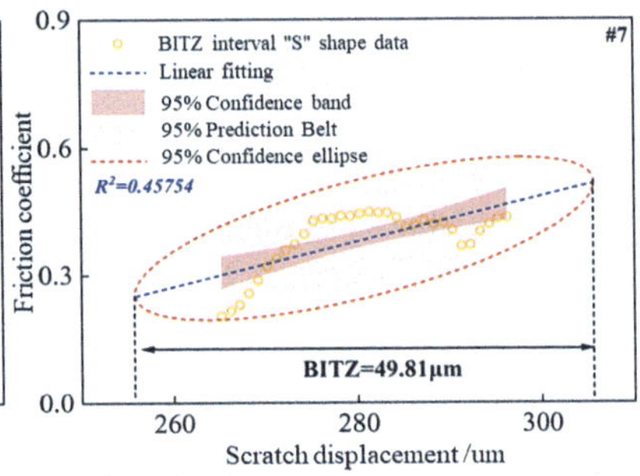

(f) #7BITZ quantitative results of samples

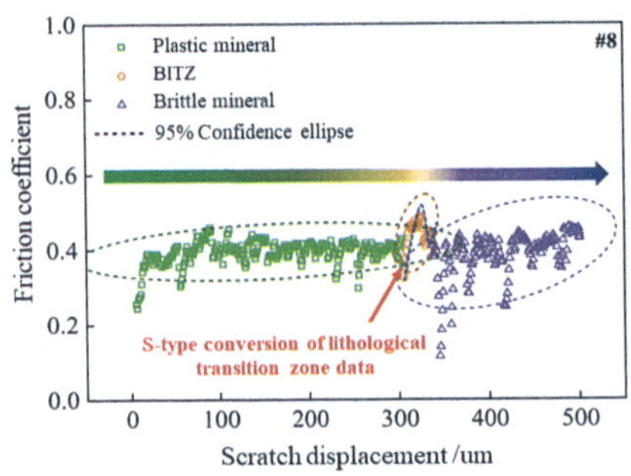

(g) #8Sample friction coefficient curve

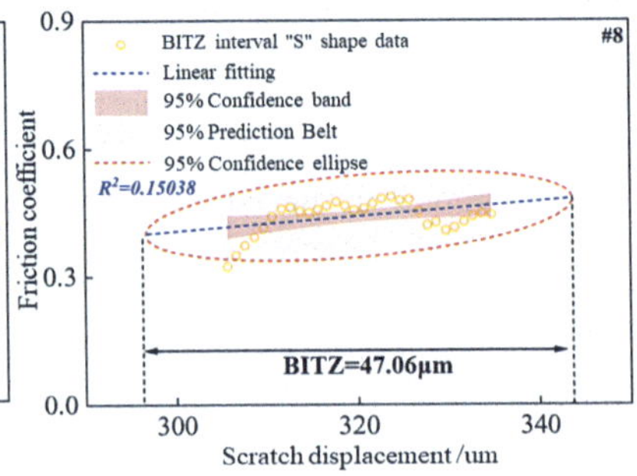

(h) #8BITZ quantitative results of samples

Fig. 3.24 Identification results of BITZ distribution length using friction coefficient method. (**a**) #3 Sample friction coefficient curve. (**b**) #3 BITZ quantitative results of samples. (**c**) #4 Sample friction

The BITZ thickness of #7 and #8 shale samples is quantified by the fracture toughness curve. It is found that the fracture toughness keeps a constant low value in the distribution range of illite minerals. With the change in the fracture resistance of the mineral under the indenter, the fracture toughness suddenly and sharply increases within a short displacement, and the data in the BITZ interval show an obvious upward trend of "J" shape. After the indenter passes through the interfacial transition zone, the fracture toughness has a short rapid decline interval, which indicates that the indenter enters the brittle mineral distribution region after passing through the interfacial transition zone. The BITZ thickness of felsic striated shale is 45.78 μm, and the correlation coefficient $R^2 = 0.79725$. The BITZ thickness of Ostraco-striated shale is 44.31 μm, and the correlation coefficient $R^2 = 0.81981$. The fracture toughness method evaluates the length of the BITZ distribution based on the ability of the mineral to resist fracture, avoiding the error caused by sample roughness. By comparing the curves of fracture toughness and friction coefficient, it is found that the data stability of BITZ section of fracture toughness curve is improved. Since only the mechanical properties of fracture toughness were considered in the evaluation process of this method, the data curve of this method still has a high degree of dispersion and low fitting confidence. As a highly heterogeneous material, shale requires a comprehensive method combining friction coefficient and deformation resistance to quantify BITZ distribution length.

3.4.4 Dimensionless Method

Dimensional Analysis and Dimensionless Parameter Combination Upgrade

During the scratch experiment, the work done by the indenter is absorbed by the mineral volume deformation and the corresponding surface damage. In the process of plastic mineral scratch experiment, most of the energy is consumed in the volume deformation, and the scratch depth increases significantly, thus forming a deep groove. The depth of scratch is a more important parameter when determining ITZ thickness. During the scratch test, the load is constant, and the hardness and fracture resistance of the mineral in the lithologic transition zone change with the displacement of the indenter. When the indenter passes through the laminar interface transition zone, the lithologic change will lead to changes in the rate and length of crack growth under the indenter. Therefore, when quantifying ITZ thickness of shale laminae, parameters such as friction coefficient, fracture toughness, hardness, and

Fig. 3.24 (continued) coefficient curve. (**d**) #4 BITZ quantitative results of samples. (**e**) #7 Sample friction coefficient curve. (**f**) #7 BITZ quantitative results of samples. (**g**) #8 Sample friction coefficient curve. (**h**) #8 BITZ quantitative results of samples

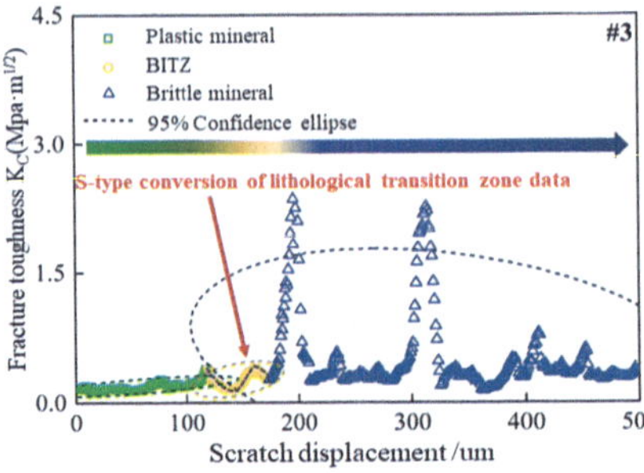

(a) #3Sample fracture toughness curve

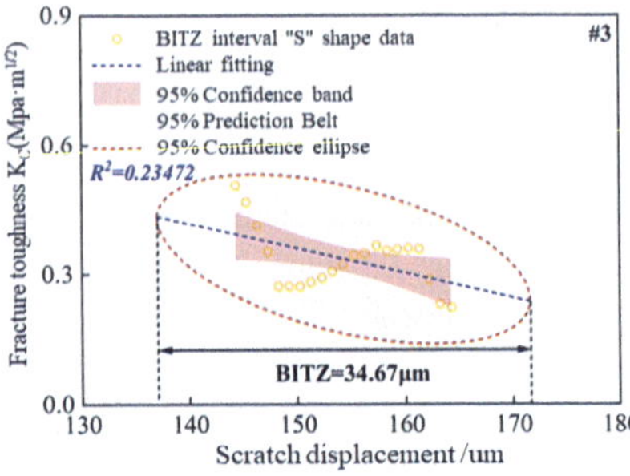

(b) #3BITZ quantitative results of samples

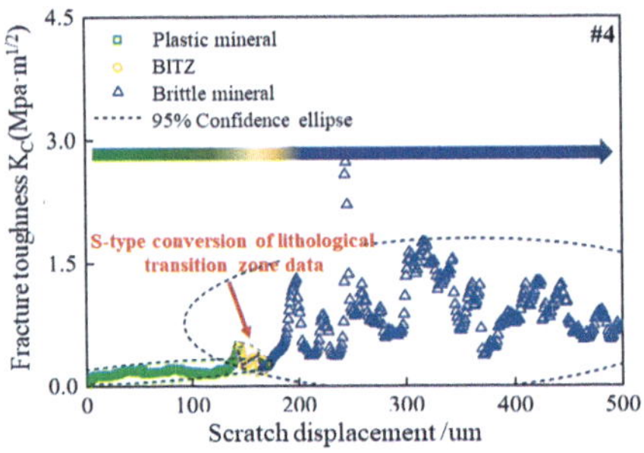

(c) #4Sample fracture toughness curve

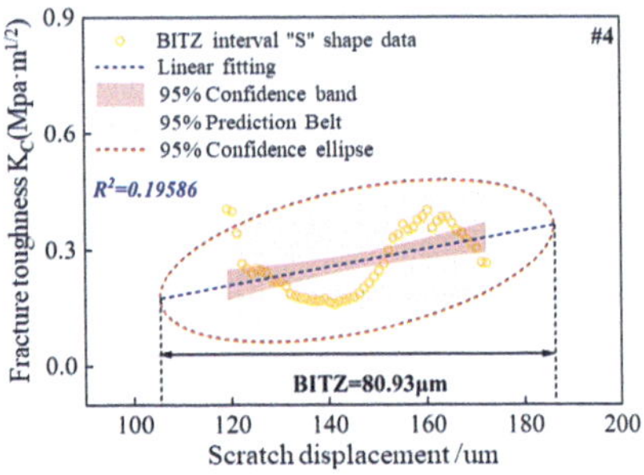

(b) #4BITZ quantitative results of samples

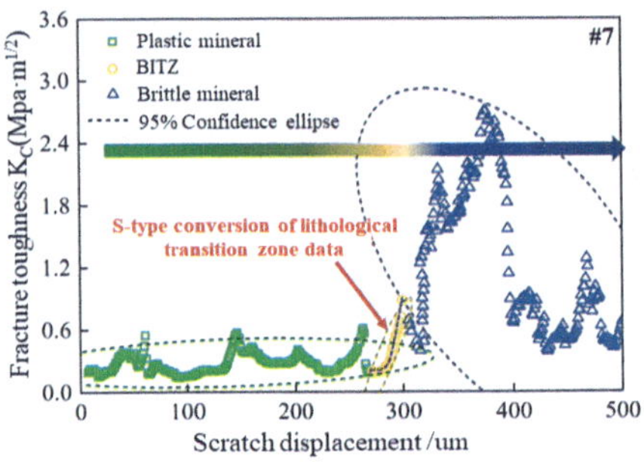

(e) #7Sample fracture toughness curve

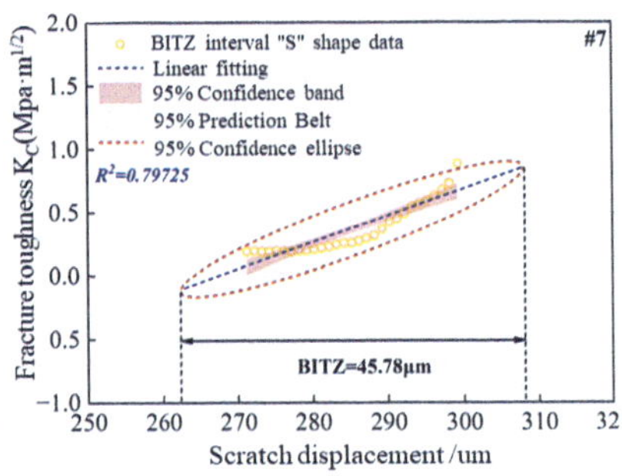

(f) #7BITZ quantitative results of samples

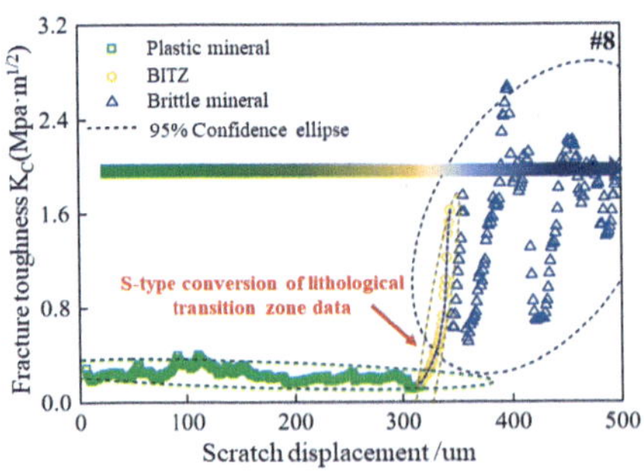

(g) #8Sample fracture toughness curve

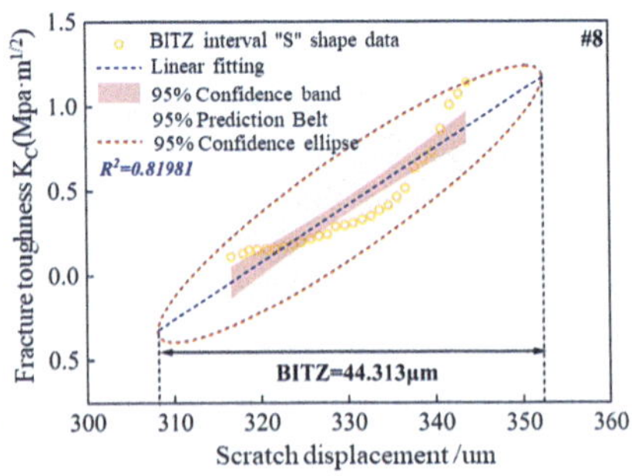

(h) #8BITZ quantitative results of samples

Fig. 3.25 Identification results of BITZ distribution length using fracture toughness method. (**a**) #3 Sample fracture toughness curve. (**b**) #3 BITZ quantitative results of samples. (**c**) #4 Sample scratch depth should be comprehensively considered. The following is a dimensional analysis of the scratch experiment process.

In the experimental scheme proposed in this paper, the scratch velocity is small enough to ignore the rate dependence of material generation caused by sample bottom fixation, and Hollomon equation is used to describe the mechanical behavior of material deformation caused by indenter.

$$\sigma = \begin{cases} E\varepsilon, & \text{for} \quad \varepsilon \le \varepsilon_y \\ K\varepsilon^n, & \text{for} \quad \varepsilon > \varepsilon_y \end{cases} \tag{3.19}$$

In Eq. (4.1), σ is the total stress, E is the elastic modulus, $K = E_{\varepsilon y}{}^{1-n}$ is the strength coefficient, σ_y is the yield stress, $\varepsilon_y = \sigma_y/E$ is the yield strain, n is the strain hardening index, and $\varepsilon = \varepsilon_e + \varepsilon_p$ is the total strain, including the elastic strain ε_e and the ductile strain ε_p.

Based on dimensional analysis, the loading force collected at the tip of the indenter:

$$F_n = F_1(d, E, v, Y, u_s) \tag{3.20}$$

Tangential force collected at the tip of the indenter:

$$F_T = F_2(d, E, v, Y, u_s) \tag{3.21}$$

In Eqs. (4.2) and (4.3), Y, E, v, μ_s, and d denote the yield strength, elastic modulus, Poisson's ratio, friction coefficient, and scratch depth, respectively.

Select Y and d as independent variables and apply the dimensional homogenization π theorem:

$$\begin{aligned} \Pi_1 = F_T Y^{-1} d^2, \Pi_2 = AYd^{-2}, \Pi_3 = F_n Y^{-1} h^{-2} \\ \Pi_4 = YE^{-1} d^0, \Pi_5 = Y^0 d^0 u_s \end{aligned} \tag{3.22}$$

According to the research of Abdullah et al., the influence of Poisson's ratio v on the experimental results of scratches can be negligible, and the following relationship is obtained:

$$\Pi_6 = F_3(\Pi_2, \Pi_3, \Pi_5) \tag{3.23}$$

Fig. 3.25 (continued) fracture toughness curve. (**d**) #4 BITZ quantitative results of samples. (**e**) #7 Sample fracture toughness curve. (**f**) #7 BITZ quantitative results of samples. (**g**) #8 Sample fracture toughness curve. (**h**) #8 BITZ quantitative results of samples

In summary, the following dimensionless relation is obtained, and the dimensionless parameter γ_0 is proposed.

$$\gamma_0 = F_3\left(\frac{F_T}{F_n}, \frac{A}{d^2}, u_s\right) = \frac{u_s K_c}{H\sqrt{d}} \tag{3.24}$$

Identification Results of ITZ Thickness of Shale Without Dimensionality Parameter

According to Eq. (4.6), dimensionless parameters γ_0 comprehensively consider material roughness, fracture resistance, penetration resistance, and deformation resistance under constant load. After dimensionless transformation of mechanical parameters collected by indenter, the discretization of dimensionless parameter curve is significantly reduced, as shown in Fig. 3.26. The BITZ distribution data of samples #3 and #4 still showed an S-shaped fluctuation pattern, with distribution lengths of 31.84 μm and 76.66 μm, respectively, and regression confidence level R^2 of 0.66794 and 0.65313, respectively. The dimensionless parameters proposed in this paper effectively weaken the discreteness of BITZ distribution length estimation and improve the confidence of BITZ distribution length estimation.

After the dimensionless transformation of mechanical parameters, the dispersion of dimensionless parameter curves of long quartz and ostracoidal laminae is further weakened, and the dimensionless parameter curves of illite minerals are almost horizontal. There is an obvious increasing trend in the ITZ data range, making the ITZ data segment more obvious. ITZ thickness of felsic laminated shale is 43.41 μm. The confidence level $R^2 = 0.85489$ indicates that the ITZ thickness of the ostraco-laminated shale is 36.56 μm. Confidence level $R^2 = 0.87232$.

3.4.5 *Comparison and Discussion*

The evaluation results of ITZ thickness by SEM, friction coefficient method, fracture toughness method, and dimensionless parameter method are summarized in Fig. 3.27. The scanning electron microscope method can observe the mineral interface in a large range, but this method does not consider the change of mechanical properties, only observe the surface morphology, and the ITZ thickness of the identified laminae is small, and the results are greatly disturbed by human factors. Due to many factors such as grinding and polishing equipment and technology, it is usually difficult to obtain a low roughness surface, which brings uncertainty to the quantification of ITZ thickness by the friction coefficient method. Due to various factors such as human factors, grinding and polishing equipment and processes, it is often difficult to obtain a low roughness surface, which brings uncertainty to the quantification of BITZ thickness using the friction coefficient method.

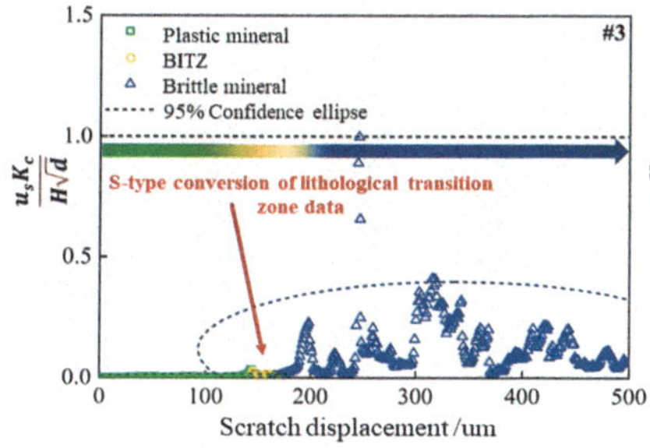

(a) #3Dimensionless parametric curve of the sample

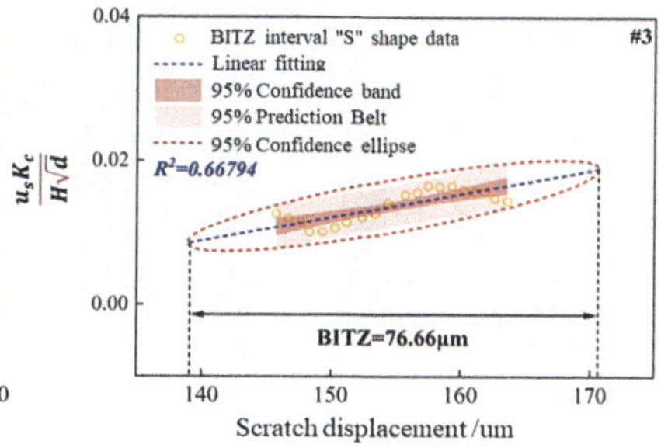

(b) #3BITZ quantitative results of samples

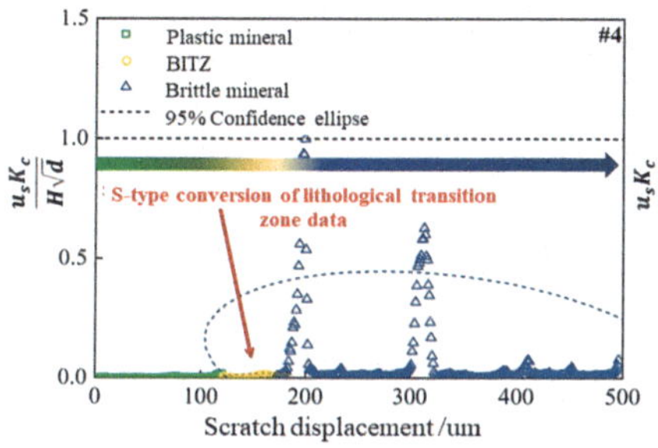

(c) #4Dimensionless parametric curve of the sample

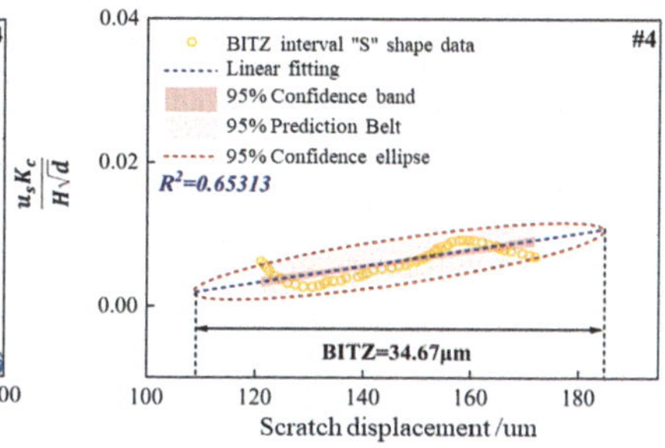

(d) #4BITZ quantitative results of samples

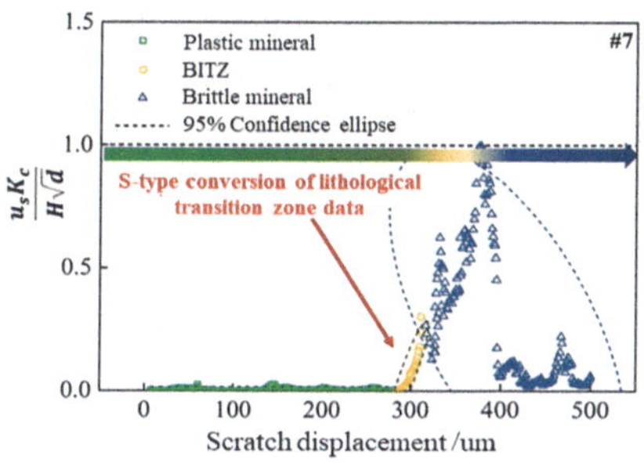

(e) #7Dimensionless parametric curve of the sample

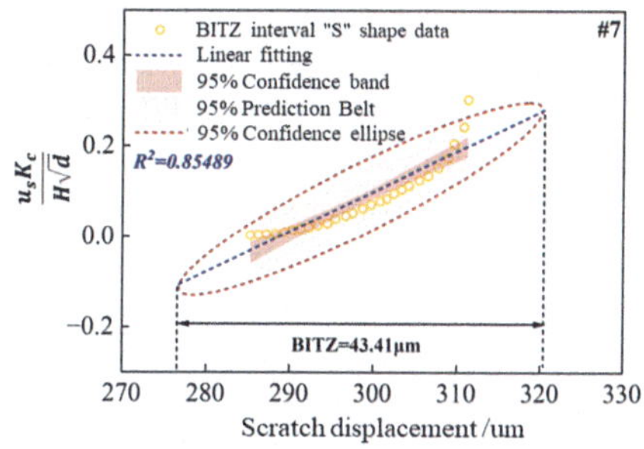

(f) #7BITZ quantitative results of samples

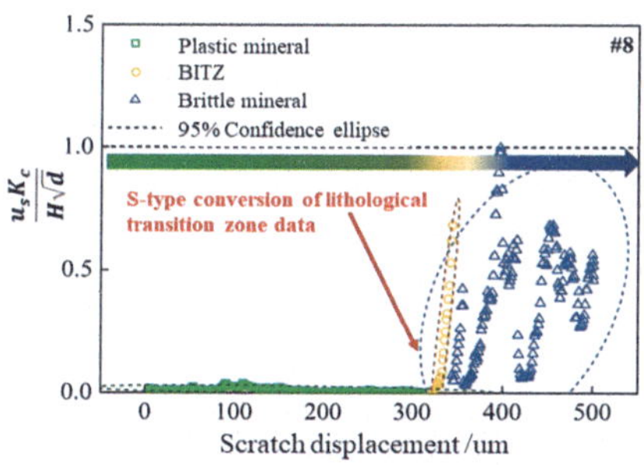

(g) #8Dimensionless parametric curve of the sample

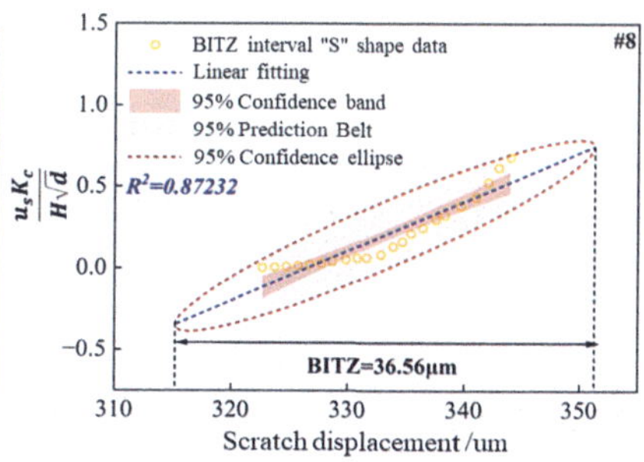

(h) #8BITZ quantitative results of samples

Fig. 3.26 Identification results of BITZ distribution length using dimensionless parameter method. (**a**) #3 Dimensionless parametric curve of the sample. (**b**) #3 BITZ quantitative results of samples.

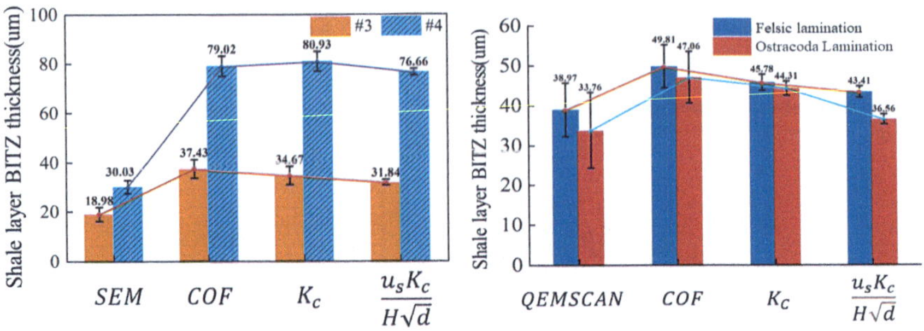

(a) Dolomite nodule transition area (b) Felsic and ostracoid transition zone

Fig. 3.27 Quantitative results of BITZ thickness in shale of Songliao Basin. (**a**) Dolomite nodule transition area. (**b**) Felsic and ostracoid transition zone

Due to the instability of surface roughness caused by the lithology transformation in the interface transition area, the friction coefficient in the BITZ range fluctuates sharply up and down, and the discreteness of the data affects the fitting area of the confidence ellipse. As a result, the BITZ thickness identified by the friction coefficient method is greater than the quantitative results obtained by the fracture toughness method and the dimensionless parameter method. Fracture toughness method can well increase the continuity of data, but it does not take into account the influence of deformation resistance and surface roughness of minerals on BITZ thickness identification, but only considers the ability of different minerals to resist fracture. Compared with the BITZ thickness quantified by the fracture toughness method, the dimensionless parameter decreases, which is because the dimensionless parameter curve comprehensively considers the influence of multiple mechanical parameters, the data dispersion is low, the fitting area of the confidence ellipse is further reduced, and the fitting correlation coefficient increases, which further improves the accuracy of the shale BITZ thickness assessment.

Compared with concrete materials, in the natural deposition process of rocks such as shale, plastic minerals and brittle minerals blend with each other, and the thickness value of the interface transition zone is larger. Wei et al. [5] used friction coefficient and fracture toughness to identify BITZ thickness in concrete materials, and the quantitative results are shown in Fig. 3.28. The BITZ thickness distribution interval between aggregate and cement determined by friction coefficient method is 0.9–0.95 μm, while the BITZ thickness distribution interval identified by fracture toughness is 0.85–0.9 μm. The quantitative results of friction coefficient method are also greater than the identification results of fracture toughness. This also indirectly

Fig. 3.26 (continued) (**c**) #4 Dimensionless parametric curve of the sample. (**d**) #4 BITZ quantitative results of samples. (**e**) #7 Dimensionless parametric curve of the sample. (**f**) #7 BITZ quantitative results of samples. (**g**) #8 Dimensionless parametric curve of the sample. (**h**) #8 BITZ quantitative results of samples

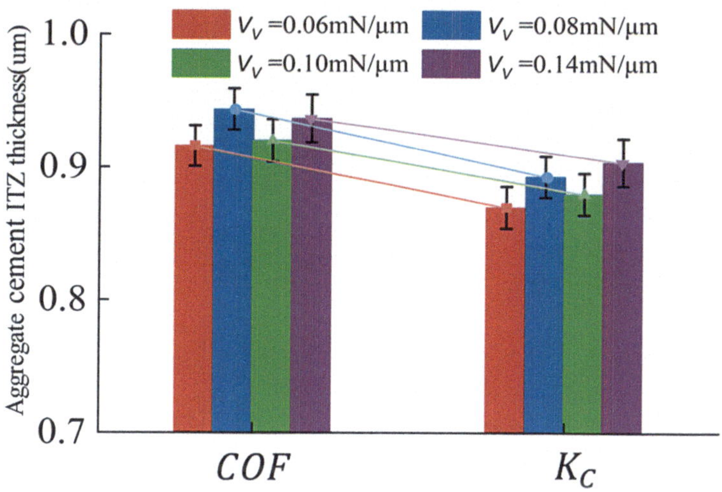

Fig. 3.28 Quantitative results of ITZ thickness for concrete materials

proves the validity of the quantitative method of shale BITZ distribution length proposed in this chapter.

References

1. Guo H, Jia W, Peng P, et al. The composition and its impact on the methane sorption of lacustrine shales from the upper Triassic Yanchang formation, Ordos Basin, China. Mar Pet Geol. 2014;57: 509–20.
2. Jarvie DM, Hill RJ, Ruble TE, et al. Unconventional shale-gas systems: the Mississippian Barnett shale of north-Central Texas as one model for thermogenic shale-gas assessment. AAPG Bull. 2007;91(4):475–99.
3. Rickman R, Mullen MJ, Petre JE, et al. A practical use of shale Petrophysics for stimulation design optimization: all shale plays are not clones of the Barnett shale. In: Paper Presented at the SPE Annual Technical Conference and Exhibition, Denver, Colorado, USA, September 2008; 2008.
4. Wang FP, Gale JF. Screening criteria for shale-gas systems. Gulf Coast Assoc Geol Soc Trans. 2009;59:779–93.
5. Dong G, Chen P, Fu J, et al. Investigation on the evaluation method and influencing factors of cross-scale mechanical properties of shale based on indentation experiment. J Petrol Sci Eng. 2022;208:109509.

Chapter 4
Scale Effect of Nanoindentation-Scratch

Shale reservoirs are characterized by a wide variety of mineral types and significant non-homogeneity, making it extremely difficult to characterize their mechanical properties. In addition, due to the non-homogeneous nature of shale, conventional core-based tests cannot fully elucidate the microscopic changes induced by pores and grain layers at all scales, making it necessary to further investigate the microscopic mechanical properties of the rock [1]. Shale, as a composite material, can be regarded as a multiscale material system from its mechanical properties, and a large amount of research work has been carried out by the previous researchers for the scale upgrading method of nanoindentation, and less research has been done on the variability of nano-centimeter scratch experimental results and scale upgrading methods. In this chapter, starting from the macro/micromechanical analysis, a systematic study is carried out on the response law of mechanical properties of textured shale at different scales by using the indentation-scratch experiments, and the fine characterization of the multiscale mechanical properties and the scale-up method are established.

4.1 Nanoindentation Upscaling Method

4.1.1 Coupled Statistical-Clustering Method

To demonstrate the response relationship between mineral content and mechanical properties, a coupled statistical-clustering analysis method is proposed. Based on statistical analysis, the mineral species of rocks are divided into different phase groups, and these phase groups are used to create physical rock models. According to the correlation between the Mohs hardness of different mineral species, the mineral species are divided into different mineral phases. If the difference between the Mohs hardness of two mineral species with good correlation is smaller than that of any other two mineral species, they are considered to have a good correlation and

L. Yang, *Multiscale Scratch Mechanics of Laminated Shale*,
https://doi.org/10.1007/978-981-95-7041-6_4

are assigned to the same mineral phase. The corresponding mineral proportions are added together, and the average of the two mineral species' hardness is taken for the merged mineral species. The mineral species contained in each mineral phase and their proportions are then statistically analyzed [2].

By analyzing the load-displacement curves of each indentation point and using Eqs. (4.1) and (4.3), the mechanical parameters of the rock, such as hardness and elastic modulus, are calculated. Based on machine learning clustering algorithms, the mechanical parameters are analyzed with different numbers of clusters, and the resulting mechanical parameters are divided into different mechanical phases. The difference coefficient D is used to quantify the difference between the statistically obtained mineral phases and the clustered mechanical phases (shown in Eq. 4.5). A smaller value of the difference coefficient D indicates a smaller difference between the mineral phases and mechanical phases, indicating better coupling. This further illustrates the relationship between the mineral composition and mechanical properties of shale at each indentation point, determines an equivalent physical model of shale, and optimizes the microscale mechanical upgrading model of shale.

$$D = \sum_{i=1}^{n} |A_i - B_i| \tag{4.1}$$

$$S(i) = \frac{b(i) - a(i)}{\max[a(i), b(i)]} \tag{4.2}$$

$$a(i) = \frac{1}{n-1} \sum_{j \neq i}^{n} \text{distance}\,(i,j) \tag{4.3}$$

$$S(i) = \begin{cases} 1 - \dfrac{a(i)}{b(i)} \ldots\ldots\ldots a(i) < b(i) \\ 0 \ldots\ldots\ldots\ldots\ldots a(i) = b(i) \\ \dfrac{b(i)}{a(i)} - 1 \ldots\ldots\ldots a(i) > b(i) \end{cases} \tag{4.4}$$

In the equation, A_i is the proportion of each type of mineral, B_i is the proportion of each cluster component of hardness and elastic modulus of the indentation, $\alpha(i)$ is the cohesion of the sample point, j is other sample points in the same cluster as sample i, and distance is the distance between i and j.

4.1.2 Upscaling Method

As shown in Fig. 4.1, traditional scale-up models generally consist of three steps. Step one involves using optical microscopy to analyze magnified micrographs and measure the microstructural parameters of the material (such as volume fraction,

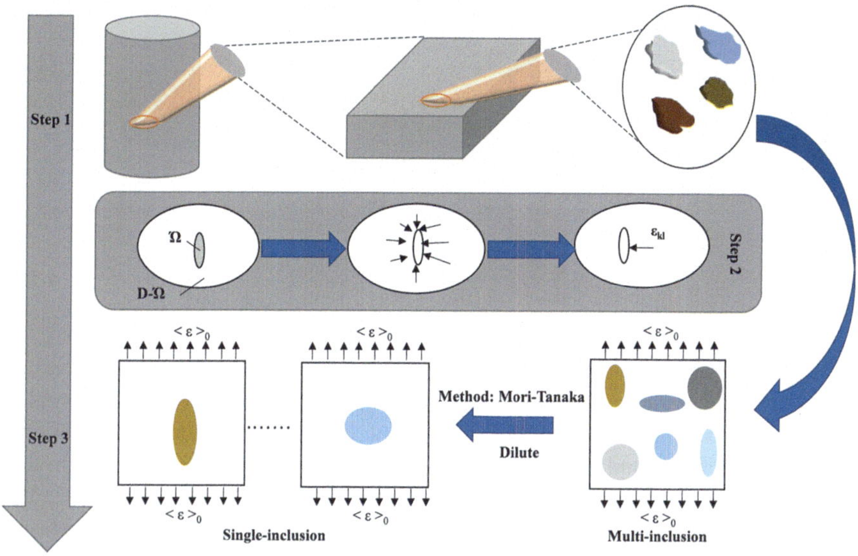

Fig. 4.1 A multiscale structure model

shape, and orientation of inclusions) that describe the microscopic features of the rock. Step two involves establishing the relationship between the microscale stress or strain and the macroscopic load acting on each component material under uniform macroscopic boundary conditions [3]. Step three involves replacing a heterogeneous material unit with an equivalent homogeneous material unit that has the same macroscopic properties by averaging the local stress and strain over the entire representative unit cell. Currently, commonly used scale analysis methods include the Mori-Tanaka method and the dilute method [2]. The details are as shown in the following equation:

$$k_{\mathrm{r}} = \frac{E_{\mathrm{r}}}{3(1 - 2v_{\mathrm{r}})} \tag{4.5}$$

$$u_{\mathrm{r}} = \frac{E_{\mathrm{r}}}{2(1 + 2v_{\mathrm{r}})} \tag{4.6}$$

$$K_{\mathrm{m}} = \frac{\sum_{r=0}^{n} f_{\mathrm{r}} \frac{k_{\mathrm{r}}}{3k_{\mathrm{r}} + 4u_0}}{\sum_{r=0}^{n} \frac{f_{\mathrm{r}}}{3k_{\mathrm{r}} + 4u_0}} \tag{4.7}$$

$$G_{\mathrm{m}} = \frac{\sum_{r=0}^{n} \frac{f_{\mathrm{r}} u_{\mathrm{r}}}{u_0(9k_0+8u_0)+6u_{\mathrm{r}}(k_0+2u_0)}}{\sum_{r=0}^{n} \frac{f_r}{u_0(9u_0+8k_0)+6u_{\mathrm{r}}(k_0+2u_0)}} \tag{4.8}$$

$$K_{\mathrm{d}} = K_0 + \sum_{r=1}^{n} f_{\mathrm{r}} \frac{(k_{\mathrm{r}} - k_0)(3k_0 + 4u_0)}{3k_{\mathrm{r}} + 4u_0} \tag{4.9}$$

$$G_{\mathrm{d}} = u_0 + \sum_{r=1}^{n} f_{\mathrm{r}} \frac{5u_0(u_{\mathrm{r}} - u_0)(3k_0 + 4u_0)}{u_0(9k_0 + 8u_0) + 6u_{\mathrm{r}}(k_0 + 2u_0)} \tag{4.10}$$

For the mineral content f_{r} of each phase defined by Mori-Tanaka and dilute upgrading method, there are limitations in describing shale minerals of complex compositions due to the extreme complexity of shale's mineral compositions, shapes, orientations, and the interactions of various forces among minerals. In this paper, based on the coupled statistical-cluster analysis method, the mineral content f_{r} of each phase is optimized to the percentage share P_{r} of each phase for cluster analysis, and the optimized scale upgrading model is defined as Z-Mori, as shown in the following equation:

$$K_{\mathrm{Z}} = \frac{\sum_{r=0}^{n} P_{\mathrm{r}} \frac{k_{\mathrm{r}}}{3k_{\mathrm{r}}+4u_0}}{\sum_{r=0}^{n} \frac{P_{\mathrm{r}}}{3k_{\mathrm{r}}+4u_0}} \tag{4.11}$$

$$G_{\mathrm{Z}} = \frac{\sum_{r=0}^{n} \frac{P_{\mathrm{r}} u_{\mathrm{r}}}{u_0(9k_0+8u_0)+6u_{\mathrm{r}}(k_0+2u_0)}}{\sum_{r=0}^{n} \frac{P_r}{u_0(9u_0+8k_0)+6u_{\mathrm{r}}(k_0+2u_0)}} \tag{4.12}$$

where k_{r} and u_{r} are the volumetric modulus and shear modulus of the r-th phase, respectively, which can be obtained using the Young's modulus E_{r} and the Poisson's ratio v_{r}. k_0 and u_0 are the volumetric modulus and shear modulus of the shale matrix, respectively. K_{m} and G_{m} are the macroscopic equivalent shear modulus and volumetric modulus, respectively, which were obtained from the Mori-Tanaka method. K_{d} and G_{d} are the macroscopic equivalent shear modulus and volumetric modulus, respectively, which were obtained from the dilute method. K_{z} and G_{z} are the macroscopic equivalent shear modulus and volumetric modulus, respectively, which were obtained from the Z-Mori method.

After obtaining the equivalent mechanical parameters of the three-phase medium, the elastic modulus of the rock at the centimeter scale can be obtained using the following equation:

$$E_{HOM} = \frac{9K_M G_M}{3K_M + G_M} \tag{4.13}$$

where K_M and G_M denote the bulk and shear moduli after homogenization, respectively.

It can be observed from Eqs. 4.15 and 4.16 that the optimized Z-Mori model requires variables that are no longer dependent on the variable f_r obtained through scanning electron microscopy in step one. Instead, relevant variable parameters for shale scale upgrading can be obtained through coupled cluster-indentation tests, thus significantly reducing the influence of mineralogical parameters on mechanical parameters in shale scale upgrading analysis.

4.1.3 Composition Analysis

Microstructure and composition analyses were conducted on samples from the Qingshankou Formation by scanning electron microscopy (SEM) at different magnifications and energy-dispersive X-ray spectroscopy (EDS). As shown in Fig. 4.2a, SEM images at 500× magnification revealed well-cemented mineral particles and good compaction, with large-sized pores barely developed. As shown in Fig. 4.2b, SEM images at 2000× magnification around the 24th indentation point showed the existence of micropores on the rock surface with complex porosity, where pore openings were smaller than ten lm. And rocks have flat and smooth surfaces as well as rough and rugged surfaces. At 10,000× magnifications, as shown in Fig. 4.2c, thin flaky minerals with a diameter much smaller than five lm were found at the indentation site. These flaky minerals were clay minerals, such as montmorillonite and kaolinite.

EDS analyses were performed to reveal the elemental contents and distributions of the Qingshankou Formation samples, providing information on the mineralogical composition of the samples. As shown in Fig. 4.2c, four analysis points around the indentation point were selected for element analyses. The EDS mapping only displayed the distribution of the most significant elements, as shown in Fig. 4.2d, revealing a large amount of calcium, silicon, magnesium, oxygen, and aluminum in the Qingshankou Formation samples. The presence of silicon and oxygen indicated the presence of quartz, while the presence of potassium, silicon, oxygen, and trace amounts of aluminum confirmed the position of feldspar. The presence of iron confirmed the existence of the mineral pyrite, while the presence of calcium, magnesium, oxygen, and carbon confirmed the existence of calcite and dolomite. And the presence of a large number of carbon atoms confirmed the existence of organic matter. Finally, by using different ratios of aluminum and silicon, it was possible to identify regions rich in clay minerals and regions rich in quartz.

As shown in Fig. 4.3, the measured content of each element varies among the four analyzed points. Specifically, Points 1 and 4 exhibit the highest content of Ca and O, suggesting a prevalence of calcite with multiple coexisting minerals. And Points

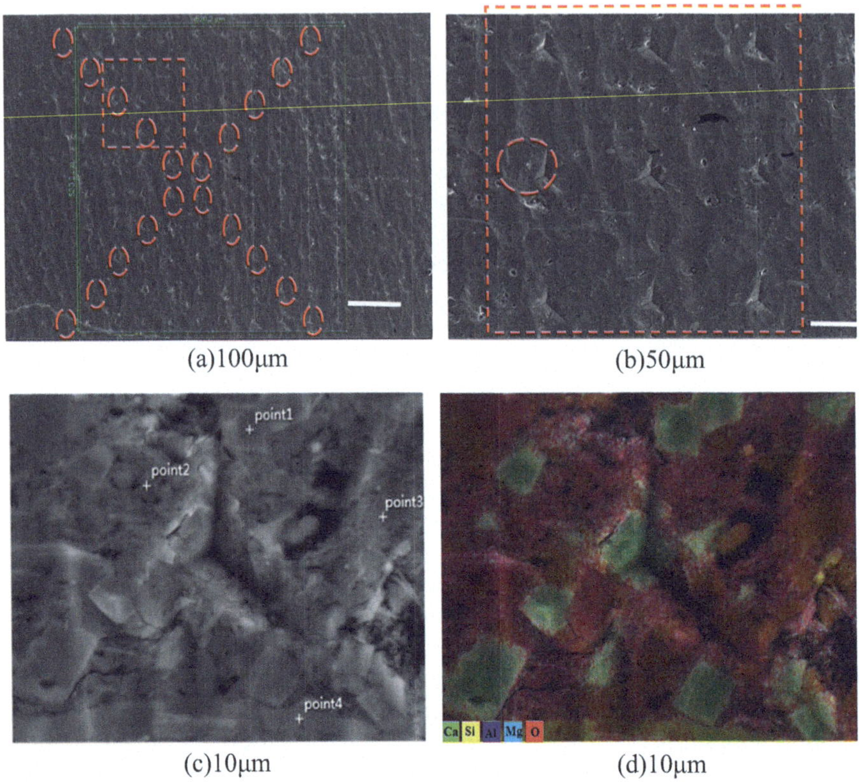

Fig. 4.2 SEM analysis of sample 1. (**a**) 100 μm. (**b**) 50 μm. (**c**) 10 μm. (**d**) 10 μm

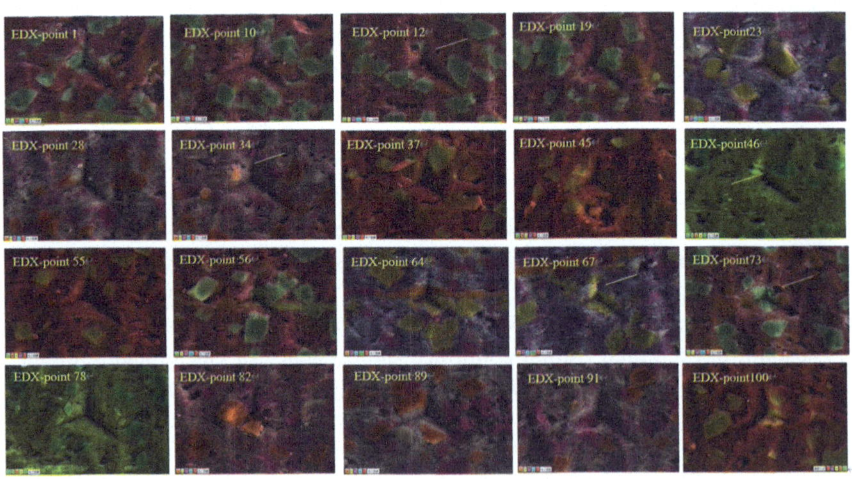

Fig. 4.3 EDX analysis of sample 1

Fig. 4.4 QEMSCAN analysis of sample 1

Table 4.1 Mineral composition and proportions of Qingshankou Formation shale sample

Major mineral types	Mineral types	Mineral proportion/%	Mohs hardness
Quartz mineral	Quartz	28.47	7
Feldspathic minerals	Potassium feldspar	0.71	6
	Albite	12.1	
	Andesine	0.02	
Clay minerals	Illite	36.15	2
	Montmorillonite	0.96	
	Kaolinite	0.03	
	Chlorite	1.2	
Dolomite minerals	Dolomite	9.84	3.5
	Ferrodolomite	0.03	
Calcite mineral	Calcite	3.71	3
Mica minerals	Muscovite	0.49	2.5
	Biotite	3.94	

2 and 3 present the highest content of O and Si, indicating a prevalence of quartz with other coexisting minerals.

Furthermore, 20 indentation points were selected along the diagonal of the indentation grid for SEM-EDX observation. The result is shown in Fig. 4.4. The results show a combination of multiple elements, suggesting the presence of multimineral inclusions at these indentation points. It is difficult to obtain homogeneous single-crystal minerals through indentation tests. Additionally (Table 4.1), microcracks were observed at indentation points 12, 34, 46, 67, and 73, indicating that their existence may affect the results of the indentation tests.

QEMSCAN tests were conducted on the grain layer samples to obtain information on the mineral content, mineral species, and distribution of minerals on the overall surface of the experimental area of the samples (Fig. 4.4). From the viewpoint of mineral surface distribution, the mineral distribution on the surface of the sample is disorganized, with a variety of minerals arranged in a disorderly manner, and the minerals are closely arranged with tiny fissures between them. From the viewpoint of mineral composition, the sample consists of a large area of illite and other clay minerals to form the matrix, and the shale has a wide variety of minerals.

To further quantify the composition of minerals, this paper divides the different mineral information into different major mineral classes based on the correlation relationship between Mohs hardness and composition, and the statistical results are shown in Table 4.1. The main mineral composition of the sample consists of 28.47% quartz minerals (quartzite), 12.84% feldspar minerals (potassium feldspar and sodium feldspar in feldspar), 9.87% mica minerals (dolomite and ferrodolomite), 38.34% clay minerals (illite, kaolinite, and chlorite), 3.71% salvations minerals (calcite), and 4.43% mica (leuco-macro-macro-macro). It also contains a small amount of monazite, pyrite, and amorphous materials that are difficult to distinguish from other minerals.

4.1.4 Phase Clustering Analysis

As shown in Fig. 4.5, the information on mechanical properties related to the mechanical phase of different samples was obtained by nanoindentation test. Sample 1 has a hardness range of about 0.035–0.253 GPa, a modulus of elasticity range of about 2.15–13 GPa, and an average modulus of elasticity of 6.57 GPa. Sample 2 has a hardness range of about 0.07–2.08 GPa, and a modulus of elasticity range of about 7.8–37.2 GPa, with an average modulus of elasticity of 23.73 GPa. Sample 3 had a hardness range of approximately 0.109–0.437 GPa, a modulus of elasticity range of approximately 3.08–10.62 GPa, and an average modulus of elasticity of 5.97 GPa. Sample 4 had a hardness range of approximately 0.28–0.54 GPa, with a range of elastic modulus of approximately 9.6–14.3 GPa and an average elastic modulus of 9.71 GPa.

Statistical analysis of the nanoindentation test results shows that the mechanical properties of different shale samples are highly variable. As shown in Fig. 4.5, there is an obvious positive correlation between the elastic modulus and hardness of the four groups of samples, but there is a large dispersion in the test results of the same sample. Combined with the conclusion that the indentation points are multimineral inclusions discussed above, this paper suggests that the main reason for the large dispersion of the indentation test results is the difference in the percentage of inclusion minerals and the composition of microfractures at each indentation point. The indentation data points with higher modulus of elasticity in Fig. 4.5 consist of harder minerals (quartz, etc.). The indentation data points with smaller modulus of elasticity are composed of softer minerals (clay minerals, etc.). The linear fit R^2 for the four groups of samples is 0.87, 0.59, 0.77, and 0.68, respectively, which indicates that there are differences in the dispersion of the test results for the different types of shale, but the range of the differences is not large, and the dispersion results are relatively close to each other.

In addition, map-like distribution plots were created to visualize the distribution of Young's modulus and hardness of the samples. As shown in Fig. 4.6, the Young's

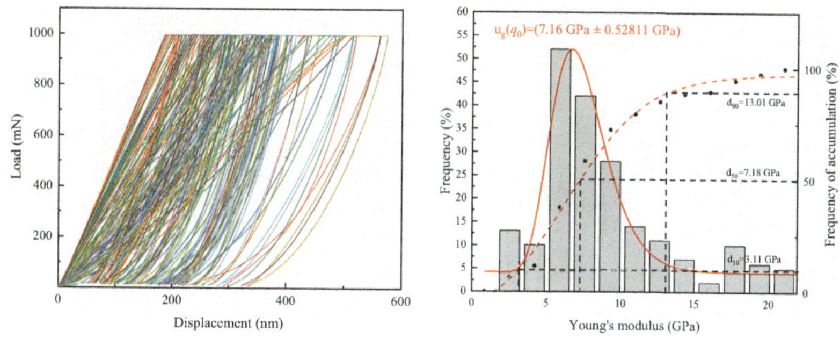

(a) sample 1

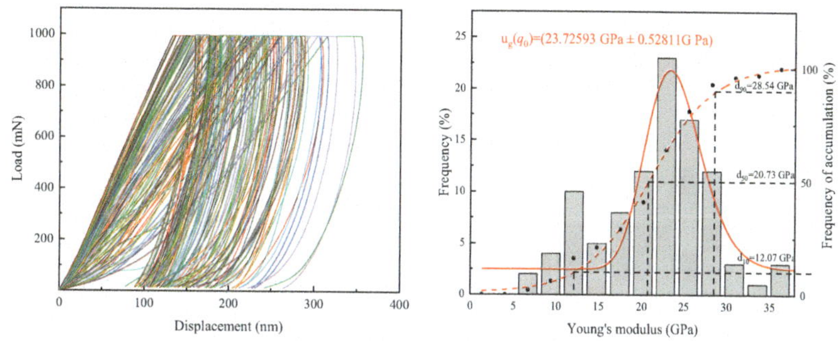

(b) sample 2

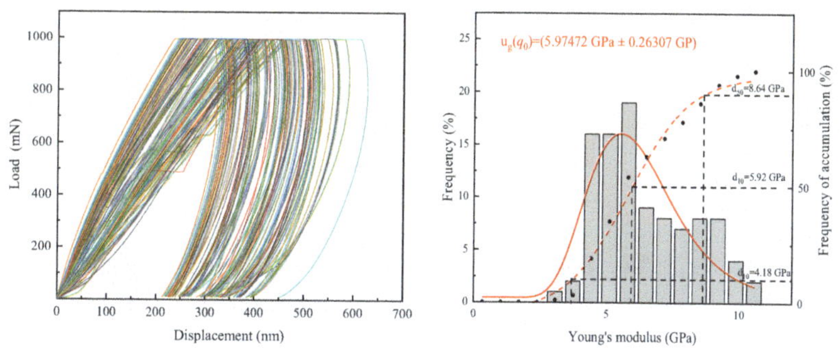

(c) sample 3

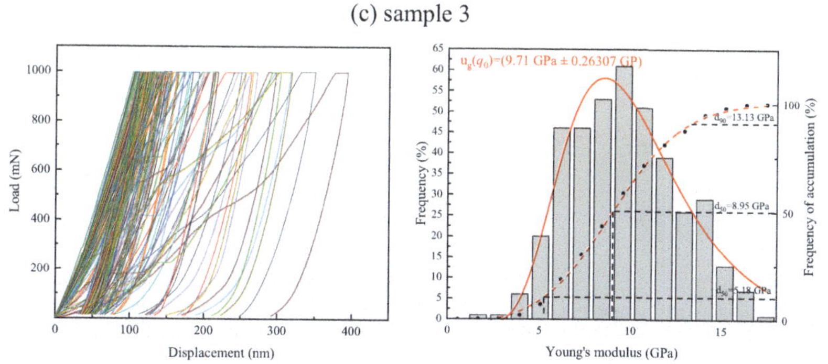

(d) sample 4

Fig. 4.5 Indentation results of different samples. (**a**) Sample 1. (**b**) Sample 2. (**c**) Sample 3. (**d**) Sample 4

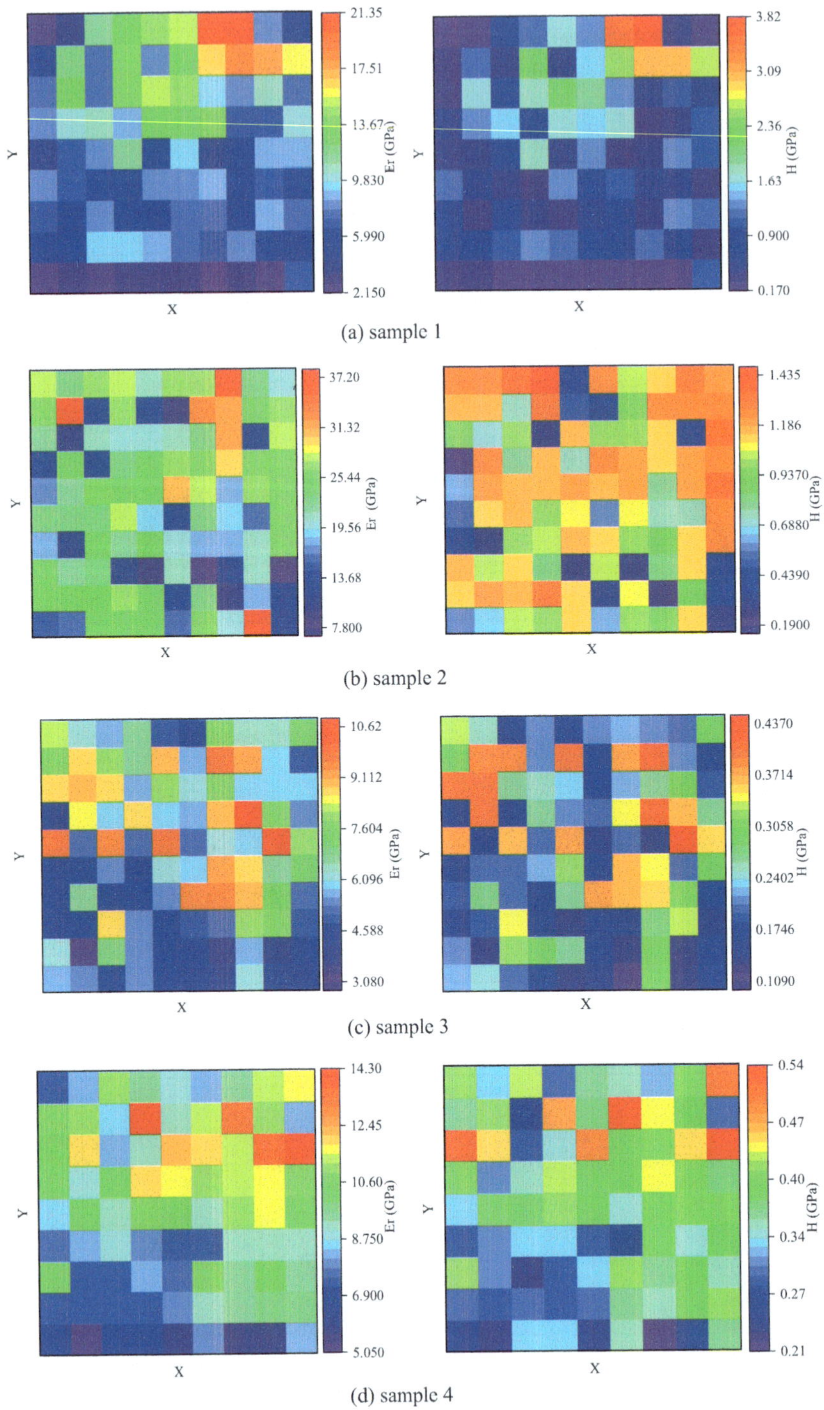

Fig. 4.6 Young's modulus and hardness analysis of sample. (**a**) Sample 1. (**b**) Sample 2. (**c**) Sample 3. (**d**) Sample 4

modulus and hardness distributions of the samples are highly discrete, demonstrating the strong heterogeneity of the shale samples. The Young's modulus and the hardness of the samples show a strong linear correlation between them on the distribution plots. The overall small values of the distribution of mechanical properties for the indentation points of Sample 1, Sample 3, and Sample 4 are mainly due to the fact that the indentation points are composed of soft minerals, such as clay minerals, as the main constituents, and the other minerals are disordered and interspersed with the clay minerals. The overall distribution of mechanical properties of the indentation sites in sample 2 is large, mainly due to the fact that the indentation sites are mainly composed of hard minerals, such as quartz minerals, and other minerals are disordered and intermingled with quartz minerals. In addition, the presence of single-crystal minerals with grain sizes larger than 20 nm in the indentation point pressure leads to the presence of extremely large values of Young's modulus in the test results.

The mechanical parameters obtained above were analyzed using the clustering algorithm in machine learning with different number of clusters and divided into different mechanical phases based on the different number of clusters. As shown in Fig. 4.7, the mechanical phases were divided into 2–6 classes based on the results of elastic modulus and hardness of the indentation test, respectively. As shown in Fig. 4.7, there are 2–6 mechanical phases based on the elastic modulus and hardness results of the indentation test. As shown in Fig. 4.7, there are 2 categories of mechanical phases, with 72% of phase 1 and 27% of phase 2; there are 3 categories of mechanical phases, with 19% of phase 1, 37% of phase 2, and 43% of phase 3; there are 3 categories of mechanical phases, with 33% of phase 1, 28% of phase 2, 29% of phase 3, and 9% of phase 4; there are 5 categories of mechanical phases, with 8% of phase 1, 8% of phase 2, and 8% of phase 3; there are 6 categories of mechanical phases. As shown in Fig. 4.7, the clustering is 5 categories, with 8% of phase 1, 8% of phase 2, 13% of phase 3, 38% of phase 4, and 32% of phase 5; the clustering is 6 categories, with 3% of phase 1, 8% of phase 2, 31% of phase 3, 28% of phase 4, 28% of phase 5, and 1% of phase 6. The clustering results of each phase are summarized in Table 4.3.

Based on the machine learning clustering algorithm to determine the optimal number of clusters classification, the introduction of the profile coefficient to evaluate the clustering effect is good or bad, the value of the profile coefficient range is [0, 1], the larger the profile coefficient, the better the clustering effect. As shown in Fig. 4.7f, when the number of clusters is 3, the contour coefficient is 0.8, which is closer to 1 than the contour coefficients of other clustered data, indicating that individuals within the same category have higher homogeneity, while the heterogeneity between categories is higher.

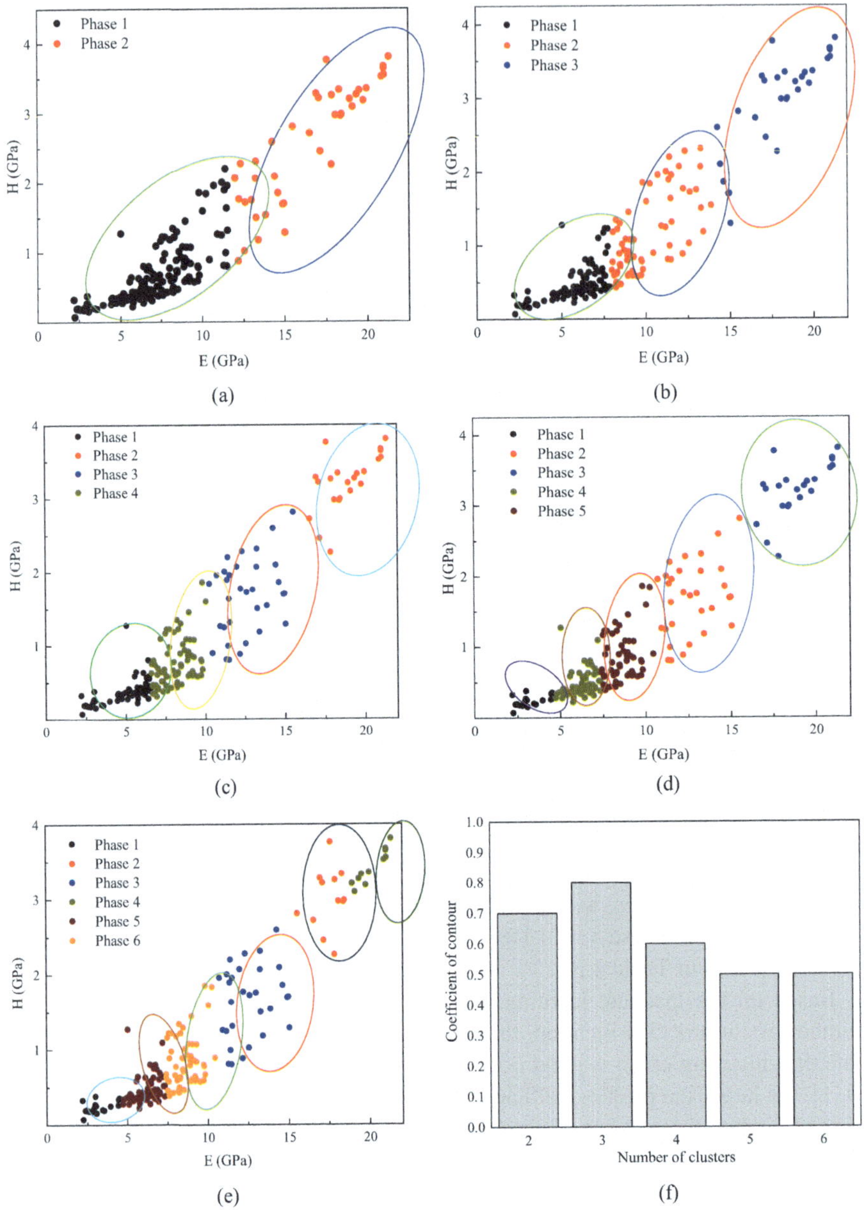

Fig. 4.7 Plot of number of clusters results

4.1.5 Coupled Statistical-Cluster Analysis

Different phase group models were established by analyzing the statistical occupancy of mineral phases and the clustering occupancy of mechanical phases. As shown in Table 4.2, the two-phase model consists of phase 1 (feldspars and quartzes) and phase 2 (clays, clasts, mica, and detritus); the three-phase model consists of phase 1 (clays), phase 2 (feldspars and quartzes), and phase 3 (clasts, mica, and detritus); the four-phase model consists of phase 1 (clays), phase 2 (quartzes), phase 3 (feldspars), and phase 4 (mica, detritus), five-phase model consists of phase 1 (clay), phase 2 (quartz), phase 3 (feldspar), phase 4 (mica, detrital), and phase 5 (mica); six-phase model consists of phase 1 (clay), phase 2 (quartz), phase 3 (feldspar), phase 4 (mica), phase 5 (mica), and phase 6 (detrital).

Through analysis of the coefficient of variation for each phase, the coefficient of variation for the two-phase model was 31.07%, for the three-phase model was 2.91%, for the four-phase model was 30.04%, for the five-phase model was 13.18%, and for the six-phase model was 28.98%. When the phase composition model was three-phase, the coefficient of variation was the smallest at 2.91%, which

Table 4.2 Mineral compositions and proportions of rippled shale samples

Phase number	Mineral types	Mineral proportion/%	Clustering percentage/%	Variation coefficient
Two-phase	Phase 1 (feldspar and quartz)	42.42	27	31.07
	Phase 2 (clay, dolomite, mica, and calcite)	56.35	72	
Three-phase	Phase 1 (clay)	38.34	37	2.91
	Phase 2 (feldspar and quartz)	42.42	43	
	Phase 3 (dolomite, mica, and calcite)	18.01	19	
Four-phase	Phase 1 (clay)	38.34	33	30.04
	Phase 2 (quartz)	28.47	29	
	Phase 3 (feldspar)	12.84	28	
	Phase 4 (dolomite, mica, and calcite)	18.01	9	
Five-phase	Phase 1 (clay)	38.34	38	13.18
	Phase 2 (quartz)	28.47	32	
	Phase 3 (feldspar)	12.84	13	
	Phase 4 (dolomite and calcite)	13.58	8	
	Phase 5 (mica)	4.43	8	
Six-phase	Phase 1 (clay)	38.34	31	28.98
	Phase 2 (quartz)	28.47	28	
	Phase 3 (feldspar)	12.84	28	
	Phase 4 (dolomite)	9.87	8	
	Phase 5 (mica)	4.43	3	
	Phase 6 (calcite)	3.71	1	

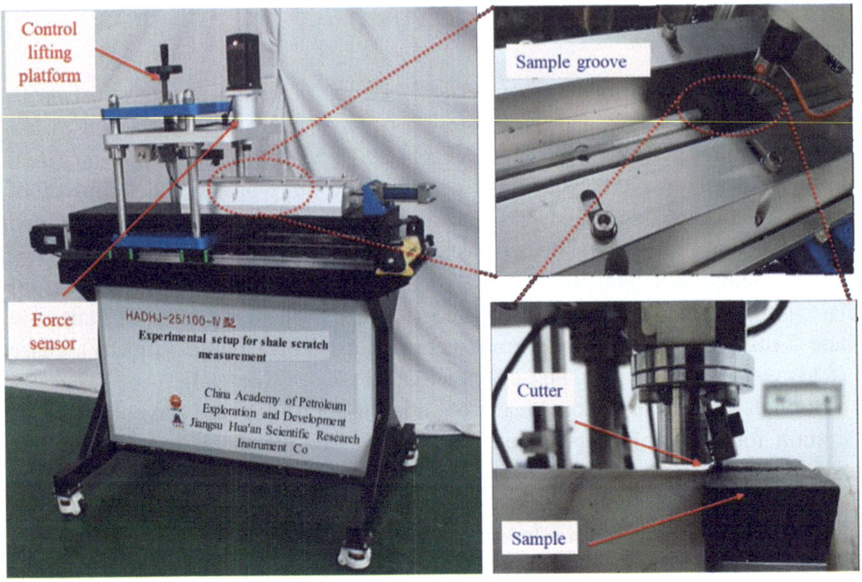

Fig. 4.8 Scratch test machine

was much lower than the other phase compositions. Based on the silhouette coefficient, the optimal number of phases for the samples from the Qingshankou Formation was determined to be three-phase. After simplifying the shale into a three-phase model, as shown in Fig. 4.7, the elastic moduli of the clustering centers for each phase of the Qingshankou Formation were 2.3, 3.24, and 4.82 GPa, respectively. The proportion of the elastic modulus Er for each phase was represented by P_r, which were 19, 37, and 43%, respectively. The volume and shear moduli of the different phases were calculated based on the elastic moduli of each phase's clustering center.

Through the use of a coupled statistical-clustering analysis method, four sets of samples were analyzed. The optimal number of phases for samples 1, 3, and 4 was 3, while the optimal number for sample 2 was two-phase. The optimal number of phases obtained from different types of shale varied, but the values were concentrated around 3. As shown in Fig. 4.9, the shale was ultimately simplified into a three-phase physical model for further analysis of its microscopic mechanical properties.

4.1.6 Analysis of Scaling Results

Uniaxial compression experiments were carried out using a hydraulic servo rigid material experimental machine to obtain the experimental results of rock stress-strain under uniaxial conditions in shale, and the macroscopic elastic moduli of the four

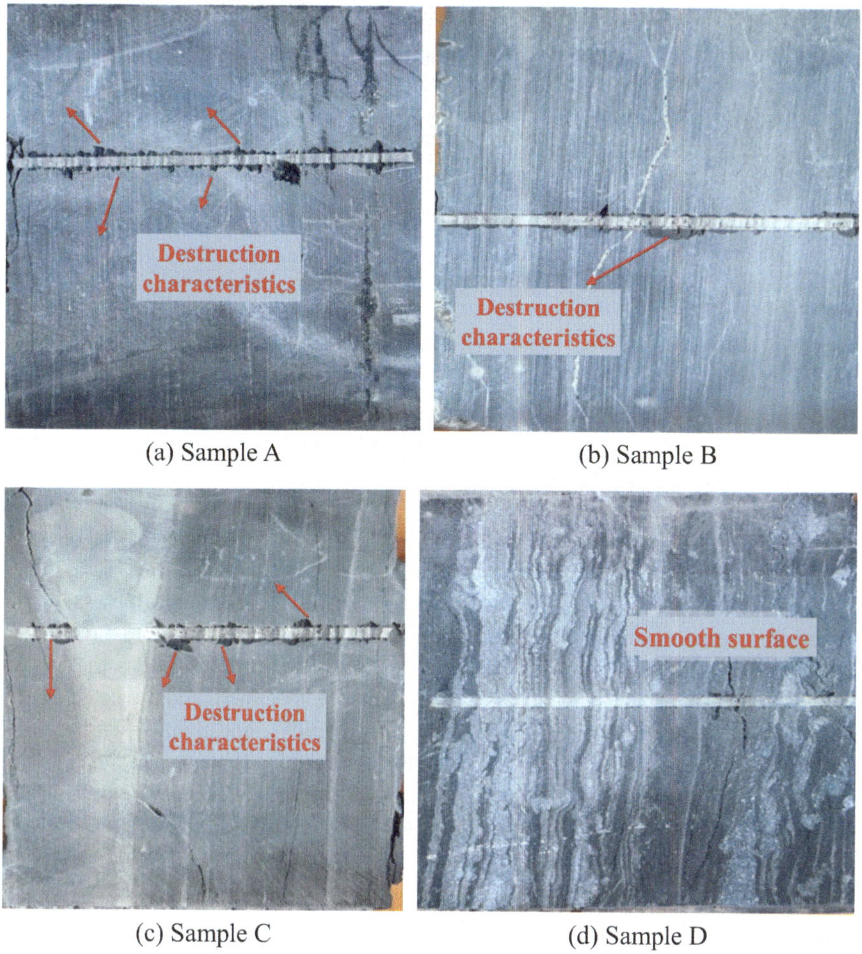

Fig. 4.9 Sample after scratch test. (**a**) Sample A. (**b**) Sample B. (**c**) Sample C. (**d**) Sample D

groups of samples were calculated using the Mori-Tanaka method, the dilute method, and the modified Z-Mori method, respectively. The macroscopic elastic modulus of sample 1 is 10.41 GPa, and the microscopic elastic modulus calculated by Z-Mori is 8.76 GPa; the macroscopic elastic modulus of sample 2 is 21.42 GPa, and the microscopic elastic modulus calculated by Z-Mori is 19.25 GPa; the macroscopic elastic modulus of sample 3 is 13.47 Gpa, and the microscopic elastic modulus calculated by Z-Mori is 7.61 GPa, and the modulus of elasticity 7.61 GPa; sample 4 has a macroscopic modulus of elasticity 12.35 GPa and a microscopic modulus of elasticity 9.34 GPa calculated by using Z-Mori.

From the calculation results of the four groups of samples, the macroscopic elastic modulus calculated using the Z-Mori upgraded model is in good agreement with the macroscopic elastic modulus in uniaxial compression, and the macroscopic elastic

Table 4.3 Results of scale upgrade elastic modulus

Sample	Modulus of elasticity obtained from the scale-up model, GPa			Statistical elasticity Modulus mean/GPa	Macroscale elastic modulus, GPa
	Mori-Tanaka	Dilute	Z-Mori		
Sample 1	6.83	5.74	8.76	7.16	10.41
Sample 2	17.94	15.93	19.25	23.73	21.42
Sample 3	5.277	4.84	7.61	5.97	13.46
Sample 4	8.66	7.95	9.34	9.71	12.35

modulus calculated using the Z-Mori upgraded model is smaller than the force macroscopic elastic modulus in uniaxial compression. This shows that the optimized Z-Mori upgraded model has good applicability, and the mechanical properties obtained by it deviate from the actual values within a certain range. The reasons for the deviation of the elastic modulus are manifold, one of which is that the homogenization process is a process of estimation of ideal assumptions, and the setting of its parameters also has a certain influence on the deviation of the results; the second is that the larger the scale of the experimental specimen, the larger the number of micropores and cracks contained in the specimen, and the more easily the rock extends from the micropores to the microcracks, and then expands to the macroscopic cracks when it is subjected to the external loads.

The advantages and disadvantages of the three upgrading methods were further compared and analyzed as shown in Table 4.3. All four groups of samples showed that the macroscopic modulus of elasticity calculated by the Z-Mori method was larger than that calculated by the Mori-Tanaka and dilute methods, and the difference between the mechanical parameters calculated by the Z-Mori method and the core-scale modulus of elasticity was the smallest. The difference between the mechanical parameters calculated by the dilute method and the core-scale modulus of elasticity is the largest, and the mean values of the statistical modulus of elasticity from the indentation test results are not consistent with the core-scale uniaxial compression. It can be seen that, in the process of upgrading shale from microscale to macro-scale, the upgrading effect of Mori-Tanaka method is better than that of dilute method, while the results of the optimized Z-Mori method are closer to the uniaxial elastic modulus of macro-scale test than the two, and the upgrading effect is more significant.

4.2 Nano- and Centimeter Scratch Properties

4.2.1 Centimeter Scratch

Experimental Equipment and Methods

In order to clarify the mechanical characteristics of the grain shale, the HADHJ-25/100-IV shale scratch measurement experimental device was used to determine the longitudinal and continuous distribution characteristics of the mechanical properties of the shale samples and to systematically study the mechanical properties and damage mechanism under the control of the reservoir characteristics of the grain shale as well as the strong non-homogeneity characteristics. Prior to the scratch test, a leveling procedure to make the surface flat was required. Then, the scratch test was performed along the core (oriented perpendicular to the laminar line) under the conditions of constant depth of cut and constant velocity. The depth of cut for this study was in the range of 0.5 mm. In order to obtain reliable data, the scratch length should be equal to at least a few centimeters. The width of the tool was 2 mm, and the force acting on the cutting surface at each depth of cut was recorded continuously.

The working principle of HADHJ-25/100-IV shale scratch measurement experimental device is that the scratch cutter head is driven by computer control to scribe the rock surface at a constant rate and speed. The computer will collect and record the displacement and force of the cutter head in real time and then substitute these data into the test theoretical model to calculate the mechanical parameters of the rock. Scratch testing has to meet the following four basic technical requirements: (1) when performing the scratch, it must be ensured that no brittle damage occurs, that is, the scratch has to be performed with plastic damage; (2) in order to ensure that it is possible to accurately measure the overall strength of the rock sample, the required minimum scratch depth must exceed the size of the rock particles rather than just measuring the strength of a particular mineral composition; (3) the choice of the speed of the scratch is closely related to the nature of the rock. The speed of scratching should be less than 10 mm/s; (4) it is important to ensure that the blade remains clean and undamaged.

Figure 4.9 shows the surface characteristics of the shale samples after scratching in the centimeter-scratch experiment. After the surface of Sample A was scratched, high-frequency thin interlayer structures were clearly observed across the scratched section. Numerous damaged areas developed on both sides of the scratch, forming a fishbone-like structure. This is due to the high clay content in sample A, and the clay layer is easy to peel off. From the surface of sample B, it can be seen that different damage patterns occur when the scratch passes through different grain layers. At the two ends of the sample, the damage characteristics of the scratch are not obvious, and the surface of the sample is relatively smooth, while obvious damage characteristics can be seen in the middle part of the sample. This is due to the fact that there is more hard material at the ends of the sample and more soft material in the middle. As can be seen from sample C, the damage characteristics of scratches are more obvious at both ends of the sample, while the damage characteristics are not obvious in the middle part of the sample, and the surface of the sample is smoother. The scratch-induced damage on the surface of Sample D was not obvious. The sample surface remained relatively smooth. The results show that the scratch experiment can well characterize the distribution of strongly inhomogeneous shale and identify the mechanically weak surface.

Fundamentals

During the scratch test, the depth of cut and the speed of the cutter were preset, and then the amount of force acting on the cutter was recorded. Figure 4.10 shows a schematic of the scratch test, where F_{T} is the horizontal force acting on the plane and F_{V} is the vertical force acting on the plane. A scratching tool is used to create a continuous groove in the rock surface. The depth of tool penetration (d) and the velocity of the tool relative to the rock are kept constant. The horizontal force (F_{T}) is parallel to the tool velocity and the vertical force (F_{V}) is perpendicular to the tool velocity. The intrinsic specific energy (ε) is the energy required to break a unit volume of rock. ε is considered a rock property characteristic directly related to the unconfined compressive strength (UCS) of the rock. The scraping tool-rock interaction model combining pure cutting and frictional contact processes and its application for estimating rock strength are shown in Eqs. 4.14–4.19 (for rectangular tools) as follows:

$$F_{\mathrm{T}} = \varepsilon(1 - \mu\xi)wd + F_{\mathrm{V}} \tag{4.14}$$

$$\mu = \tan\phi \tag{4.15}$$

$$E = E_0 + \mu S \tag{4.16}$$

$$E_0 = \varepsilon(1 - \mu\xi) \tag{4.17}$$

$$E = \frac{F_{\mathrm{T}}}{wd} \tag{4.18}$$

$$S = \frac{F_{\mathrm{V}}}{wd} \tag{4.19}$$

where w is the width of the cutter; d is the depth of tool penetration; ε is the intrinsic specific energy; ξ is the inclination of the mean force acting on the tool end face; μ is the friction coefficient at the wear surface/rock interface; φ is the angle of internal friction of the rock; S is the drilling strength of the tool; and E is the specific energy.

The hardness of the material can be determined as:

$$H = \frac{F_{\mathrm{V}}}{A} \tag{4.20}$$

where A is the load-bearing area and can be derived as $A = wd$.

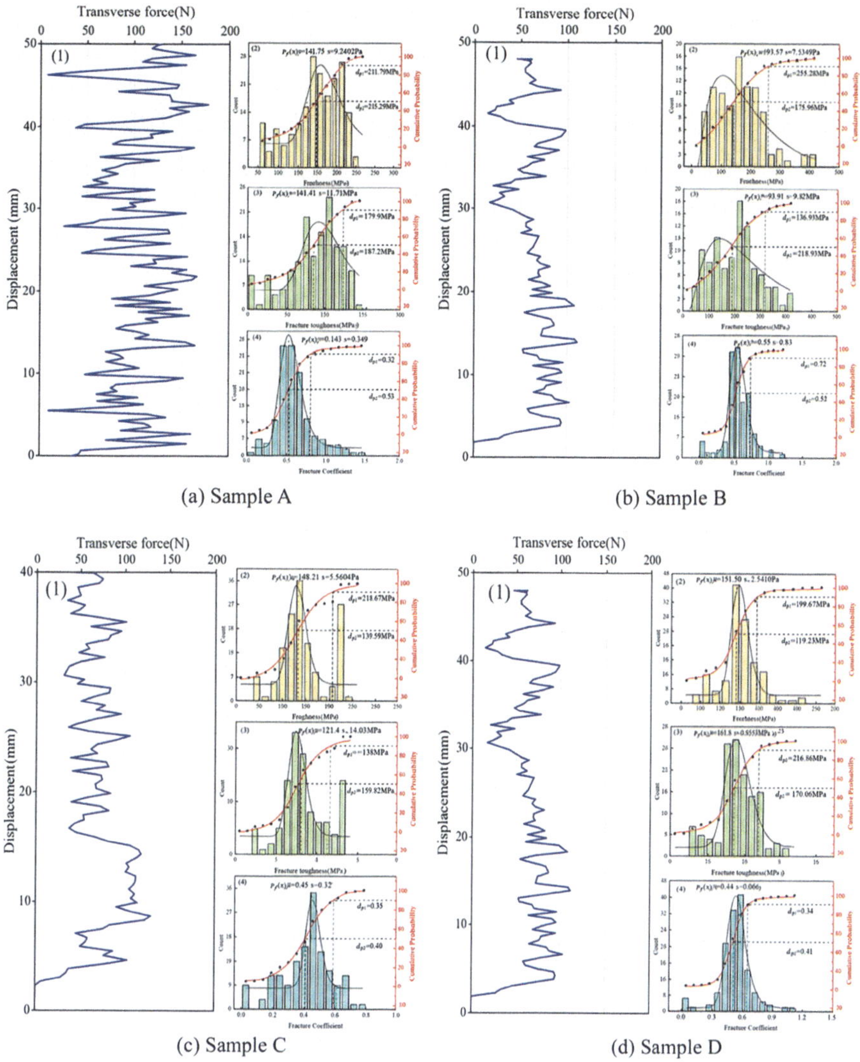

(a) Sample A (b) Sample B

(c) Sample C (d) Sample D

Fig. 4.10 Scratch results: (**a**) curve of lateral force with displacement, (**b**) hardness histogram, (**c**) fracture toughness histogram, and (**d**) friction coefficient histogram

To some extent, the modulus of elasticity and specific energy are correlated. However, this correlation is not very strong and may exist because strength and stiffness are correlated to some extent. Scratch signals are usually quite complex and obviously contain more information than the specific energy calculations, which represent only the average value of the tangential force. Assuming that the force fluctuations in a scratch test are a series of small stress accumulations similar to those in a uniaxial test, the derivative of the force can be

$$\chi = \frac{{}^{d}F_{\mathrm{T}}}{dL} \tag{4.21}$$

It is closely related to the rock stiffness. L is the displacement of the tool along the rock surface.

The average of the positive peaks of the material $dF_{\mathrm{T}}/d_{\mathrm{L}}$ curve correlates well with the modulus of elasticity modulus, defined as:

$$E = \frac{\frac{{}^{d}F_{\mathrm{T}}}{dL}}{wd} \tag{4.22}$$

Scratch fracture mechanics modeling, under plane strain conditions, the energy release rate can be expressed relationally as a function of scratch force and scratch probe geometry:

$$G = \frac{1 - V^2}{E} \frac{F_{\mathrm{eq}}^2}{2\rho A} \tag{4.23}$$

where E is the modulus of elasticity, ν is Poisson's ratio, and F_{eq} is the equivalent force considering the lateral force F_{T} and the vertical force F_{V}.

To mark the onset of cracking, a threshold criterion applicable to the energy release rate is utilized: Crack extension occurs once $G = G_{\mathrm{f}}$, where the threshold is the fracture energy G_f, assumed to be a material constant. From Eq. (2.11), it is possible to relate the fracture energy G_f to the scratch toughness K_s, where E is the elastic modulus and ν is Poisson's ratio.

$$G_{\mathrm{f}} = \frac{1 - V^2}{E} K_{\mathrm{s}}^2 \tag{4.24}$$

Therefore, the linear elastic fracture mechanics (LEFM) of anisotropic materials in estimating the relationship between rock scratch toughness (K_s) and fracture energy release rate is

$$K_{\mathrm{s}} = \frac{F_{\mathrm{eq}}}{\sqrt{2\rho A}} \left[\mathrm{MPa} \cdot \sqrt{m}\right] \tag{4.25}$$

where A is the contact area of the rectangular tool under horizontal force in the direction of the scratch, ρ is the perimeter of the fracture surface at the maximum depth of cut (d), and $2\rho A$ is the shape function of the tool, where F_{eq} is the equivalent force considering the lateral force F_{T} and the vertical force F_{V}.

$$F_{\mathrm{eq}} = \sqrt{F_{\mathrm{T}}^2 + \frac{3}{5} F_{\mathrm{V}}^2} \tag{4.26}$$

The relationship between fracture toughness (K_{IC}) and scratch toughness (K_s) can be expressed as:

$$K_s\left(F_{eq}, d\right) = K_{IC} \tag{4.27}$$

Experimental Results

The left-hand panel of Fig. 4.10 shows an example of the variation in transverse force recorded when the surface of a shale sample was incised in a centimeter-scratch experiment. The curve can be based on the fact that the average values of the transverse force are all subdivided into several different regions, which can be caused by variations in mineral composition. In Fig. 4.10, the right-hand side plot distribution shows the probability density function and cumulative probability density function plots for hardness, fracture toughness, and friction coefficient over the four samples. 50% probability (d50) were 137.75 MPa and 1.81 MPa m. The histograms show that Sample B had the highest mean scratch hardness and fracture toughness. The corresponding values were 195.57 ± 73.54 MPa and 2.01 ± 0.92 MPa·m$^{1/2}$, respectively. At the 90% probability level (d90), the scratch hardness and fracture toughness were 255.27 MPa and 3.13 MPa·m$^{1/2}$, respectively. At the 50% probability level (d50), they were 137.75 MPa and 1.81 MPa·m$^{1/2}$, respectively. In contrast, Sample A showed the lowest mean scratch hardness and fracture toughness, with values of 141.78 ± 9.44 MPa and 1.41 ± 0.11 MPa·m$^{1/2}$, respectively. The fracture toughness value and hardness of sample B were approximately about one and a half times higher than that of sample A. The fracture toughness and hardness of sample B were approximately 1.5 times those of sample A. For samples C and D, the fracture toughness values were 1.75 ± 0.04 and 1.61 ± 0.05 MPa m$^{1/2}$, respectively, while the corresponding hardness values were 145.21 ± 5.54 and 151.50 ± 2.54 MPa, respectively. The coefficients of friction were 0.54, 0.55, 0.45, and 0.44 for samples A, B, C, and D, respectively. This indicates that sample B and sample A have greater surface roughness than that of sample C and sample D. As can be seen from sample C, there is an obvious mineral separation interface, and the values of transverse force in the 0-5 mm test section and the 16-40 mm test section are significantly smaller than those in the 5–15 mm test section. The value of transverse force in the 5–15 mm test section is about 100 N, and the value of transverse force in the 0–5 mm test section and in the 16–40 test section is about 55 N.

Figure 4.11 shows that hardness and fracture toughness show the same trend along the length of the scratch in the centimeter-scratch experiment. Sites with large hardness values tend to have larger fracture toughness values. Figure 4.12 similarly shows that the fracture toughness of the sample in the centimeter-scratch experiment can be estimated from the hardness using the following equation:

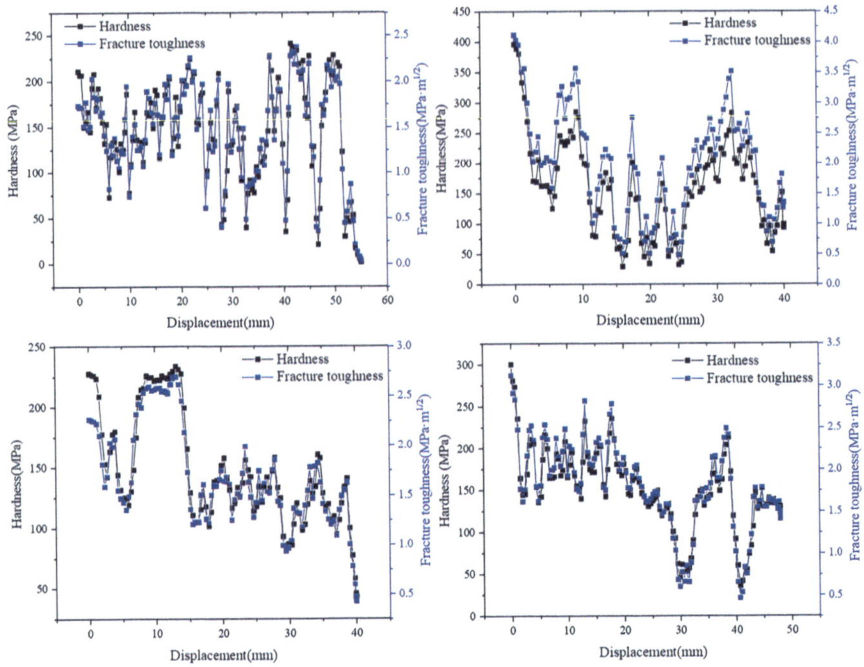

Fig. 4.11 Variation of hardness and fracture toughness along the scratch length

$$K_c = c \times H + d \tag{4.28}$$

where c and d are the fitting coefficients. For the studied striated shale samples, c ranged between 0.088 and 0.104 in the centimeter-scratch experiments.

4.2.2 Nanoscratch

In order to further obtain the micromechanical property characteristics of the striated shale samples and explore the micro-nonhomogeneity of the mineral distribution on the shale surfaces, small pieces of processed samples were selected from the samples after completing the centimeter-scratch experiments, and the mechanical properties of the sample surfaces were quantitatively evaluated and analyzed by the scratch experiments at the nanoscale. Figure 4.13 demonstrates the surface characteristics of the shale samples after scoring in the nanoscratch experiment. As can be seen from the figure, after scratching the surface of sample A, it is found that the whole section of the scratched surface can be seen as a clear thin interlayer structural feature, and a large number of damage surfaces appear on both sides of the scratch. From the surface of sample B, it can be seen that when the scratch passes through the harder

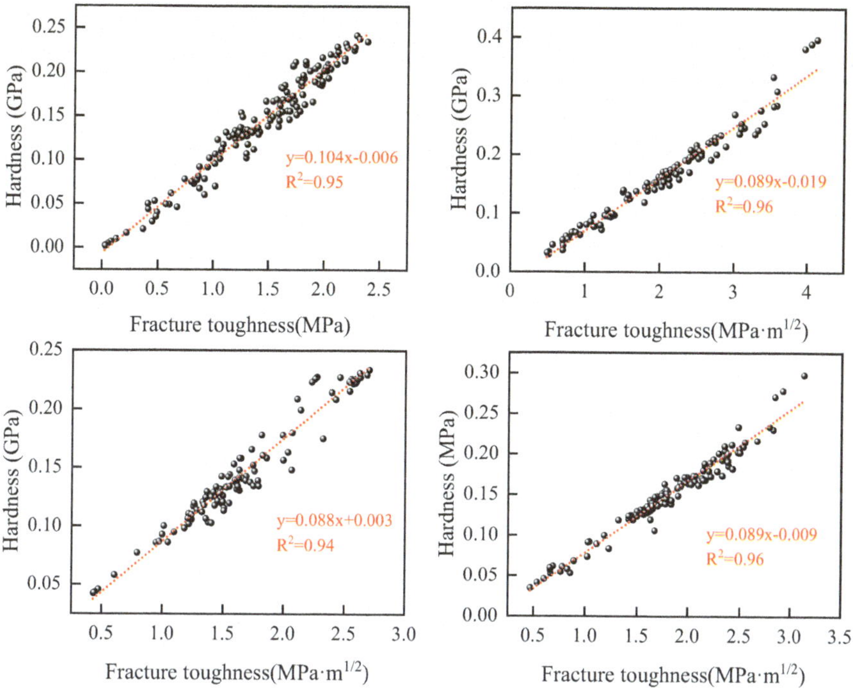

Fig. 4.12 Correlations between the hardness and fracture toughness

grain layer, the damage characteristics are not obvious at both ends of the sample, and the surface of the sample is relatively smooth. Sample C shows that the surface of the sample is relatively homogeneous, the damage characteristics of the scratch at both ends are not obvious, and the surface of the sample is relatively smooth. Sample D shows that the surface of the sample is more non-homogeneous, with more minerals, and the scratch damage characteristics are more obvious.

The results are shown in Fig. 4.14. Figure 4.14a demonstrates the variation of transverse forces recorded while scoring the surface of the shale samples. Figure 4.14 (b), (c), and (d) demonstrate the distributions of hardness, fracture toughness, and coefficient of friction values for the four samples. From the histograms, it can be seen that sample C has the largest mean value of scratch hardness and fracture toughness in the nanoscratch experiment with the maximum values of 3.55 GPa and 3.41 MPa $m^{1/2}$, respectively, whereas sample A has the smallest mean value of scratch hardness and fracture toughness in the nanoscratch experiment with the minimum values of 1.92 GPa and 1.45 MPa $m^{1/2}$, respectively. Sample B has the highest mean values of scratch, the mean values of scratch hardness and fracture toughness for sample B in the nanoscratch experiment were 3.42 GPa and 2.43 MPa $m^{1/2}$, respectively, while the mean minimum values of scratch hardness and fracture toughness for sample D in the nanoscratch experiment were 2.32 GPa and 2.21 MPa $m^{1/2}$, respectively.

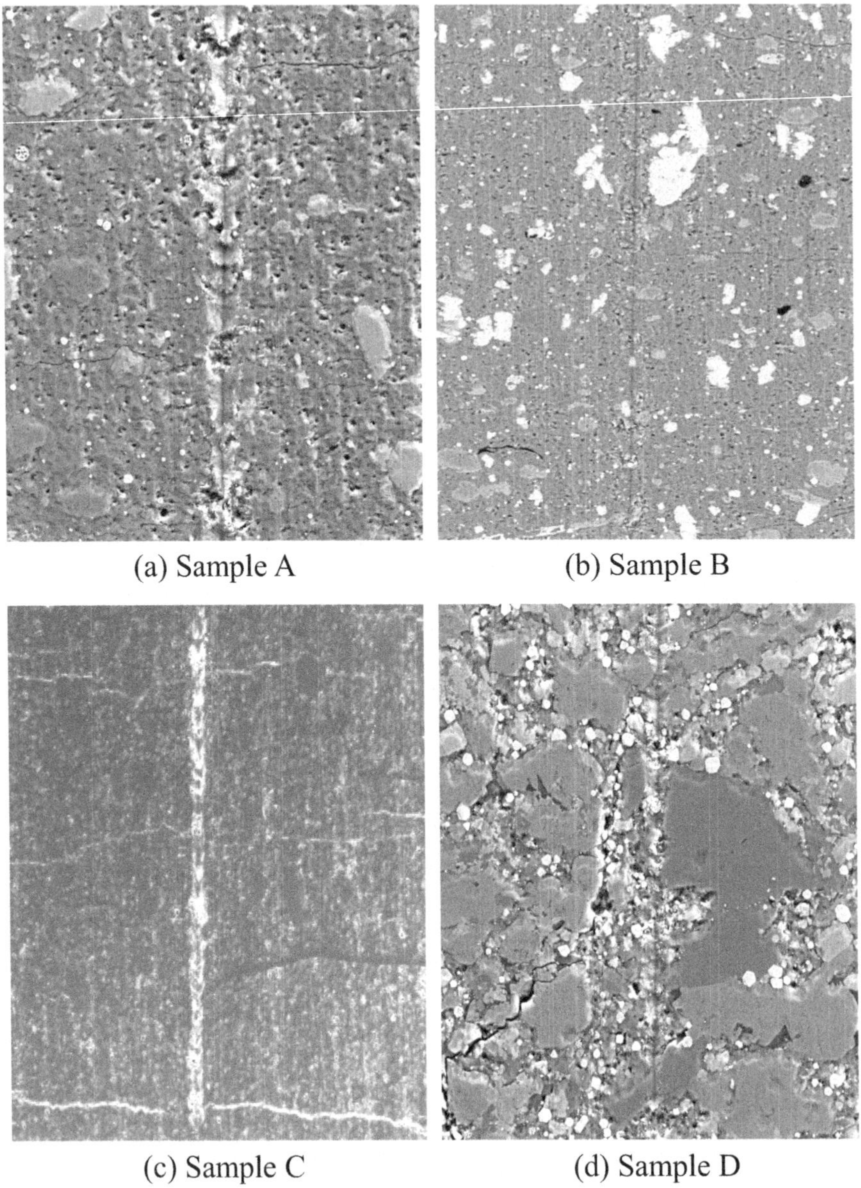

Fig. 4.13 Sample after scratch test. (a) Sample A. (b) Sample B. (c) Sample C. (d) Sample D

Figure 4.15 shows a similar relationship in the nanoscratch experiment as in the centimeter scale. Hardness and fracture toughness likewise show the same trend along the length of the scratch. Sites with large hardness values tend to have larger fracture toughness values. We further compared the hardness and fracture toughness.

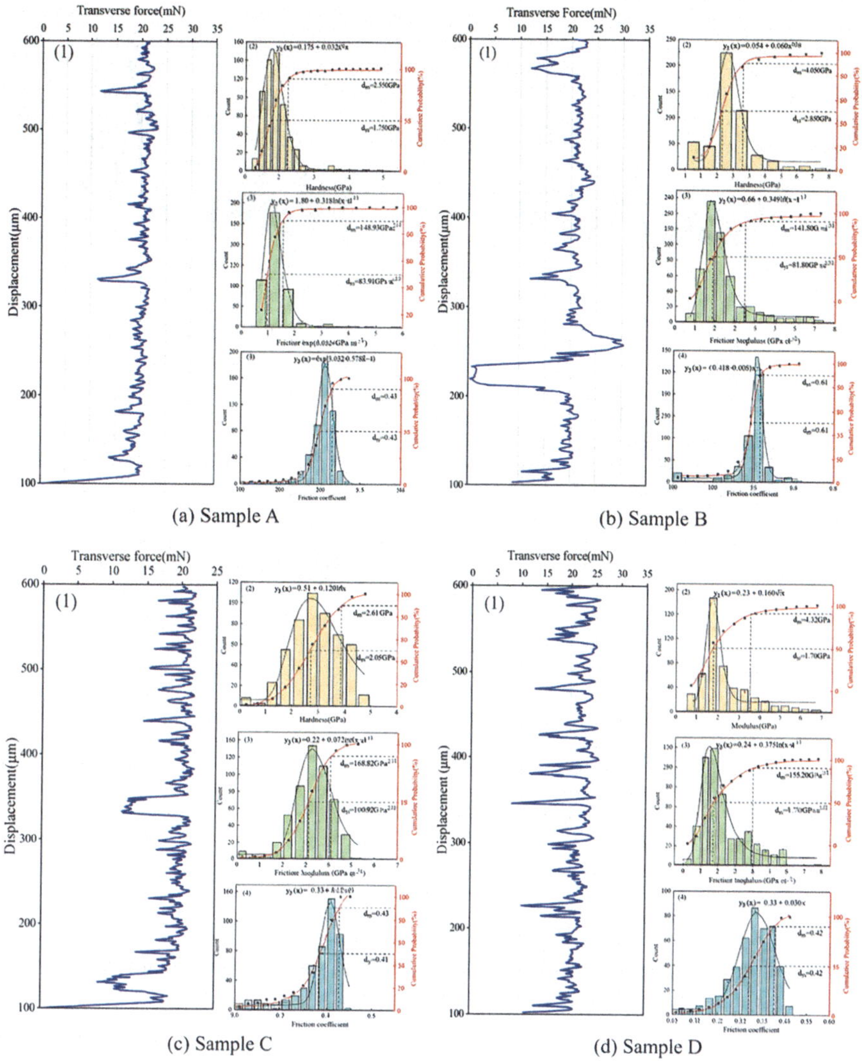

(a) Sample A (b) Sample B (c) Sample C (d) Sample D

Fig. 4.14 Scratch results: (**a**) Curve of lateral force with displacement, (**b**) hardness histogram, (**c**) fracture toughness histogram, and (**d**) friction coefficient histogram

As shown in Fig. 4.16, it indicates that there is some linear relationship between the fracture toughness and hardness of the samples, which we can estimate from the hardness using the following equation:

$$k_c = a \times H + b \tag{4.29}$$

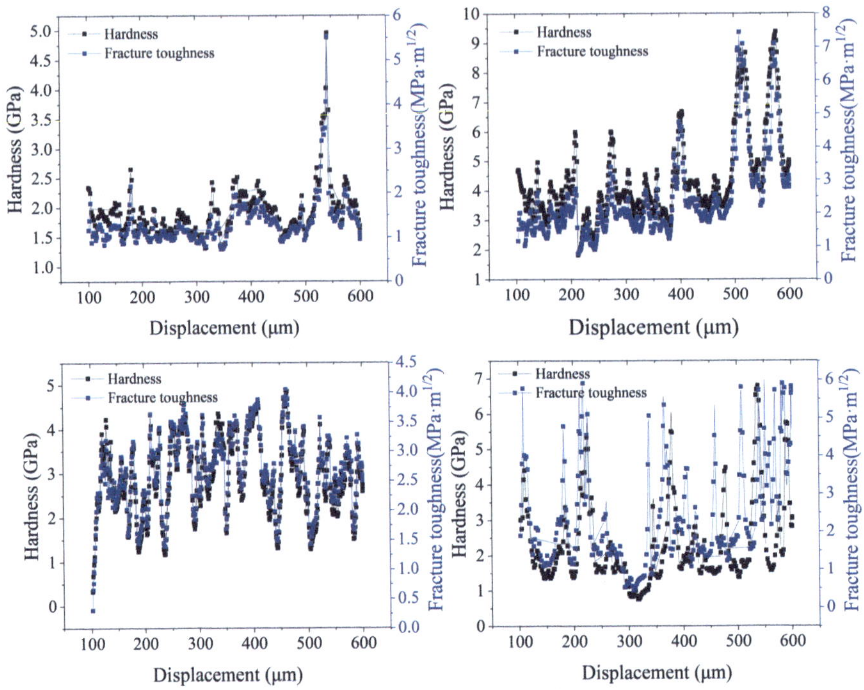

Fig. 4.15 Variation of hardness and fracture toughness along the scratch length

where a and b are the fitting coefficients. For the studied grain shale samples, a is between 0.712 and 1.885 in the nanoscratch experiments. However, the slope values may be different for samples of different types and regions, which need to be further confirmed in future experiments.

As shown in Figs. 4.17 and 4.18, there is a clear correlation between mineral composition and fracture toughness despite the limited number of samples. As the quartz + feldspar mineral content increases, the scratch fracture toughness and hardness values show an increasing trend. However, both scratch fracture toughness and hardness values decreased with increasing clay content. This is due to the lower hardness of clay minerals and the higher hardness of quartz and feldspar minerals.

4.2.3 Difference Analysis

A comparison of the fracture toughness obtained by two different methods and two different scales, the nanoscratch method and the centimeter-scratch method, is shown in Fig. 4.19. As can be seen from the figure, the fracture toughness at the centimeter scale is significantly lower than the value at the nanoscale, and the fracture toughness values derived from nanoscratching are one to three times higher

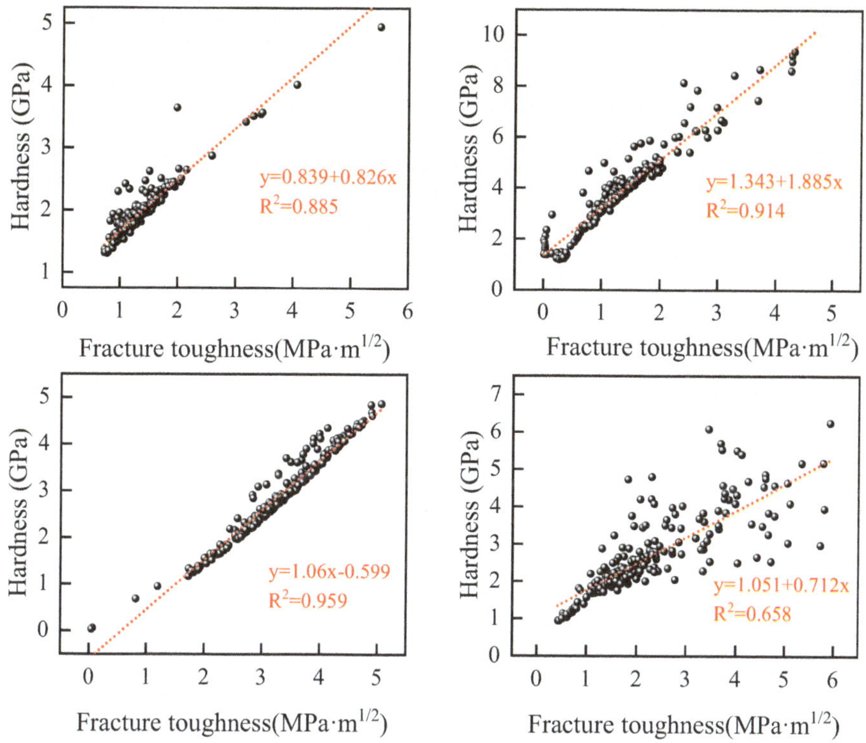

Fig. 4.16 Correlations between the hardness and fracture toughness

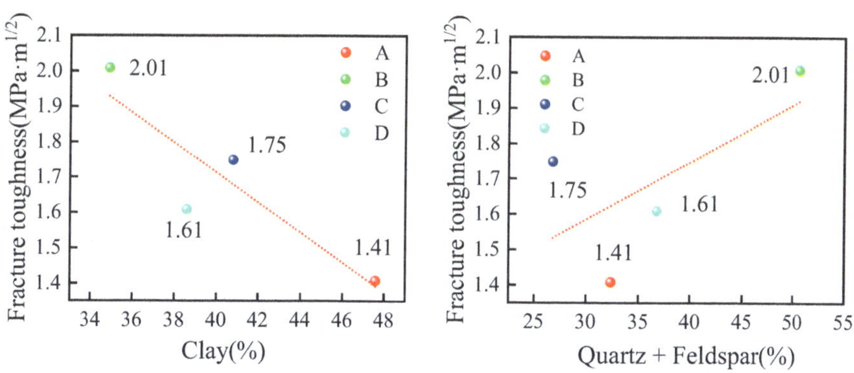

Fig. 4.17 Impact of the mineral compositions on the fracture toughness

than those derived from centimeter scratching. This is because the two methods are not only based on different types of measurements but also on different scales.

Figure 4.20 shows the correlation between the fracture toughness obtained from both the nanoscratch and centimeter-scratch experiments. A polynomial regression

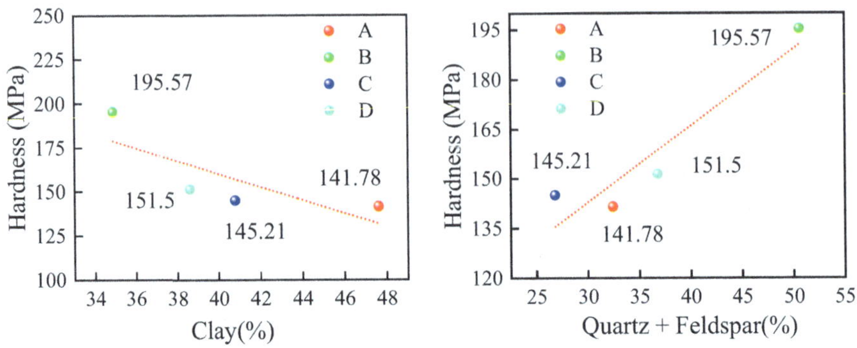

Fig. 4.18 Impact of the mineral compositions on the hardness

Fig. 4.19 Comparison of fracture toughness between nanoscratch and centimeter scratch

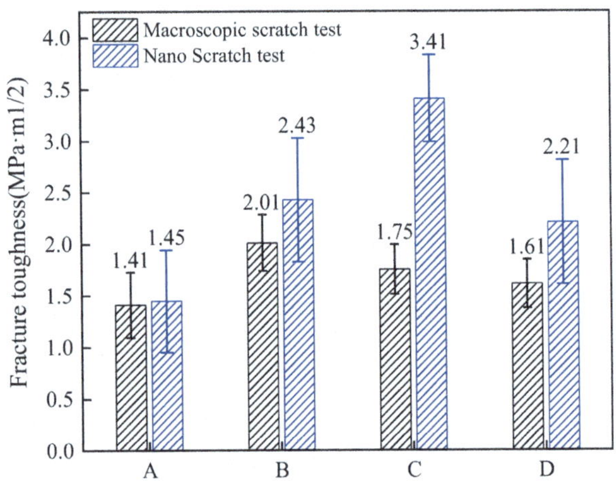

was fitted to the fracture toughness values obtained at the two different scales, and the curve fitting results show that the values of nanoscratch fracture toughness and centimeter-scratch fracture toughness are correlated. The fitting effect is not the same for shales of different lithologies. The polynomial fitting equations are different for samples of different lithologies. As can be seen in Fig. 4.21, there are errors in the experimental and predicted data for different samples because the two methods are not only based on different types of measurements but also have different scales. So there may be some errors. For sample A and sample B with high clay content, the error is the largest, which is related to the development of shale grain layer of sample A and sample B. Sample C has the smallest error and sample D has the second highest error.

The results in Fig. 4.21 show that after data preprocessing the scratch data at both scales fit better and the correlation between the experimental and predicted data is high. The above results show that the small number of samples and the two different scales lead to some error margins in the results. However, as the number of

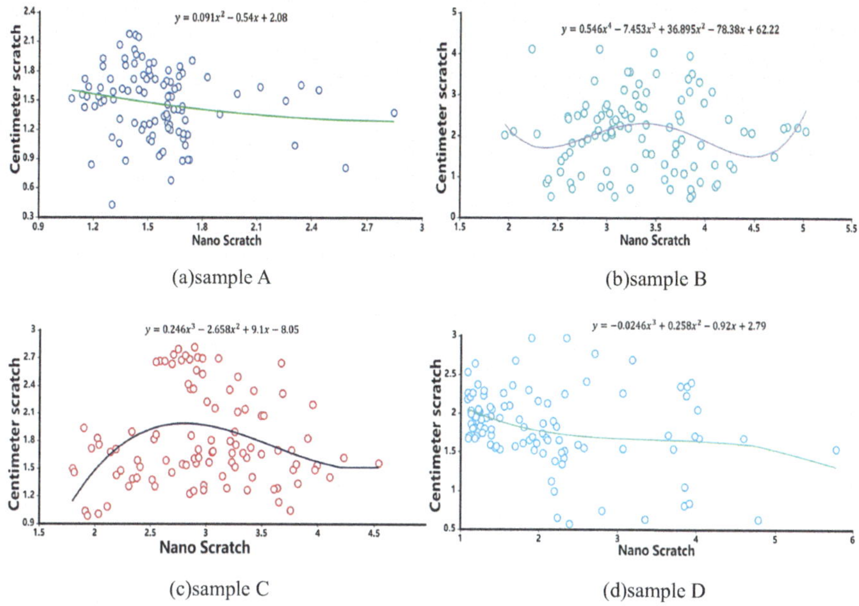

Fig. 4.20 The experimental data of nanoscratch and centimeter scratch were fitted. (**a**) Sample A. (**b**) Sample B. (**c**) Sample C. (**d**) Sample D

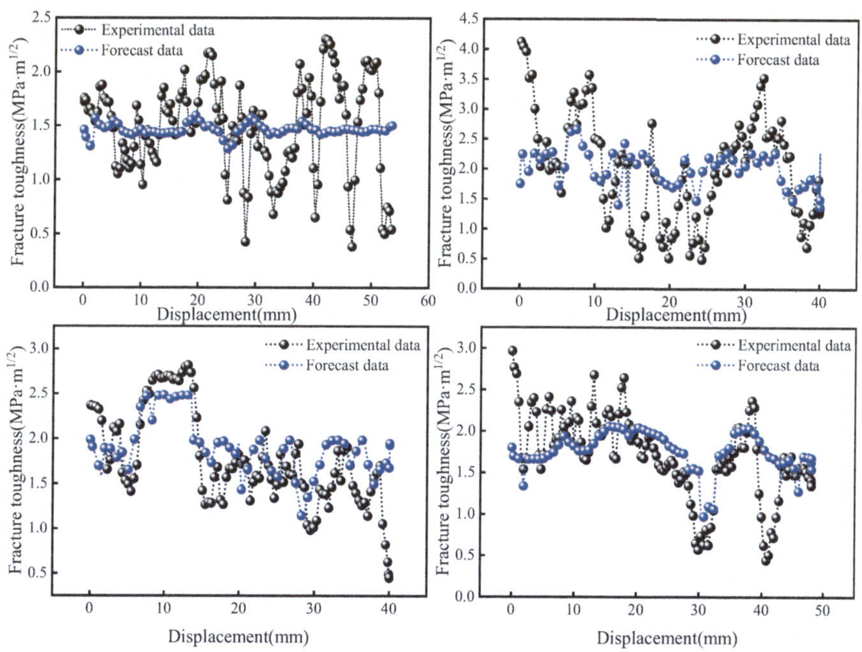

Fig. 4.21 Experimental data and predicted data of shale fracture toughness

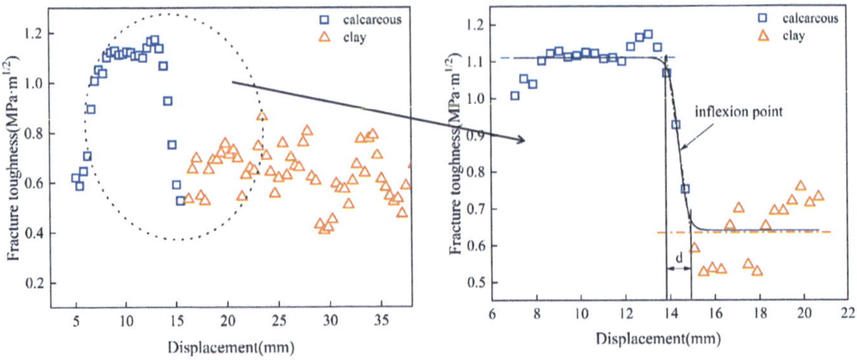

Fig. 4.22 Study on the interface thickness of sample C

experiments increases, a more accurate calculation method may be derived to convert the fracture toughness from one method to the other. Shale drill cores usually consist of a number of intact and incomplete sections, which presents a sampling challenge. The nanoscratch method can utilize small samples to obtain fracture toughness, but requires relatively large experimental time and sample preparation effort. The centimeter-scratch test, on the contrary, is very fast but requires a relatively intact core section. The experimental results show that the fracture toughness values obtained at both scales can be converted by a polynomial relationship, and that the combination of the nanoscratch test and the centimeter-scratch test can provide a general tool for evaluating the fracture toughness of a range of shale segments, including both intact and non-intact core samples. In this study, only the fracture toughness of shale was compared by the nanoscratch test and the centimeter-scratch test. In the future, the fracture toughness obtained from the three-point bending experiment will also be compared with the fracture toughness obtained from the nanoscratch and centimeter-scratch tests.

For shale, grinding and polishing processes are necessary to prepare a smooth surface for scratch testing. However, it is difficult to obtain surfaces with low roughness due to many factors, such as human factors, grinding, and polishing materials and processes. As a result, the COF method cannot accurately quantify the interfacial thickness due to the effect of the COF measured on the surface roughness of the shale samples. This introduces uncertainty to the quantification of interfacial thickness by the COF method. Previous scratch test results have revealed that sample C has a distinct mineral interface and that sample C has a carbonate content of 32.4% and that the carbonate mineral is predominantly composed of dolomite. The scratch fracture toughness curve divides this sample into two distinct regions, with a clear differential change in fracture toughness between the dolomite and the clayey material. Therefore, fracture toughness curves derived from scratch experiments are proposed to be used for interfacial studies. As shown in Fig. 4.22, d denotes the interface thickness. The fracture toughness data were first fitted by S-shape analysis. The inflection point on the fitted curve was then identified as the tangent point. A tangent line through this point was drawn and intersected with the extension of the two baseline lines. The region between the two intersections was identified as the interface. The thickness of the interface was derived to be 1.05 mm. The use of fracture toughness showed superiority in terms of interface thickness with a fit R^2 of 0.93.

References

1. Diao Z, Huo F, Li P. Characterization of shale pore heterogeneity and its controlling factors: a case study of the Longmaxi formation in Western Hubei, China. Energy Sci Eng. 2024;12(7): 2899–917.
2. Manjunath GL, Jha B. Geomechanical characterization of Gondwana shale across nano-micro-meso scales. Int J Rock Mech Min Sci. 2019;119:35–45.
3. Sun C, Li G, Elgharib GM, et al. Meso-scale mechanical properties of mudstone investigated by nanoindentation. Eng Fract Mech. 2020;238:107245.

Chapter 5
Finite Element Simulation of Nanoindentation-Scratch

In the previous chapters, nanoindentation experiments and theoretical derivation were carried out to study the factors affecting the work energy of laminated shale when soft and hard minerals mixed under the striated development conditions. Combined with indentation energy analysis, the ratio U_e/U_t of elastic energy to total energy of laminated shale is affected by dimensionless parameters $\nu_1 + \nu_2$, $Y_1 + Y_2$ and $E_1 + E_2$ of laminated minerals. When studying the mechanical properties of shale, it is necessary to consider the composition and structure of the lamination comprehensively, such as the characteristics of different lamination Poisson's ratio, lamination thickness, and different lamination dip angle on the microscale, instead of only considering a single mechanical parameter. In order to further deduce the specific relationship between mechanical parameters in the process of shale indentation and scratch experiment, commercial software ABAQUS is used to analyze and simulate several influential factors in the indentation and scratch experiment. Using finite elements to simulate the scratch characteristics of the material surface, not only can we obtain the stress distribution of the material, but also we can predict the tendency of crack generation and extension of the material under loading [1]. In addition, based on the magnitude analysis and nanoscratch finite element analysis, the plastic parameters such as yield stress, strain hardening index, and interfacial friction coefficient of the material can be combined dimensionlessly and fitted with the required coefficients [2]. Based on the heterogeneous characteristics, a two-dimensional plane model is established under the mixed conditions of soft and hard minerals, and the lamination process of rich laminated shale with different lamination parameters is simulated. The effects of different scratch depth and scratch displacement on stress transfer and failure modes are discussed based on 3D nanoscratch simulation. The curve fluctuation patterns of ITZ interval of continuous indenter measurement are analyzed.

L. Yang, *Multiscale Scratch Mechanics of Laminated Shale*,
https://doi.org/10.1007/978-981-95-7041-6_5

5.1 Finite Element Model

5.1.1 Dimensional Systems

ABAQUS does not have its own unit system. When building the model, it is necessary to determine the dimension adopted. The commonly used consistency dimension system is shown in Table 5.1.

Common dimensional systems are SI, SI (mm), US Unit (ft), and US Unit (inct). The sample thickness and particle size prepared in this paper are in the micro- and nano-level, so the international SI (mm) unit system is selected. In order to reduce the calculation time, the finite element simulation of nanoindentation is simplified as a plane strain problem, and a three-dimensional model is used for nanoscratches. When establishing the simulation model, the pressing depth should be less than 0.1 times of the total height of the material. Set the material height to 5 μm, the pressing depth to 500 nm, and the model width to 10 μm. This can greatly reduce the workload and time of analysis and calculation on the basis of ensuring correctness. Three components are established, from top to bottom are a flexible elastic thin layers, a rigid body indenter, and a base, as shown in Table 5.2.

5.1.2 Basic Parameter Setting

The core of ABAQUS software lies in two built-in solvers, namely ABAQUS/Standard and ABAQUS/Explicit. Explicit is an explicit analysis solver, an analysis

Table 5.1 ABAQUS dimensional system

Quantifier	SI	SI (mm)	US unit (ft)	US unit (inct)
Length	m	mm	ft	in
Load	N	N	lbf	lbf
Quality	kg	Tonne	slug	$lbfs^2/in$
Time	s	s	s	s
Stress	Pa (N/m^2)	MPa (N/mm^2)	lbf/ft^2	psi
Energy	J	mJ	ft lbf	in lbf
Density	kg/m^3	$tonne/mm^3$	$slug/ft^3$	$lbf\ s^2/in^4$

Table 5.2 Types of components of model

Part name	Nanoindentation components	Nanoscratch components
Compressible elastic thin layer	Two-dimensional variable body	Three-dimensional variable body
Base	Two-dimensional analytic rigid body	Three-dimensional analytic rigid body
Conical indenter	Two-dimensional analytic rigid body	Three-dimensional analytic rigid body

module for special purposes, which uses explicit integration of time to solve dynamic finite element equations. Explicit analyzers are suitable for dealing with dynamic and high-speed shocks.

Since mechanical behavior at the nanoscale is usually slow and stable, static analysis is suitable. Static analysis can accurately predict the deformation and stress distribution of materials in nanoindentation experiments, which provides an important reference for the study of mechanical properties of nanomaterials. In addition, for small-scale models, more accurate and stable solutions can be obtained using the ABAQUS implicit analyzer, and convergence can be achieved faster. Combined with the micro-nano scale model studied, ABAQUS/Standard solver is selected for numerical simulation in this chapter, and the specific settings are shown in Fig. 5.1.

The contact algorithm uses a simple master-slave contact calculation method, and the contact mode between the indenter and the measured material is chosen as the face-surface contact mode. A node on the slave surface cannot invade any part of the master surface, and the master surface can invade the slave surface between nodes on the slave surface. As shown in Fig. 5.2, combined with the actual nanoindentation experiment, the indenter was set as the main plane, the measured material was set as the slave plane, and the normal contact was set as "hard contact." When using the finite sliding contact formula, ABAQUS will often determine the contact state of the area on the main surface with each node on the slave surface. When the contact pressure becomes zero or negative, the contact surface is separated, the constraint is

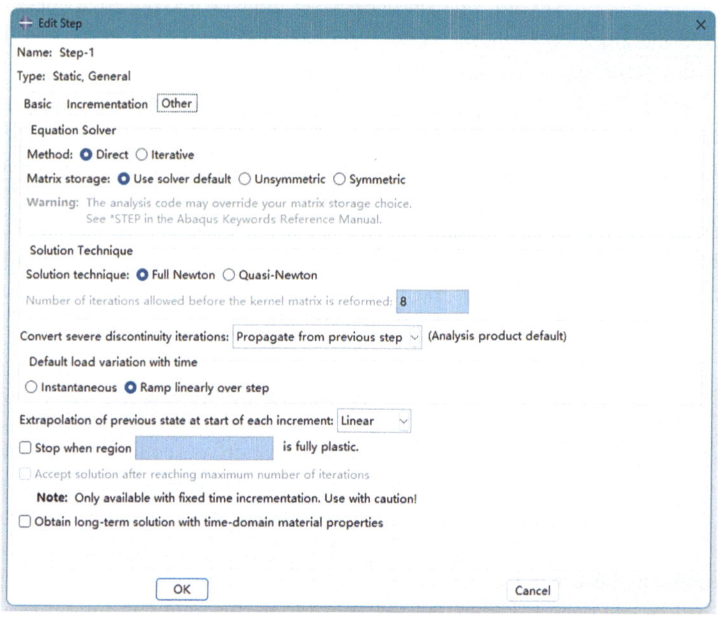

Fig. 5.1 Solver settings

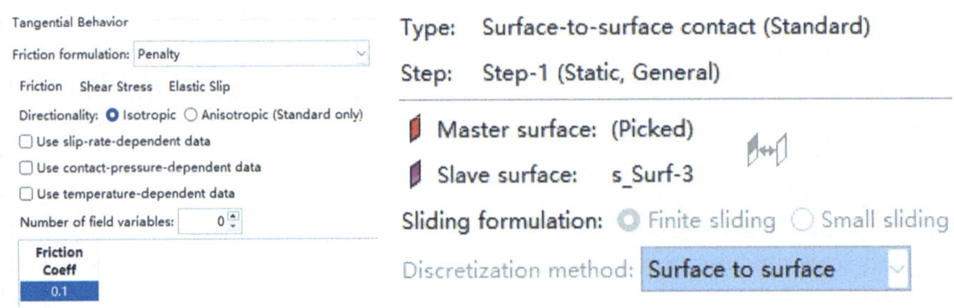

Fig. 5.2 Friction coefficient and contact condition settings

withdrawn, the tangential contact attribute is limited slip, set as "penalty contact," and the friction coefficient is set as 0.1 for easy calculation.

5.1.3 Indenter and Material Modeling

Modeling of Two-Dimensional Indenter for Nanoindentation

For the axisymmetric structure of the model, the two-dimensional finite element simulation has a relatively close calculation accuracy to the three-dimensional simulation. In order to save calculation time, the two-dimensional modeling of the nanoindentation is carried out. The angle between the side and the adjacent surface is 140.6°, and the angle between the vertical line and the surface is 70.3°, which has a high elastic modulus and hardness. The elastic modulus of indenter is set at 1140 GPa and Poisson's ratio is 0.07, which is not prone to deformation and can avoid the coupling effect of indenter on specimen measurement. The Berkovich indenters are set to maintain self-similarity in small scales, thus reducing the influence of friction and ensuring suitability for small-scale indentation experiments. The nanoindentation tip simulation two-dimensional geometry model is shown in Fig. 5.3. The half angle of the indenter in the model is set to 70.3°, which has the same projected area as the real Berkovich indenter.

Modeling of Three-Dimensional Indenter for Nanoscratches

In nanoscratch finite element simulation, the residual cross-sectional profile, friction coefficient, and residual stress along the scratching direction on the sample surface change considerably when the sample surface is damaged by a larger indenter angle [3]. The amount of sample damage buildup and the friction coefficient decrease with the increase in the tip radius of the indenter, and the desired scratch profile and surface can be obtained by adjusting the appropriate tip radius in the finite element simulation of the scratch test [4]. It is essential to predict and measure the effect of

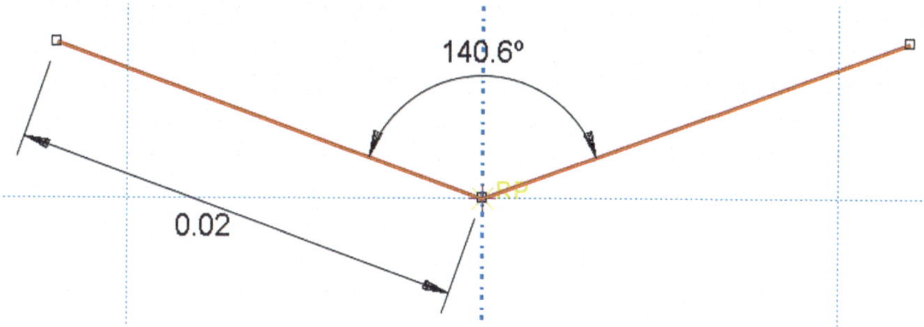

Fig. 5.3 Simulation geometric model of nanoindentation tip

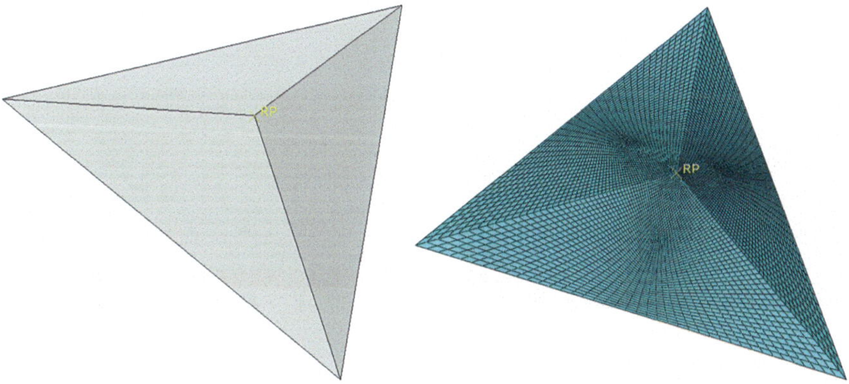

Fig. 5.4 Three-dimensional geometric model of nanoscratch indenter

the indenter tip radius during the scratch test; otherwise, the accuracy of the nanoscratch finite element simulation results can be seriously affected [5].

During nanoscratch experiments, the shape and geometric characteristics of the indenter and the material under test may be non-axisymmetric. Compared with 2D modeling, 3D modeling can more truly reflect the actual situation and accurately describe the deformation and stress distribution during the scratch process. In addition, compared with two-dimensional modeling, three-dimensional modeling can provide more comprehensive and accurate information about scratch morphology, which is helpful for in-depth understanding of scratch mechanism. To sum up, three-dimensional modeling is adopted in the finite element simulation of nanoscratches, and the three-dimensional model of the indenter is shown in Fig. 5.4.

Two-Dimensional Planar Modeling of Nanoindentation

Since the indentation test is a local test method, the overall size of the specimen is affected by the shape of the indenter, the depth of the indenter, and the properties of the material. In general, the dimensional size of the sample is about 10 times the depth of the compression to ensure that the sample is not affected by the constraint change of the outer interface of the material. A plane model of 100 × 50 μm was established considering the depth and stress transfer range of the experiment, and the real stress and plastic strain data of laminated shale were selected as material parameters to be input into the finite element model for calculation. In order to more closely simulate the lamellar shale pressing process, the laminate is located above the substrate rigid material. As shown in Figs. 5.5, 5.6 and 5.7, the total layer thickness is set to 250 nm. The lamination inclination is set to 0°, 30°, 45°, and 90°, and the lamination thickness is 250 nm, 500 nm, 1000 nm, and 2000 nm, and the distribution of plastic minerals and brittle minerals is overlapping. In the finite element simulation process, the indentation is completely pressed into the two layers of minerals, and the pressure and energy transfer range of the indenter are far enough

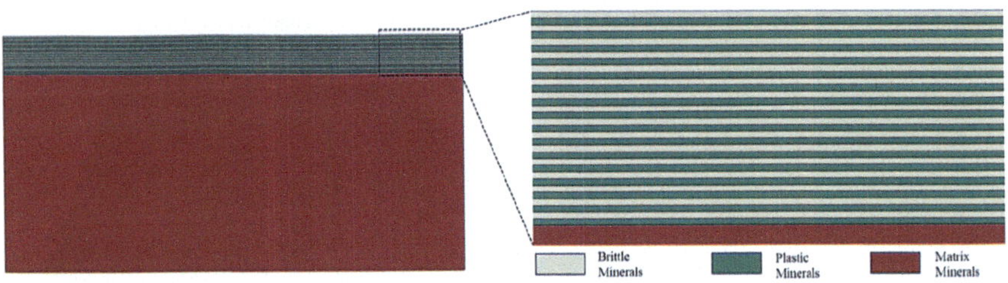

Fig. 5.5 Modeling of horizontal layered shale materials

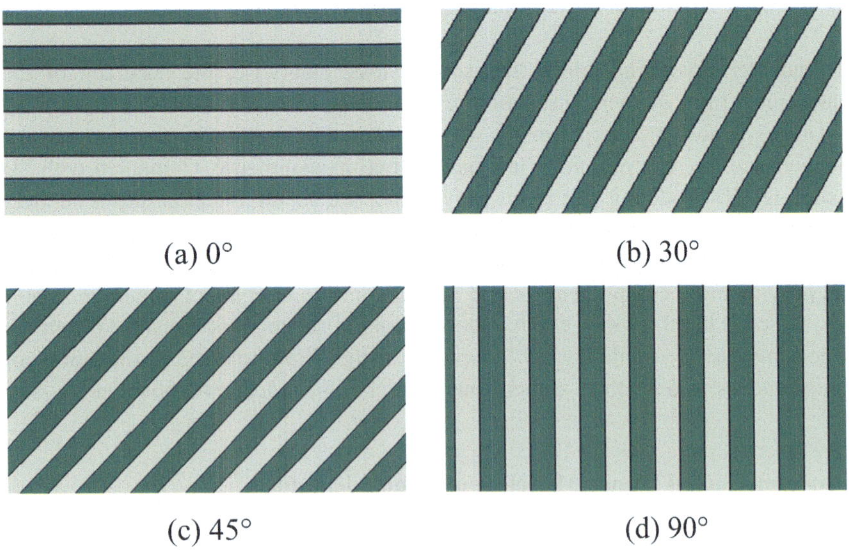

Fig. 5.6 The inclination angle of the two-dimensional model of nanoindentation. (**a**) 0°. (**b**) 30°. (**c**) 45°. (**d**) 90°

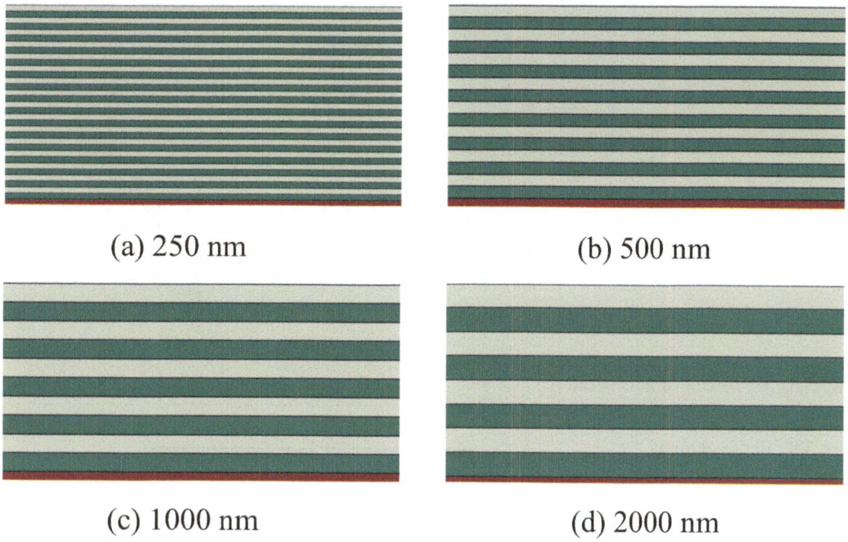

Fig. 5.7 Layer thickness of two-dimensional nanoindentation model. (**a**) 250 nm. (**b**) 500 nm. (**c**) 1000 nm. (**d**) 2000 nm

from the substrate material to avoid the influence of the collected mechanical parameters on the substrate material.

Nanoscratch 3D Cube Modeling

In nanoscratches, numerical simulation focuses on collecting mechanical property parameters of materials across regions, and three-dimensional modeling can better consider the volume effect of materials, so as to more accurately predict the mechanical response of scratches and material properties. The three-dimensional morphology of materials after destruction has an important impact on understanding deformation behavior and mechanical properties. The calculation time of nanoscratch test is slow because of the large number of mesh of three-dimensional model. The model should be built to ensure that the dimensional size is about 10 times the depth of compression, to ensure that it is not affected by the constraint change of the outer interface of the material. As shown in Fig. 5.8, a cube model with a length, width, and height of 700 × 400 × 100 μm is established, and a height of 30 μm is selected for partitioning for local encryption. The mechanical properties obtained by nanoindentation of laminated continental shale are selected as material parameters to be input into the finite element model for calculation. The stress and energy transfer range of the indenter is far enough from the substrate material to avoid the influence of the collected mechanical parameters on the substrate material.

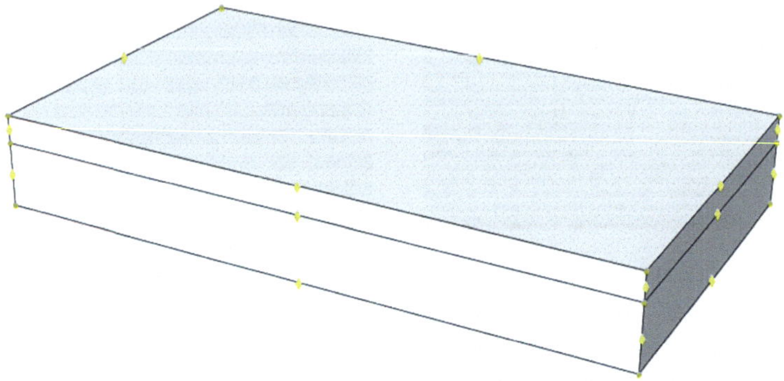

Fig. 5.8 3D model of nanoscratch

Fig. 5.9 Finite element model of layered shale

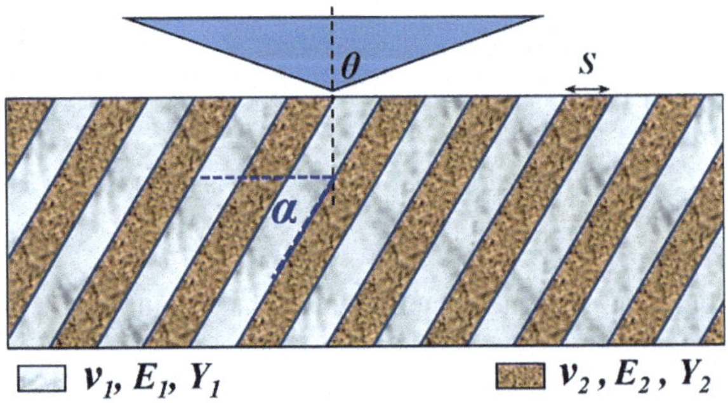

The Nanoindentation Model Gives Attributes

The finite element simulation model of nanoindentation is shown in Fig. 5.9. Poisson's ratio, elastic modulus, and yield strength of brittle minerals are ν_1、 E_1, and Y_1, respectively; Poisson's ratio, elastic modulus, and yield strength of plastic minerals are ν_2, E_2, and Y_2, respectively; layer thickness is S and interlayer inclination is α. Based on the elastic modulus and hardness of laminated shale obtained by nanoindentation experiment and considering the extreme situation of pure plastic or pure brittle minerals under the indenter, the material properties of each part in the setting model of each mineral are shown in Table 5.3.

The Nanoscratch 3D Model Gives Attributes

In ABAQUS, Maxe damage is a common constitutive relation used to describe the damage behavior of a material during stress. Maxe damage model is based on the linear elastic theory and takes into account the elastic and plastic damage deformation of materials in the process of stress, which is used to describe the damage behavior of various materials, including metals, composite materials, and rocks. In

Table 5.3 Mechanical parameters of minerals in finite element model

E_1 (Gpa)	Y_1 (Mpa)	E_2 (Gpa)	Y_2 (Mpa)	v	S (nm)	α (°)
60	130	20	50	0.1	250	0
80	170	30	70	0.15	500	30
100	210	40	90	0.2	1000	45
120	250	50	120	0.3	2000	90

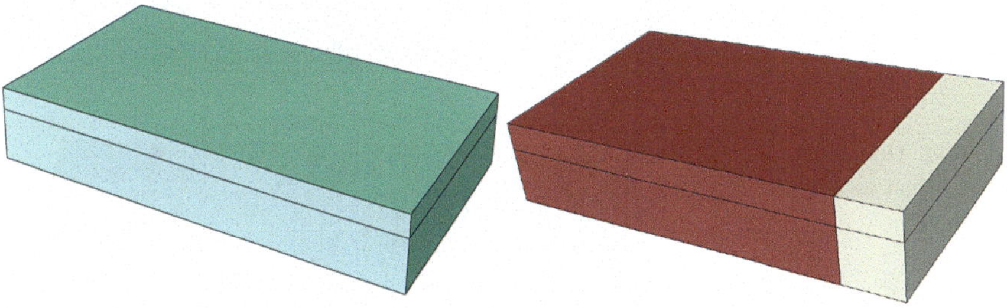

(a) Homogeneous model (b) Heterogeneous mineral transition zone model

Fig. 5.10 The nanoscratch model assigns attribute cross-sections. (**a**) Homogeneous model. (**b**) Heterogeneous mineral transition zone model

order to better observe the damage of the material when the indenting head is crossed, the material is damaged by Maxe, and the damage evolutionary fracture energy is set to 0.05, the elastic modulus is set to the density of 2.65×10^{-09}, the yield stress is set to 50 MPa, and the Poisson's ratio is set to 0.2 [98]. As shown in Fig. 5.10, the material properties in the transition area of heterogeneous minerals are set according to the elastic modulus of the hybrid lamellar shale of feldspar minerals collected by nanoindentation test. The elastic modulus of the plastic material is 6.414 GPa and that of the brittle material is 31.714 Gpa.

5.1.4 Grid Division and Boundary Conditions

Nanoindentation Two-Dimensional Model Meshing

In order to improve the computational efficiency, the mesh along the indentation section is refined. The stress and strain of the tested sample are mainly concentrated in the area near the indentation, and the far stress and strain are almost zero. Therefore, it is only necessary to build a local material model and adopt local mesh refinement, which can reduce the total number of units, improve computing efficiency, and save computing time. Using the linear reduction integral unit (CPS4R) cells suitable for large strain analysis and encrypted by local fine mesh, the sample was divided into 32,970 grids (Fig. 5.11).

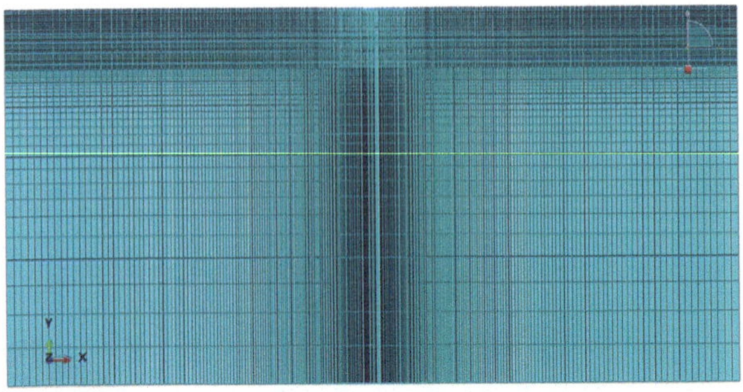

Fig. 5.11 Horizontal layered shale local grid refinement

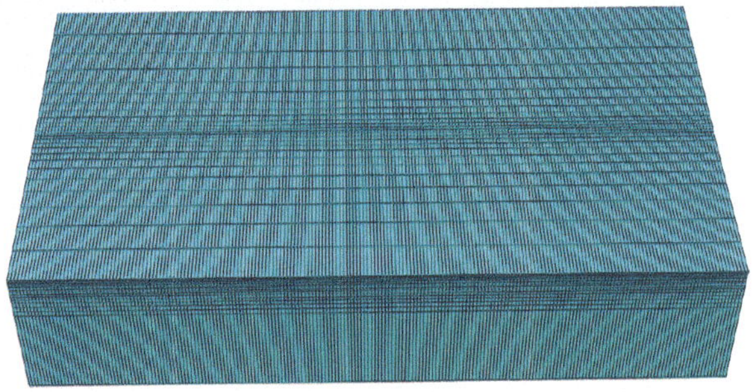

Fig. 5.12 Horizontal layered shale local grid refinement

Nanoscratch 3D Model Meshing

In order to improve the computing efficiency, the mesh along the 3D model is also locally refined to reduce the number of units, thus improving the computing efficiency and saving the computing time. Due to the large deformation of the material in the process of nanoscratch, the linear reduction integral hexahedron element C3D8R, which has high computational accuracy and is suitable for large strain analysis, was selected for mesh partitioning. The specimens were divided into 118,000 grids, and hourglass control was adopted to reduce and integrate, as shown in Fig. 5.12.

Analyze Step and Boundary Conditions

ABAQUS allows a maximum of 1000 incremental steps for analysis. Due to the complex contact and large plastic deformation in the nanoscratch simulation process, the analysis is not easy to converge. Therefore, the initial increment step is set to 10^{-8}, the minimum allowed increment step is set to 10^{-20}, and the maximum allowed increment step is set to 0.01.

In the actual nanoindentation experiment, there are three steps: loading, stabilizing load, and unloading. The purpose of the stable load is to eliminate the effect of material creep on the performance. In the indentation model, the influence of interlaminar mineral soft and hard hybrid and laminar parameters is mainly

Fig. 5.13 Boundary conditions of nanoindentation model

considered, and the material is set as ideal elastic-plastic, without considering the influence of creep. The boundary condition is an important part of finite element calculation. The horizontal displacement along the symmetry axis was set to zero. The indenter was constrained to move only in the downward direction, and the axial displacement at the bottom of the tested material was fixed at zero.

Due to the large actual size of continental shale material, the outer surface can be used as a free surface or a horizontal displacement constraint surface, and the calculated results are consistent under the two boundary conditions (Fig. 5.13). As shown in Fig. 5.14, the indenter is a rigid body, and the force and displacement are concentrated to a point named as a reference point to facilitate data output. The nanoscratch test is displacement-controlled, and the indenter moves to destroy the material after pressing into the material. The analysis steps and boundary conditions are set in accordance with the two-dimensional indentation model.

5.2 Nanoindentation Simulation

5.2.1 Poisson's Ratio Effect

A total of 64 nanoindentation simulations were performed on layered shale with a fixed inclination of 0° and a striated thickness of 500 nm to verify the influence of Poisson's ratio on the indentation results. Figure 5.15 shows a finite element simulated cloud image of nanoindentation with different Poisson's ratio parameters. The results show that the load increases gradually with the increase in the indenter depth, and the stress is concentrated in the laminates with higher elastic modulus and yield strength. The Poisson's ratio has little effect on the deformation and stress transfer process.

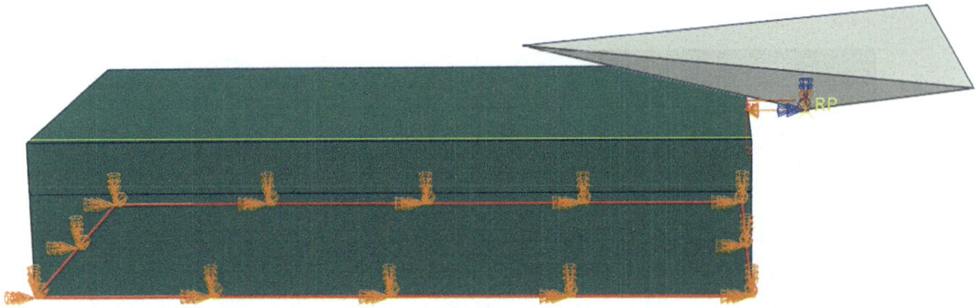

Fig. 5.14 Boundary conditions of nanoscratch model

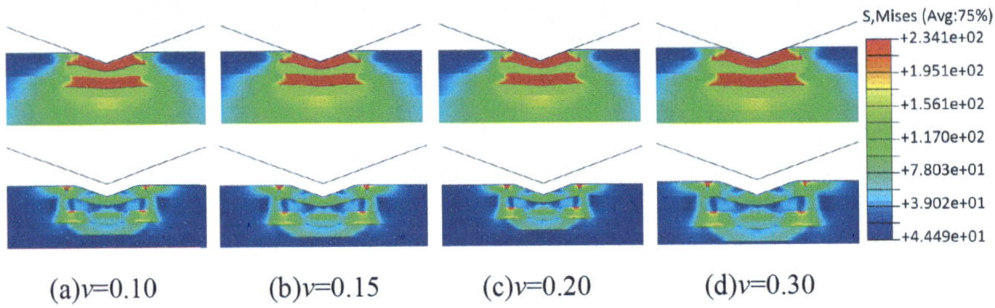

Fig. 5.15 Simulation of indentation in shale with different Poisson's ratio bedding planes. (**a**) $\nu = 0.10$. (**b**) $\nu = 0.15$. (**c**) $\nu = 0.20$. (**d**) $\nu = 0.30$

Figure 5.16a shows the dimensionless relationship of finite element simulation results for materials with different Poisson's ratios. It can be seen that the recovery rate of dimensionless indentation energy U_e/U_t increases with the growth of $Y_1/E_1 + Y_2/E_2$ data, and the trends of different Poisson's ratios are basically the same. In Fig. 5.16b, it can be seen that the influence of Poisson's ratio on dimensionless parameter U_e/U_t in laminar parameters is almost negligible. It is worth noting that when ν=0.1, the energy recovery rate will have an early growth area, which is due to the small Poisson's ratio of the laminate leading to the large transverse deformation of the material. An advance growth zone occurs when the dimensionless parameter $Y_1/E_1 + Y_2/E_2$ is greater than 0.35. In the numerical simulation with ν greater than 0.2, there is no advance growth zone, but with the increase in $Y_1/E_1 + Y_2/E_2$, energy degradation zone appears in both regions. This is because with the increase in $Y_1/E_1 + Y_2/E_2$, the deformation between hard and soft minerals is inconsistent, resulting in incomplete contact and dimensionless energy degradation.

Poisson's ratio has little effect on the pressure relief work of shale. For horizontal laminates, it is important to analyze the relationship between mechanical parameters and indentation work. In order to study the effect of dimensionless parameters $Y_1/E_1$and Y_2/E_2 on the indentation energy, the DoseResp surface function fitting model was used to fit the three-dimensional data. As shown in Fig. 5.17a, the correlation between U_e/U_t、 Y_1/E_1 and Y_2/E_2 can be determined according to the fitting equation, and the surface fitting formula is shown in Eq. (5.1). The correlation coefficient of

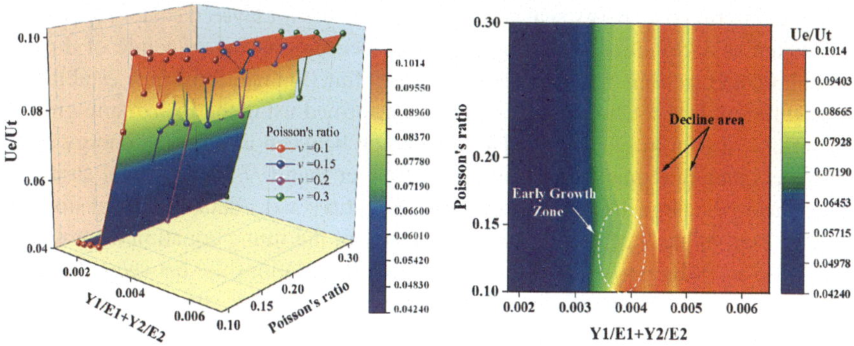

(a) Three-dimensional space surface (b) Two-dimensional planar heat map

Fig. 5.16 Energy recovery rates of different Poisson's ratio bedding planes. (**a**) Three-dimensional space surface. (**b**) Two-dimensional planar heat map

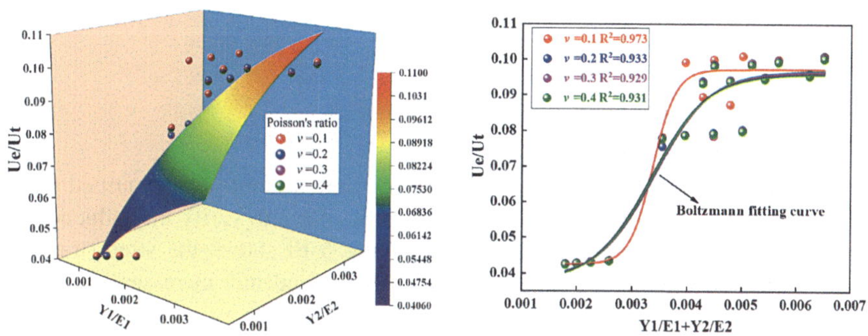

(a) DoseRespSurface function fitting (b) BoltzmannFunction fitting

Fig. 5.17 Dimensionless relationship fitting of different Poisson's ratio bedding planes. (**a**) DoseResp Surface function fitting. (**b**) Boltzmann Function fitting

dimensionless parameters is Y_1/E_1, which is much higher than that of 1.456×10^6. The results show that under different horizontal laminate Poisson's ratio, the minerals with higher yield strength and elastic modulus have a decisive influence on the indentation energy.

$$\frac{U_e}{U_t} = -2.519 + 9.487 \times \left(1/\left(1 + \left(1.456 \times 10^6 \times \frac{Y_1}{E_1}\right)^{-0.023}\right)\right) \times \left(1/\left(1 + \left(800 \times \frac{Y_2}{E_2}\right)^{-0.0283}\right)\right) \tag{5.1}$$

In order to find the energy scale relationship suitable for layered shale, $Y_1/E_1 + Y_2/E_2$ is taken as a key dimensionless parameter according to the principle of dimensional homogenization, and the energy recovery function relationship is established under different Poisson's ratios. The fit was performed using the Boltzmann growth function, as shown in Fig. 5.17b. The correlation between indentation energy scale parameters and Boltzmann function R^2 is stronger than 0.9 in horizontal stratified shale. Due to the large discreteness of Poisson's ratio $v = 0.1$ data, the initial slope of the fitting curve has a certain change. The result of the fitting equation is shown in Eq. (5.2), according to which the dimensionless relationship between the new dimensionless parameter and the indentation energy recovery rate can be obtained.

$$\frac{U_e}{U_t} = 0.04269 - 0.05466/\left(1 + \exp\left(\left(\frac{Y_1}{E_1} + \frac{Y_2}{E_2} - 0.00336\right)/(5.57 \times 10^{-4})\right)\right) \tag{5.2}$$

5.2.2 *Thickness Effect*

A total of 64 sets of nanoindentation simulations were performed on stratified shale with a fixed inclination of 0° and a Poisson's ratio of 0.3 to verify the influence of layer thickness on the indentation results. Figure 5.18 shows the stress transfer results under the indentation area. As the depth of the indenter increases gradually, the stress of the indenter increases gradually. The transfer of stress and energy between laminae mainly depends on the minerals with increased yield strength and elastic modulus. When the grain thickness is small, the compression energy propagation distance increases. In addition, the area of residual stress on the shale surface also increases with the increase in lamination thickness.

As shown in Fig. 5.19a, the energy recovery U_e/U_t shows an upward trend with the increase in $Y_1/E_1 + Y_2/E_2$ data, and the change trend of different layer thicknesses is basically the same, indicating that layer thickness has a relatively weak influence on dimensionless parameter energy recovery. In Fig. 5.19b, there is a weak early growth area at $S = 2000$ nm, which is due to the weak effect on pressure energy transfer with the increase in layer thickness, and the stress and energy cannot be transferred to the next layer quickly. In addition, at $Y_1/E_1 + Y_2/E_2 = 0.0045$ and 0.0052, dimensionless energy degradation regions appear, which is similar to the situation shown by lamination processes with different Poisson's ratios.

In order to test the influence of dimensionless Y_1/E_1 and Y_2/E_2 parameter on the energy recovery rate in models with different laminae thickness, DoseResp surface function was used to fit the finite element model, as shown in Fig. 5.20a. The surface fitting formula is Eq. (5.3), where the correlation coefficient of Y_1/E_1 is 1.757×10^7, which means that brittle minerals with high yield strength and elastic modulus have a

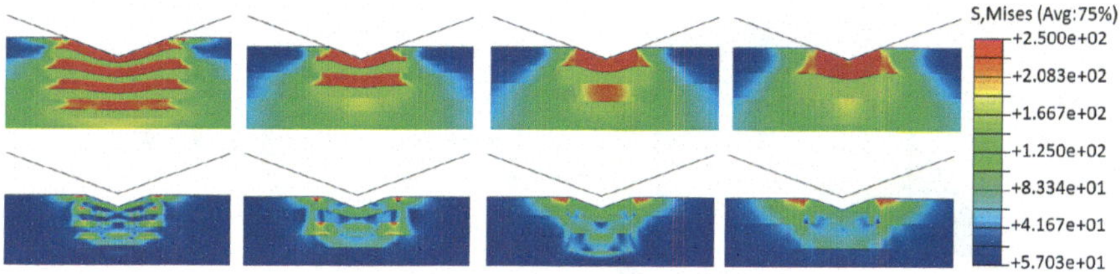

(a) S=250 nm (b) S=500 nm (c) S=1000 nm (d) S=2000 nm

Fig. 5.18 Stress cloud map of layered indentation with different thicknesses of bedding. (**a**) $S = 250$ nm. (**b**) $S = 500$ nm. (**c**) $S = 1000$ nm. (**d**) $S = 2000$ nm

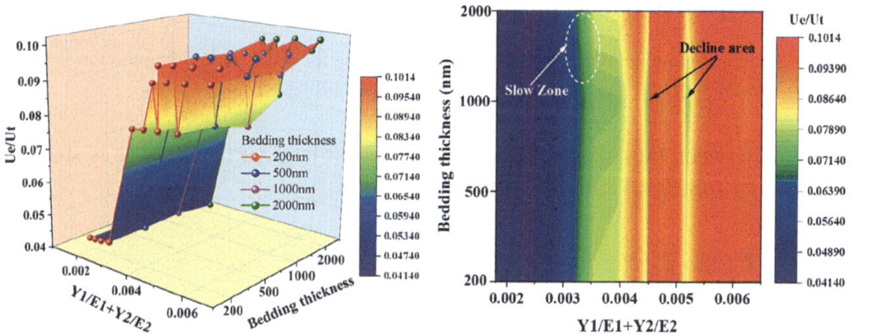

(a) Three-dimensional space surface (b) Two-dimensional planar heat map

Fig. 5.19 Energy recovery rates of different thickness bedding planes. (**a**) Three-dimensional space surface. (**b**) Two-dimensional planar heat map

significant impact on the correlation coefficient of energy recovery. In shales with different thickness layers, minerals with higher yield strength and elastic modulus still have significant effects on dimensionless energy recovery.

$$\frac{U_e}{U_t} = -1.667 + 6.228 \times \left(1/\left(1 + \left(1.757 \times 10^7 \times \frac{Y_1}{E_1}\right)^{-0.020}\right)\right) \times \left(1/\left(1 + \left(2441 \times \frac{Y_2}{E_2}\right)^{-0.0267}\right)\right) \tag{5.3}$$

The Boltzmann growth function is used to establish the fitting relationship of dimensionless parameter $Y_1/E_1 + Y_2/E_2$ and U_e/U_t, as shown in Fig. 5.20b. For laminar thicknesses below 1000 nm, the increased thickness creates stress-transfer faults and reduces dimensionless energy recovery. However, for the grain thickness above 2000 nm, the influence of the grain thickness on the indentation process is reduced, ultimately leading to a further increase in dimensionless hardness. The addition of Y_1/E_1 and Y_2/E_2 is considered to be a dimensionless parameter. The

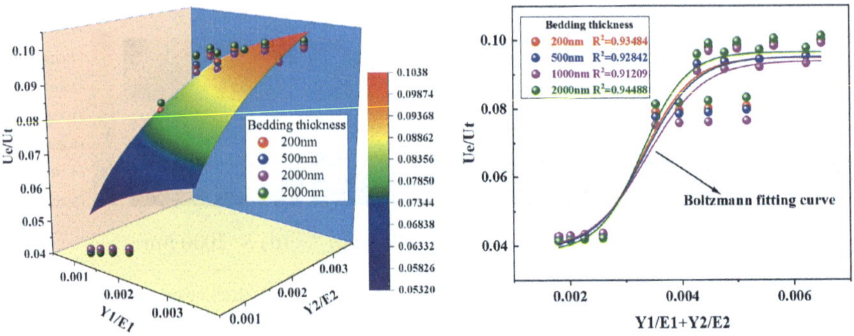

(a) DoseRespSurface function fitting (b) BoltzmannFunction fitting

Fig. 5.20 Dimensionless nondimensional relationship fitting for different bedding thicknesses. (**a**) DoseResp Surface function fitting. (**b**) Boltzmann Function fitting

correlation between these parameters under different laminar thicknesses is studied, and the fitting function is described in formula (5.4).

$$\frac{U_e}{U_t} = 0.03915 - 0.09502/\left(1 + \exp\left(\left(\frac{Y_1}{E_1} + \frac{Y_2}{E_2} - 0.00332\right)/(4.59E - 4)\right)\right) \quad (5.4)$$

5.2.3 *Angle Effect*

A fixed lamination thickness of 250 nm and a constant Poisson's ratio of 0.3 were set, and the lamination inclination α was used as the variable. The stress nephogram generated by finite element simulation of indentation of laminated shale with different inclination angles is shown in Fig. 5.21. The stress is mainly distributed along the laminated direction, and most of the energy is concentrated under the indenter to form a circular concentrated stress area.

Figure 5.22a shows the correlation between the dimensionless parameter $Y_1/E_1 + Y_2/E_2$ and the energy release rate. U_e/U_t steadily increases with the increase in $Y_1/E_1 + Y_2/E_2$. Figure 5.22b shows that when $Y_1/E_1 + Y_2/E_2$=0.0033, the indentation energy U_e/U_t of $\alpha = 30°$ shows a relatively weak advance growth area, which is because the stress distribution is more uniform due to the smaller lamination angle. In the regions $Y_1/E_1 + Y_2/E_2$=0.0045 and 0.0052, dimensionless energy degradation regions are observed, and this phenomenon is also observed in simulations with Poisson's ratio and thickness as variables. In addition, in the $Y_1/E_1 + Y_2/E_2$=0.0042

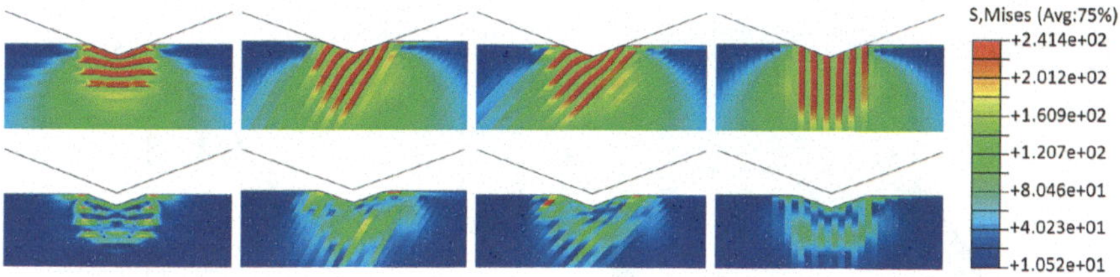

(a) S=250 nm (b) S=500 nm (c) S=1000 nm (d) S=2000 nm

Fig. 5.21 Stress cloud map of indentation simulation with different dip angle layers. (**a**) $S = 250$ nm. (**b**) $S = 500$ nm. (**c**) $S = 1000$ nm. (**d**) $S = 2000$ nm

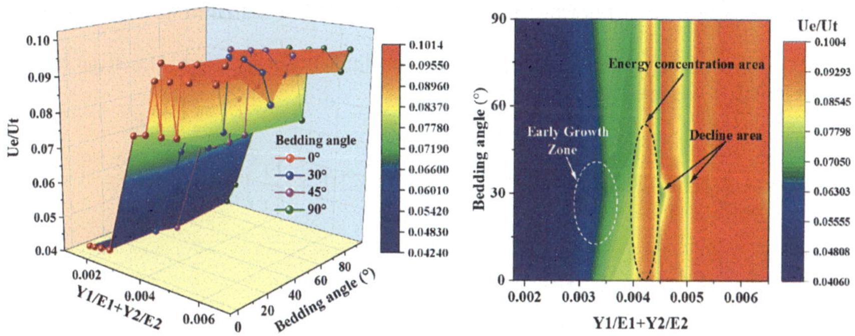

(a) Three-dimensional space surface (b) Two-dimensional planar heat map

Fig. 5.22 Energy recovery rate of different dip angle bedding planes. (**a**) Three-dimensional space surface. (**b**) Two-dimensional planar heat map

region, models with laminar angles below 60° show areas of energy accumulation due to small residual stresses.

In order to test the degree of influence of dimensionless Y_1/E_1 and Y_2/E_2 parameters on various laminar angle models, dimensionless relationships were obtained by fitting models with surface functions, as shown in Fig. 5.23a. Equation (5.5) provides an analytical expression for surface fitting. The coefficients of dimensionless parameter Y_1/E_1 and Y_2/E_2 are 1.7547×10^4 and 1.988×10^7, respectively. In the fracturing process of inclined laminated shale, minerals with low yield strength and elastic modulus will affect the fracturing energy. In the actual shale indentation test, it is important to emphasize the influence of plastic minerals such as illite and chlorite on the indentation energy.

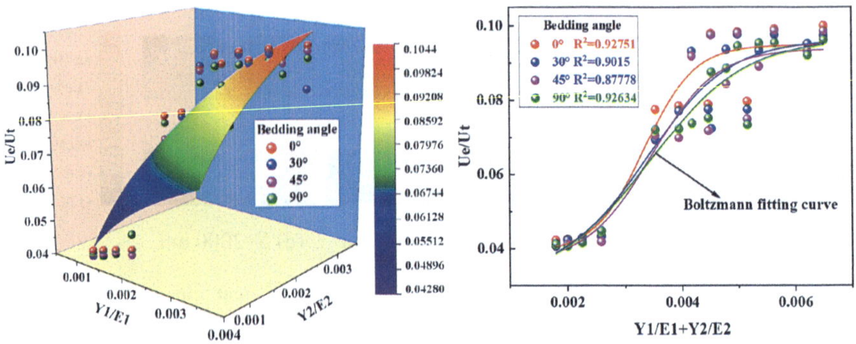

(a) DoseRespSurface function fitting (b) BoltzmannFunction fitting

Fig. 5.23 Dimensionless fitting of bedding planes with different dip angles. (**a**) DoseResp surface function fitting. (**b**) Boltzmann function fitting

$$\frac{U_e}{U_t} = -2.225 + 7.189 \times \left(1/\left(1 + \left(1.755 \times 10^4 \times \frac{Y_1}{E_1}\right)^{-0.0221}\right)\right) \times \left(1/\left(1 + \left(1.99 \times 10^7 \times \frac{Y_2}{E_2}\right)^{-0.0233}\right)\right) \tag{5.5}$$

The functional relationship between dimensionless parameters $Y_1/E_1 + Y_2/E_2$ and U_e/U_t under different laminar conditions was studied, and the Boltzmann growth function was used for fitting, as shown in Fig. 5.23b. It can be seen that the dimensionless energy reaches the highest value when the laminar inclination is 0°. Due to the influence of inclination angle on stress and energy transfer, energy recovery is incomplete, resulting in residual stress accumulation in the laminae. The dimensionless energy decreases with the increase in angle α. The growth trend of dimensionless relationship in different layers is similar. With the increase in layer angle, the data dispersion also increases significantly. The Boltzmann growth function fits the data as shown in formula (5.6), and the correlation coefficient R^2 exceeds 0.85, indicating that there is a good correlation between the fitting results.

$$\frac{U_e}{U_t} = 0.9562 - 0.06424/\left(1 + \exp\left(\left(\frac{Y_1}{E_1} + \frac{Y_2}{E_2} - 0.00342\right)/\left(2.91 \times 10^{-4}\right)\right)\right) \tag{5.6}$$

5.3 Nanoscratch Simulation

5.3.1 *Stress Distribution*

Indenter into the Stage

The indenter adopts a displacement control mode, the pressing depth is 5000 nm, the horizontal movement rate is 25 μm/s, and the total scratch displacement is 500 μm. The cloud map of stress distribution at different scratch distances is obtained, as shown in Fig. 5.24. It can be seen from the local stress nephogram that the high stress region of the matrix is mainly concentrated right below the reference point of the indenter, showing a high stress distribution. With the sliding of the indenter, the material in the initial drawing stage fails, leaving a certain residual stress between the material and the nearby matrix.

Scratch Test Metaphase

Figure 5.25 shows the stress transfer state as displacement changes in the middle stage of the scratch test. When the indenter is drawn into the material, the stress is concentrated under the indenter, and the plane drawn into the material will also bear some of the stress. With the increase in the stroke displacement, the stress is uniformly transferred, and the high stress concentration area is always kept directly below the indenter. The residual stress decreases with increasing depth in the scratch path, but uneven residual stress is left below the initial scratch stage. Therefore, it is necessary to ensure the displacement of a sufficiently long scratch in scratch tests and simulations.

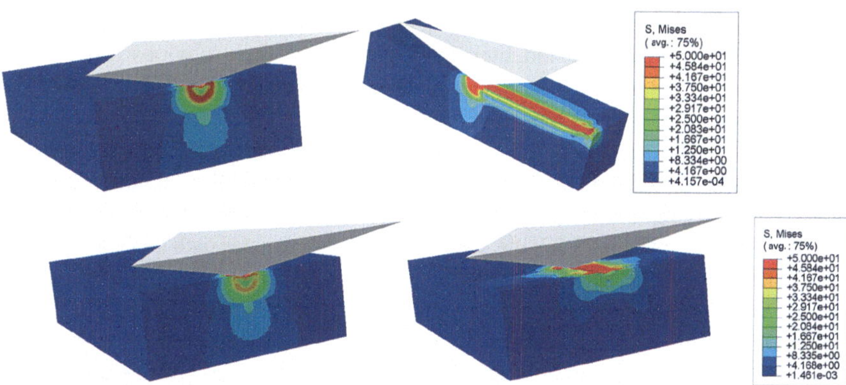

Fig. 5.24 Stress distribution at the starting surface of different scratch distances

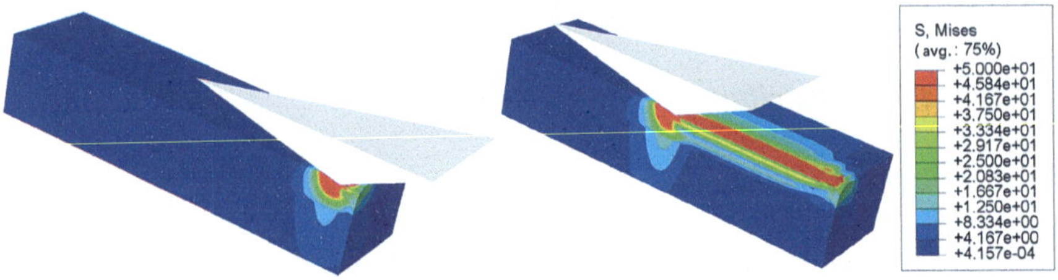

Fig. 5.25 Stress distribution at different scratch distance profiles

5.3.2 *Failure Mode*

During the nanoscratch test, it is not possible to know which side of the positive pyramidal indenter is aligned with the direction of the scratch. Although there are relevant calibration methods, the influence of the deflection angle of indenter on the damage form of the material during the scratch test is still worth further study.

Indenter into the Stage

At the initial stage of indenter cutting, the stress of Berkovich indenter without deflection at the initial stage of scratch test was axisymmetrical under the indenter, and the damage of the material was relatively uniform. The indenter with a deflection of 30° shows more serious damage, which is because the area of the indenter in contact with the material becomes larger in the initial entry stage, and the stress under the indenter is concentrated on the left side of the diagonal line of the indenter, and the material deformation is more obvious (Fig. 5.26).

Scratch Test Metaphase

With the increase in the displacement of the indenter, the damage modes of the indenter caused by different indenter deflection angles show obvious differences. The stress of the indenter without deflection angle spreads evenly to both sides during the scratch test, and the axial symmetry of the center line is present, and the stacking height of the material on both sides of the indenter is the same. For the Berkovich indenter with a certain deflection angle, the material accumulation in the scratch path is more obvious and the damage area is larger when the deflection angle is larger. As shown in Fig. 5.27, the scratch path stress transfer path of the deflecting indenter can be clearly seen at the yellow ellipse, and obvious concentrated stress damage occurs in the deflecting direction of the deflecting indenter. In addition, compared with the failure mode in Fig. 5.26, it can be found that the non-deflected indenter has a wider propagation range of upper and lower stress during the scratch

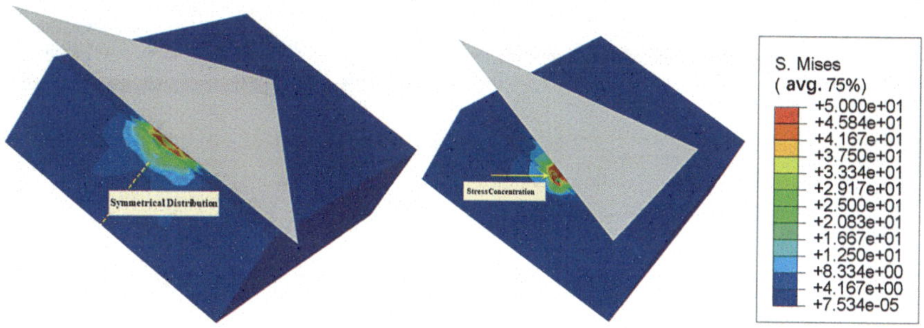

(a) No deflection of the indenter (b) The indenter deflects at 30°

Fig. 5.26 The influence of head deflection angle on initial failure. (**a**) No deflection of the indenter. (**b**) The indenter deflects at 30°

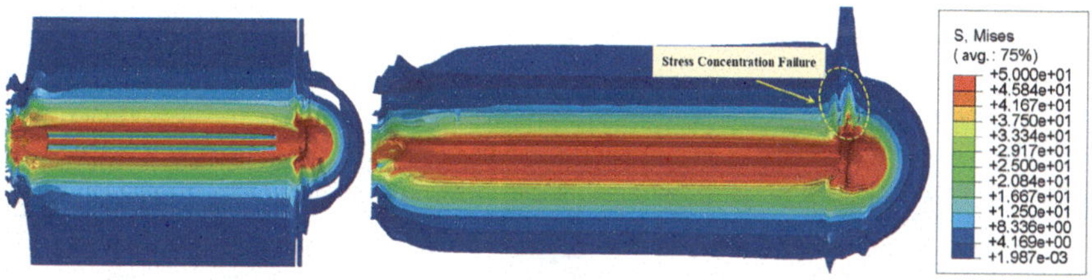

(a) No deflection of the indenter (b) The indenter deflects at 30°

Fig. 5.27 Equivalent surface of mid-term failure caused by head deflection angle. (**a**) No deflection of the indenter. (**b**) The indenter deflects at 30°

test, but leaves less residual stress. However, the indenter with a certain deflection angle causes greater residual stress in the scratch path and more obvious damage.

Tangential and normal stress curves obtained from scratch tests with different indenter deflection angles are shown in Fig. 5.28a. During the initial drawing stage of indenter, the growth rate of tangential stress of indenter with a certain deflection angle is greater than the stress value $\alpha = 0°$. With the increase in the scratch displacement, the curves of different deflection angles converge at $L = 100$ μm. In the subsequent scratch displacement, the scratch stress curve of $\alpha = 0°$ is larger than the tangential stress of $\alpha = 30°$. This phenomenon may be due to the fact that the indenter with a certain deflection angle causes a larger damage area in the drawing stage, and the residual depth of the scratch remains unchanged with the increase in the scratch displacement, but the volume of the scratch groove with a certain deflection angle is smaller. Figure 5.28b shows the friction coefficient curves of different indenter deflection angles. Since the indenter with a certain deflection angle has a larger contact material area during the scratching process, the amplitude of the friction coefficient curve is higher.

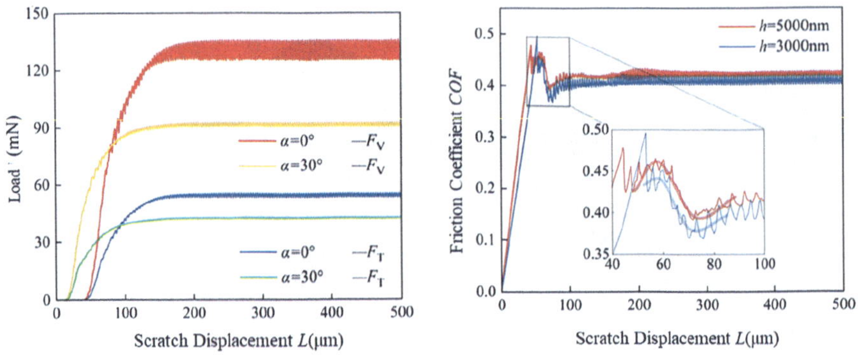

(a)Influence of deflection Angle of indenter on stress

(b) Influence of deflection Angle of indenter on friction coefficient

Fig. 5.28 Scratch curves of indenters with different deflection angles. (**a**) Influence of deflection angle of indenter on stress. (**b**) Influence of deflection angle of indenter on friction coefficient

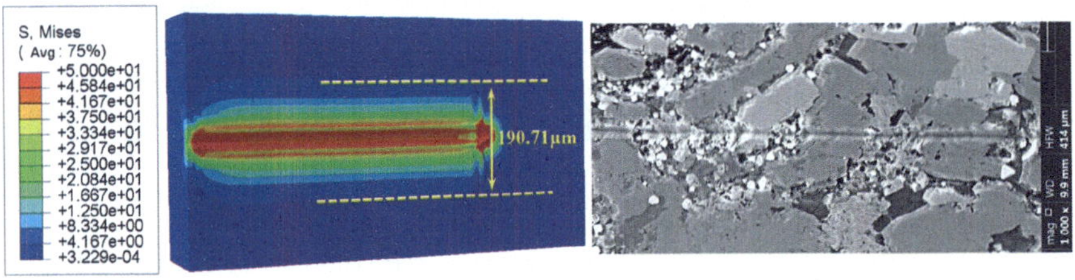

(a) Failure mode of finite element model (b) Failure modes of brittle minerals

Fig. 5.29 The failure mode at a smaller depth of penetration. (**a**) Failure mode of finite element model. (**b**) Failure modes of brittle minerals

5.3.3 Scratch Depth Effect

With the increase in the drawing depth, the stress transfer range within the material also expands, resulting in the deterioration of the material. Therefore, in the face of different materials, it is necessary to choose the appropriate scratch depth and stress to ensure that the material can maintain its stability when subjected to stress. Figure 5.29 shows the failure form of the finite element model when the depth is smaller. When the depth is 3000 nm, the stress diffusion range reaches 190.71 μm, which is close to the actual failure model of ductile minerals, such as dolomite and apatite. Brittle minerals are harder, scratch less deep, and the material does not accumulate around it, but leaves a shallower scratch path. When the indenter passes over the area with weak mineral bonding strength, the strain energy release quickly leads to rock fragmentation and separation, resulting in destruction of microscale rock fracture。 .

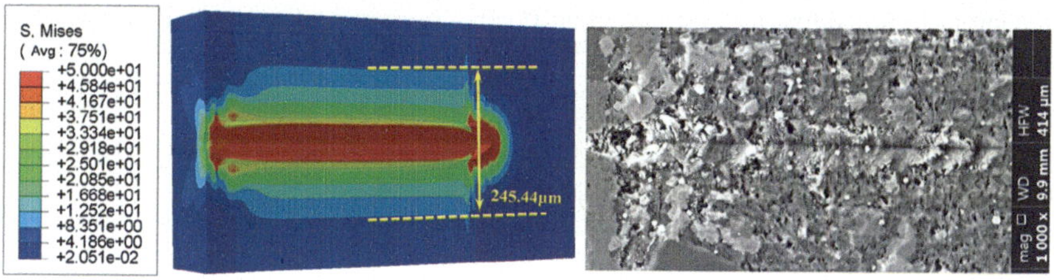

(a) Failure mode of finite element model (b) Failure mode of plastic minerals

Fig. 5.30 The failure mode at larger depth of penetration. (**a**) Failure mode of finite element model. (**b**) Failure mode of plastic minerals

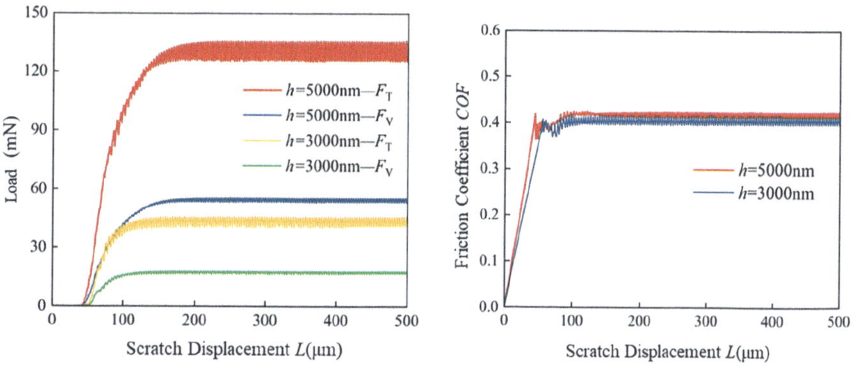

(a) Stress curve (b) Friction coefficient curve

Fig. 5.31 Scratch curves with different indentation depths. (**a**) Stress curve. (**b**) Friction coefficient curve

When the depth is 5000 nm, Fig. 5.30 shows that the stress diffusion range reaches 245.44 μm, which is close to the failure model of actual shale plastic minerals. Plastic minerals can better absorb the energy generated during the scratch test into volume deformation, showing a failure mode of debris accumulation on both sides. Because the texture is weak, the depth is large, and the stress diffusion range is wide.

Figure 5.31a shows the tangential stress and normal stress curves at different drawing depths. At the drawing depth of 3000 nm, the tangential force stabilized at 43.74 mN after the initial growth stage, and the normal stress finally stabilized at 18.06 mN. Because the tangential force dominates the material damage during the scratch experiment, the fluctuation amplitude of F_T curve is larger than that of F_V curve. At a depth of 5000 nm, the normal stress F_V stabilizes at 55.04 mN, while the tangential force F_T eventually stabilizes at 130.79 mN. The stress in the process of nanoscratch increases with the increase in scratch depth, and the tangential force in scratch test is much larger than the normal stress. Figure 5.31b shows the friction coefficient curves when the depth is 3000 nm and 5000 nm. A deeper depth will lead to a larger initial slope of the friction coefficient curve and a faster speed to reach the peak of the friction coefficient curve.

5.3.4 Interface Transition Zone

The stress distribution during the scratching process of the heterogeneous mineral transition zone is shown in Fig. 5.32. At the depth $h = 3000$ nm, the stress diffusion range at the junction of brittle minerals and plastic minerals is small, and the stress concentration of brittle minerals is more obvious than that of plastic minerals. In the middle stage of scratch test, the residual stress in the scratch path of plastic minerals is obviously smaller than that of brittle minerals. At the depth $h = 5000$ nm, the stress diffusion range at the junction of brittle minerals and plastic minerals is larger, and the stress transfer range of brittle minerals is larger than that at the depth of 3000 nm. In the middle stage of scratch test, the plane stress transfer range of brittle minerals extends almost to the whole plane, while the stress transfer range of plastic minerals is still limited to a certain range.

As shown in Fig. 5.33, the contoured failure cloud map of the shale laminar interface transition zone shows that the material at the junction of the material at the interface transition zone is not completely deformed when the depth of the material is shallow, which means that complete material information is not collected during the scratch test, resulting in inaccurate evaluation of mechanical properties. When the drawing depth is shallow, the contact between the indenter and the material in the

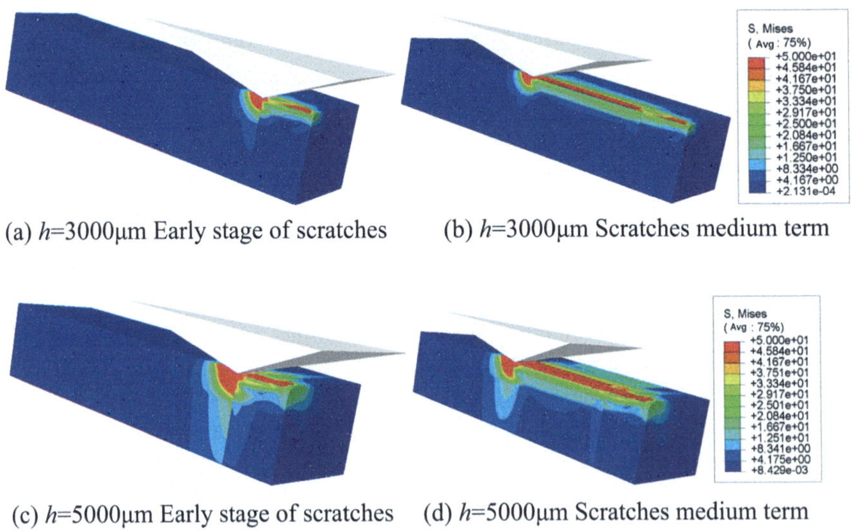

Fig. 5.32 Scratched stress distribution in the transition zone of heterogeneous minerals. (**a**) $h = 3000$ μm Early stage of scratches. (**b**) $h = 3000$ μm Scratches medium term. (**c**) $h = 5000$ μm Early stage of scratches. (**d**) $h = 5000$ μm Scratches medium term

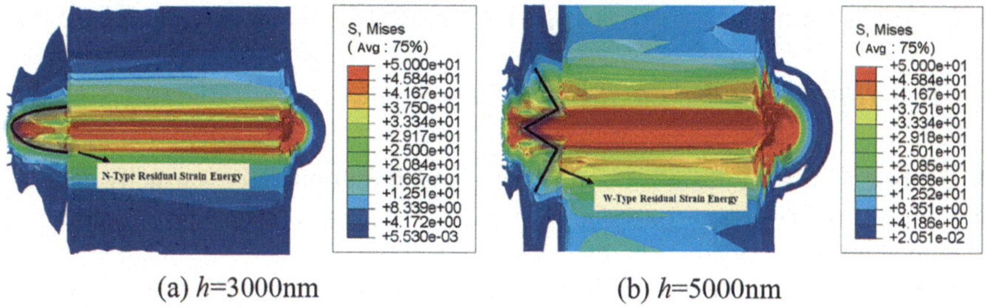

Fig. 5.33 Cloud map of contour line failure in the transition zone of interface. (**a**) $h = 3000$ nm. (**b**) $h = 5000$ nm

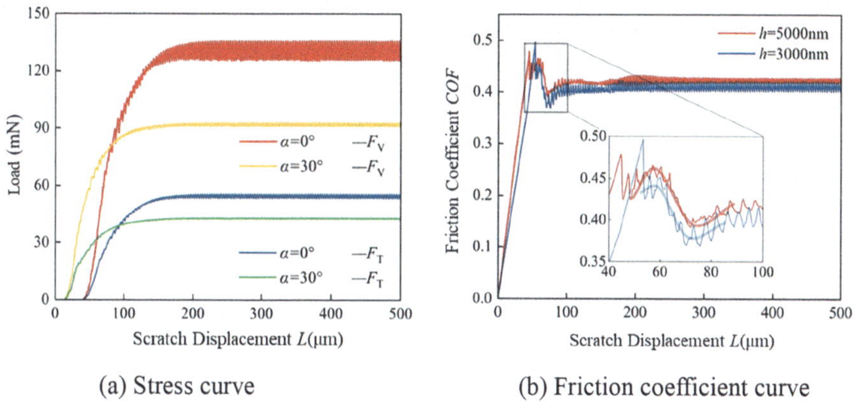

Fig. 5.34 Stress and friction coefficient curves in the interface transition zone. (**a**) Stress curve. (**b**) Friction coefficient curve

interface transition zone is insufficient, resulting in "n" type residual deformation, while when the drawing depth is large, the material in the interface transition zone is fully stressed, showing a "w" type residual deformation.

Figure 5.34a shows the tangential stress and normal stress curves of the laminar interface, and the amplitude of the curves is close to the analysis results of different subgrade depths. The normal stress curve shows a sharp amplitude fluctuation after the indenter passes through the interface transition zone, while the tangential force curve has a small fluctuation, and the curve with larger depth has a more drastic fluctuation. Figure 5.34b shows the friction coefficient curves of different zoned depths, whose amplitude is relatively close. When the indenters pass through the interfacial transition zone, the friction coefficient curves also show an "S" type fluctuation, which verifies the feasibility of quantifying the distribution length of the interfacial transition zone of laminated shale based on the characteristics of friction coefficient fluctuation.

5.4 Dimensionless Functional Relationship

5.4.1 Dimensional Analysis

When the penetration depth is constant, the scratch friction process usually consists of the following processes (see Fig. 5.35). The first is the contact process under the combined influence of constant displacement, and the second is the transient sliding friction under the combined influence of normal pressure and cutting pressure. However, the scratch area δ, scratch residual depth h_r, tangential force F_t, normal force F_n, and other scratch strain parameters are still in the process of fluctuating change at this time.

When the pressing depth is constant, a typical sliding friction process usually consists of three stages (see Fig. 5.35). The first is the contact process under constant displacement. The second is the transient sliding friction under the combined action of normal load and tangential load. At this time, the scratch response parameters such as scratch width δ, scratch residual depth h_r, tangential force F_t, and normal stress F_n are still in the fluctuation stage. When pressing into a constant depth, the scratch shape and the force on the indenter will reach a stable stage, and the scratch test process will reach a steady state. Figure 5.36 shows the schematic diagram of cross section topography in the scratch direction, including material accumulation height h_p, scratch width δ, and residual depth h_r.

For the scratch process, select tangential stress F_t, normal stress F_n, scratch width δ as variables, they can use h_r, E, Y, v, μ_s, and β dimensional combination to represent

$$F_n = f_1(h_r, Y, E, \nu, \beta, \mu_s) \tag{5.7}$$

$$\delta = f_2(h_r, Y, E, \nu, \beta, \mu_s) \tag{5.8}$$

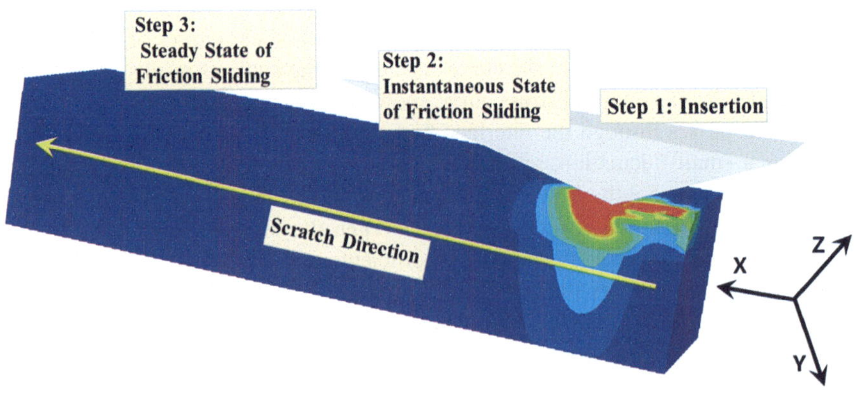

Fig. 5.35 Scratch process under constant normal load

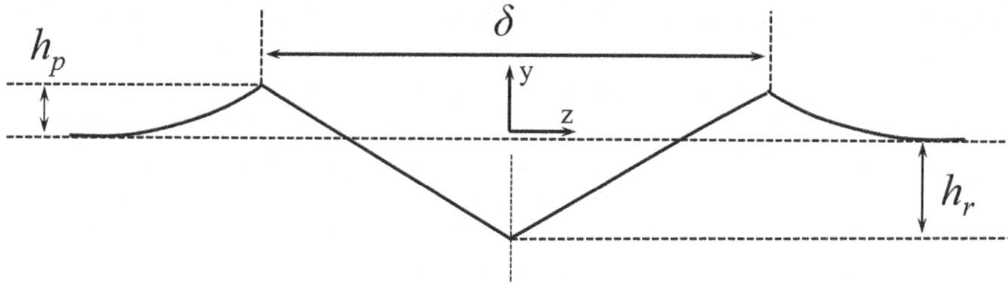

Fig. 5.36 Schematic diagram of cross-sectional morphology in the direction of scratches

$$F_t = f_3(h_r, Y, E, \nu, \beta, \mu_s) \tag{5.9}$$

Here E is Young's modulus, Y is yield stress, v is Poisson's ratio, μ_s is sliding friction coefficient, and β is scratch finite element simulation of indentor deflection angle. E and h are selected as independent variables, and other quantities can be represented by these two quantities.

$$\begin{gathered}\Pi_1 = F_n E^{-1} h_r^{-2},\ \Pi_2 = \delta E^0 h_r^{-1} = \delta F_n^{-0.5} E^{0.5},\ \Pi_3 = F_t E^{-1} h_r^{-2} = F_t F_n^{-1} \\ \Pi_4 = Y E^{-1} h_r^0,\ \Pi_5 = \beta E^0 h_r^0,\ \Pi_6 = \mu_s E^0 h_r^0,\ \Pi_7 = v E^0 h_r^0\end{gathered} \tag{5.10}$$

According to reports by Liu et al. and Zhang et al., the influence of Poisson's ratio can be ignored, and the following relationship can be obtained:

$$\Pi_1 = f_4(\Pi_4, \Pi_5, \Pi_6) \tag{5.11}$$

$$\Pi_2 = f_5(\Pi_4, \Pi_5, \Pi_6) \tag{5.12}$$

$$\Pi_3 = f_6(\Pi_4, \Pi_5, \Pi_6) \tag{5.13}$$

Applying the π theorem to the dimensionless parameter, we can obtain:

$$\Pi_1 = \frac{F_n}{E h_r^2} = f_7\left(\frac{Y}{E}, \beta, \mu_s\right) \tag{5.14}$$

$$\Pi_2 = \frac{\delta}{\sqrt{F_n/E}} = f_8\left(\frac{Y}{E}, \beta, \mu_s\right) \tag{5.15}$$

$$\Pi_3 = \frac{F_t}{F_n} = f_9\left(\frac{Y}{E}, \beta, \mu_s\right) \tag{5.16}$$

5.4.2 Dimensionless Stress Function

Previous studies have shown that the subdivision depth has little effect on F_n/Eh_r^2, and the results obtained with different friction coefficients are similar. In the process of scratch test, the normal deformation of the material is mainly affected by the elastic modulus, that is, the dimensionless equation becomes

$$\frac{F_n}{Eh_r{}^2} = \Pi_1\left(\frac{Y}{E}, \beta\right) \tag{5.17}$$

Figure 5.37a shows the influence of different elastic modulus and head deflection angle on dimensionless normal stress. When the value of *Y/E* increases, the value of F_n/Eh_r^2 rises steadily. When *Y/E* is less than 0.009, the impact of head deflection angle β on F_n/Eh_r^2 is small, while the value of *Y/E* has a significant impact on F_n/Eh_r^2. When *Y/E* is greater than 0.012, F_n/Eh_r^2 decreases first and then increases with the increase in the deflection angle of the indenter. When $\beta = 0°$, the dimensionless normal load reaches the maximum value. From Fig. 5.37b, it can be observed that F_n/Eh_r^2 shows an approximately linear rise as the values of $\sin\beta$ and *Y/E* increase. The following equation is fitted with the 3D-Plane function, where the unit of Young's modulus is MPa.

$$\frac{F_n}{Eh_r^2} = 0.0296 + 62.6843\frac{Y}{E} - 0.0832\sin\beta \tag{5.18}$$

The yield strength largely determines the tangential failure of the material. In order to improve the accuracy of the fitting formula, the relationship between

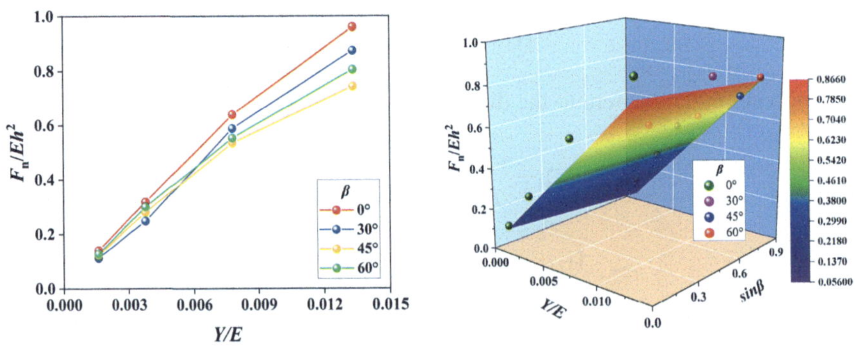

(a) Influence of Y/E on F_n/Eh_r^2 under different head deflection Angle

(b) Three-dimensional surface relationship of $sin\beta$, Y/E and F_n/Eh_r^2

Fig. 5.37 The dimensionless relationship between β, *Y/E*, and F_n/Eh_r^2. (**a**) Influence of *Y/E* on F_n/Eh_r^2 under different head deflection angles. (**b**) Three-dimensional surface relationship of $\sin\beta$, *Y/E*, and F_n/Eh_r^2

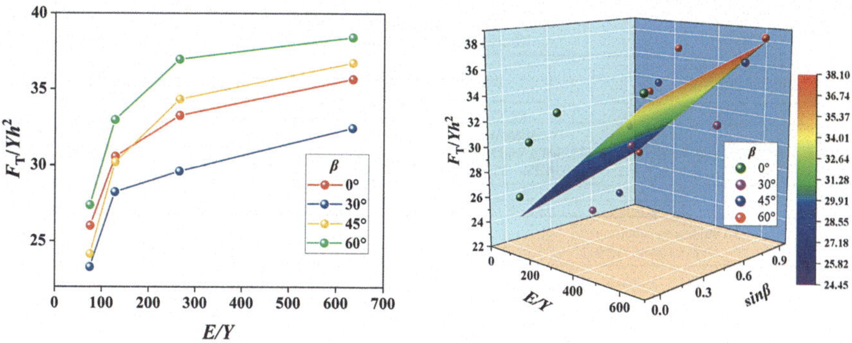

(a) Influence of E/Y on F_T/Yh_r^2 under different head deflection Angle

(b) Three-dimensional surface relationship of $sin\beta$, E/Y and F_T/Yh_r^2

Fig. 5.38 The dimensionless relationship between β, *E*/*Y*, and $F_{\mathrm{T}}/Yh_{\mathrm{r}}^2$. (**a**) Influence of *E*/*Y*on $F_{\mathrm{T}}/Yh_{\mathrm{r}}^2$ under different head deflection angles. (**b**) Three-dimensional surface relationship of *sin* β, *E*/*Y*, and $F_{\mathrm{T}}/Yh_{\mathrm{r}}^2$

dimensionless function $F_{\mathrm{T}}/Yh_{\mathrm{r}}^2$ and *E*/*Y* and head deflection angle β was analyzed (formula 5.19). It should be noted that the process always follows the principle of dimensional homogenization, that is, the dimensions of horizontal and vertical coordinates are 1.

$$\frac{F_{\mathrm{T}}}{Yh_{\mathrm{r}}^2} = \Pi_2\left(\frac{E}{Y}, \beta\right) \tag{5.19}$$

Figure 5.38a shows the relationship between different elastic modulus and head deflection angle and dimensionless tangential stress. The increase in $F_{\mathrm{n}}/Eh_{\mathrm{r}}^2$ results in a rapid increase in *E*/*Y* value. The indenter deflection angle β has a great influence on $F_{\mathrm{n}}/Eh_{\mathrm{r}}^2$. When the indenter deflection angle is 60°, $F_{\mathrm{T}}/Yh_{\mathrm{r}}^2$ reaches the maximum value. It is worth noting that the dimensionless tangential stress values are close when the head deflection angle is 0° and 45°. When *E*/*Y* is less than 120, the value of β=0° is larger. When *E*/*Y* is greater than 120, the dimensionless shear stress value of β=45° is larger. It can be observed from Fig. 5.38b that $F_{\mathrm{n}}/Eh_{\mathrm{r}}^2$ presents an approximate linear increase with the increase in sin β and *E*/*Y* values. The following Eq. (5.19) is fitted with the 3D-plane function:

$$\frac{F_{\mathrm{T}}}{Yh_{\mathrm{r}}^2} = 23.069 - 0.0187\frac{E}{Y} + 3.6657 \sin\beta \tag{5.20}$$

According to the test results of nanoscratches, F_{T}, F_{n}, h_{r}^2, and other parameters can be collected according to the mechanical parameter sensor, and the indenter deflection angle β can be set in advance. Therefore, based on the scratch results and Eqs. (5.19) and (5.20), the binary first-order equations were formed for the two

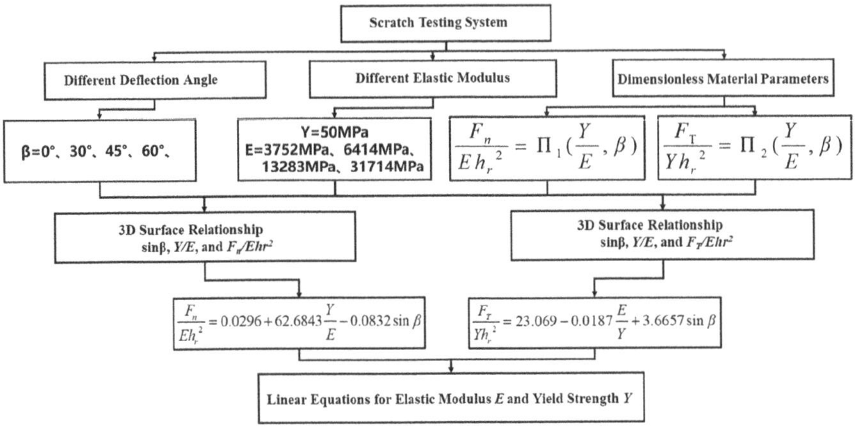

Fig. 5.39 Scratch test inversion process for elastic modulus and yield strength

unknowns of yield strength Y and elastic modulus E, and the elastic modulus and yield strength of the material were calculated based on the scratch experiment. The specific process is shown in Fig. 5.39.

The elastic modulus is a measure of how much a material deforms when stressed, and a high elastic modulus means that a rock deforms more quickly when stressed, making it easier to break. The size and distribution characteristics of the surface elastic modulus of laminated shale based on scratch test inversion can provide important reference value for the evaluation of the compressibility sweet spot of unconventional energy shale oil. According to the proposed elastic-plastic mechanical parameters, the dimensionless equations are reversed, and the elastic modulus of 8 groups of nanoscratch samples developed in Chap. 3 is backdeduced. Figures 5.40 and 5.41 show the inversion results. For #1 and #2 samples, due to the high content of illite minerals and soft texture, the elastic modulus collected in most of the scratch curves ranges from 0 to 20 GPa. Due to the presence of brittle minerals such as quartz and pyrite behind the scratch path, high elastic modulus peaks appear in a few locations.

In the process of hydraulic fracturing to exploit shale oil and gas, rocks with high elastic modulus are more susceptible to pressure. When injected with high-pressure water flow, rocks are more likely to be squeezed and broken, thus forming cracks and channels, which are conducive to releasing oil and gas in shale. Therefore, it is of great significance to invert the elastic modulus of shale and predict fracturing sweet spots. Figure 5.42a shows the elastic modulus inversion curves of a total of 6 scratch curves of samples #3 to #8. Due to the strong heterogeneity of shale, the curves show very violent fluctuations. The elastic modulus of plastic minerals such as illite is small, and the wave peak amplitude is close to the level, while the wave peak of brittle minerals is high; consequently, it is impossible to accurately evaluate the mechanical characteristics of shale from the scratch elastic modulus curve. Figure 5.42b shows the plane heat maps of elastic modulus for inversion of 6 groups

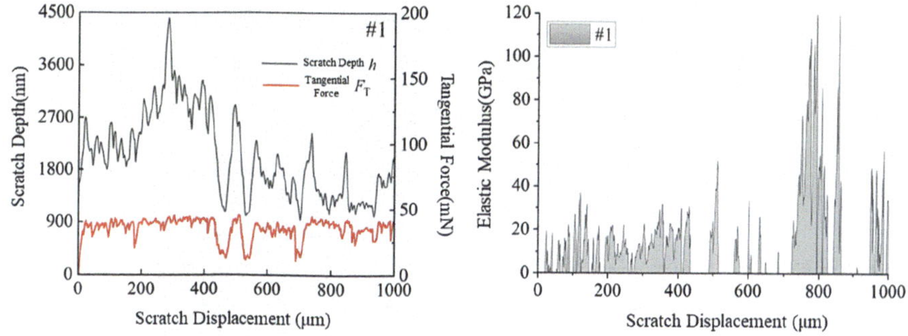

(a) Scratch depth and tangential force curve (b) Elastic modulus inversion results

Fig. 5.40 Inversion results of #1 sample elastic modulus. (**a**) Scratch depth and tangential force curve. (**b**) Elastic modulus inversion results

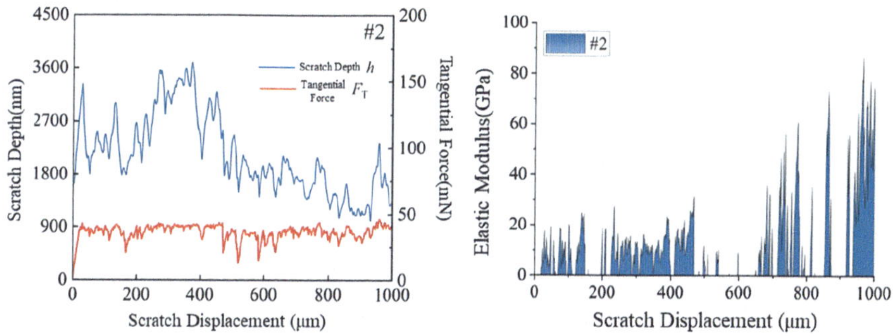

(a) Scratch depth and tangential force curve (b) Elastic modulus inversion results

Fig. 5.41 Inversion results of #1 sample elastic modulus. (**a**) Scratch depth and tangential force curve. (**b**) Elastic modulus inversion results

of scratch curves. The chromatographic interval is set to be 0–150 GPa, and data points higher than 150 GPa are shown in red on the map, so as to predict engineering sweet spots. According to the distribution of red regions in the heat map, it can be observed that samples #7 and #8 are predicted engineering sweet spots at the scratch displacement of 300 μm, which also proves the accuracy and feasibility of the proposed inversion method of mechanical parameters.

5.4.3 Dimensionless Scratch Width

In the process of scratch test, under the same yield strength and friction coefficient, the influence of pressing depth on $\delta/\sqrt{F_n/E}$ is weak, and the influence of friction coefficient is also small. Therefore, the effect of friction coefficient μ_s and

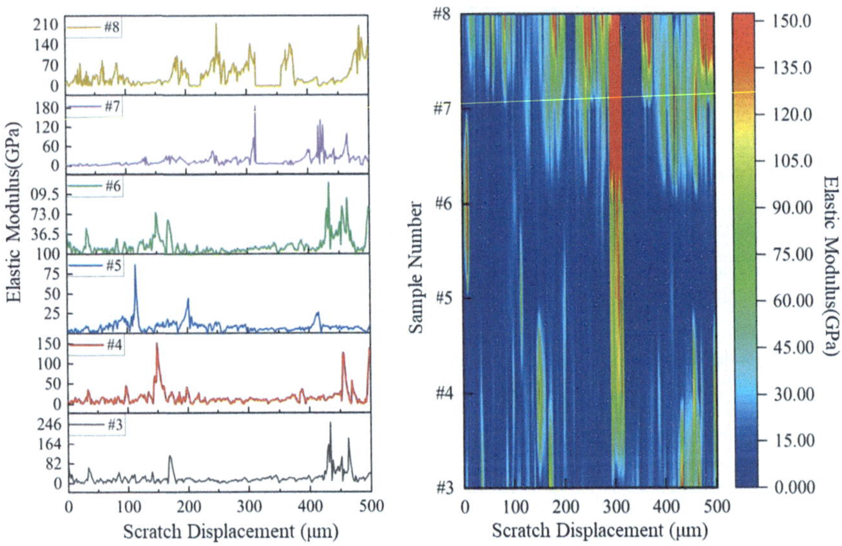

(a) Elastic modulus inversion curve (b) Project dessert selection results

Fig. 5.42 Based on the engineering sweet-spot selection results using scratch elastic modulus. (**a**) Elastic modulus inversion curve. (**b**) Project dessert selection results

indentation depth h on normalized scratch width $\delta/\sqrt{F_n/E}$ can be ignored, that is, the dimensionless equation becomes

$$\frac{\delta}{\sqrt{F_n/E}} = \Pi_3\left(\frac{Y}{E}, \beta\right) \tag{5.21}$$

Figure 5.43a shows the parameterized relationship between dimensionless scratch width $\delta/\sqrt{F_n/E}$ and Y/E, as well as sinusoidal values of indenter deflection angle β. As a general trend, with the increase in Y/E, the value of $\delta/\sqrt{F_n/E}$ shows a slow decreasing trend after a sharp decline. The change of indenter deflection angle β has no obvious influence on dimensionless scratch width $\delta/\sqrt{F_n/E}$, while Y/E has a significant influence on $\delta/\sqrt{F_n/E}$. In order to better analyze the relationship between head deflection angle and mechanical performance, the sinusoidal functions of head deflection angle sin β, Y/E, and $\delta/\sqrt{F_n/E}$ are used to form three-dimensional spatial data distribution for analysis. In Fig. 5.43b, surface fitting of dimensionless scratch width is performed with the help of three-dimensional surface function, and the analytical expression of dimensionless function is obtained:

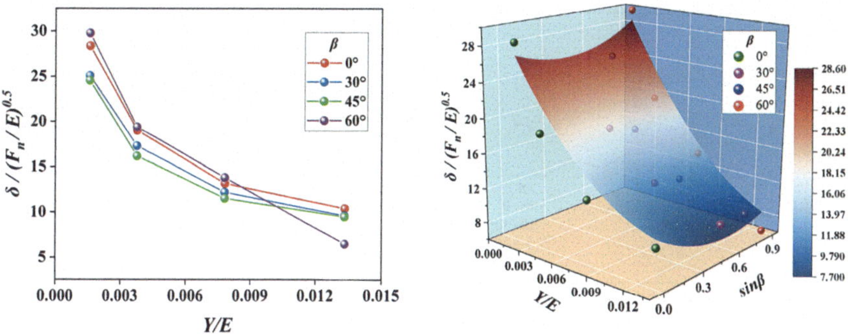

(a) Influence of Y/E on $\delta/\sqrt{F_n/E}$ under different head deflection Angle

(b) Three-dimensional surface relationship between $sin\beta$, Y/E and$\delta/\sqrt{F_n/E}$

Fig. 5.43 The dimensionless relationship between β, *Y*/*E*, and $\delta/\sqrt{F_{\rm n}/E}$. (**a**) Influence of *Y*/*E* on $\delta/\sqrt{F_{\rm n}/E}$ under different head deflection angles (**b**) Three-dimensional surface relationship between sin β, *Y*/*E*, and $\delta/\sqrt{F_{\rm n}/E}$

$$\frac{\delta}{\sqrt{F_{\rm n}/E}} = 30.96 - 2623.31\frac{Y}{E} - 7.12\sin\beta + 8.64E4\frac{Y^2}{E^2} + 10.97\sin^2\beta - 401.21\frac{Y}{E}\sin\beta \tag{5.22}$$

In order to comprehensively consider the influence of tangential force and normal stress on dimensionless scratch width, similar to Formula 5.19, the original normalized scratch width $\delta/\sqrt{F_{\rm T}/Y}$ is taken as a new dimensionless function, and its relationship with *Y*/*E* and indenter deflection angle β is analyzed, that is, the dimensionless equation becomes

$$\frac{\delta}{\sqrt{F_{\rm T}/Y}} = \Pi_3\left(\frac{Y}{E}, \beta\right) \tag{5.23}$$

In Fig. 5.44a, as the value of *Y*/*E* increases, the value of $\delta/\sqrt{F_T/Y}$ presents a relatively slow upward trend, *Y*/*E* has a significant impact on $\delta/\sqrt{F_{\rm n}/E}$, and the change of indenter deflection angle β has a significant impact on dimensionless scratch width $\delta/\sqrt{F_{\rm T}/Y}$. When β=0° and 60°, the numerical size of $\delta/\sqrt{F_{\rm T}/Y}$ is first and second, respectively. In Fig. 5.44b, surface fitting of dimensionless scratch widths is performed with the help of polynomial three-dimensional plane functions:

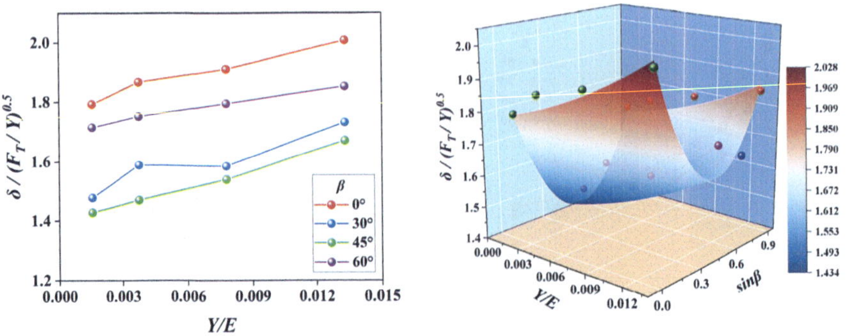

(a) Influence of Y/E on $\delta/\sqrt{F_T/Y}$ under different head deflection Angle

(b) Three-dimensional surface relationship between $sin\beta$, Y/E and $\delta/\sqrt{F_T/Y}$

Fig. 5.44 The dimensionless relationship between β, Y/E, and $\delta/\sqrt{F_T/Y}$. (**a**) Influence of Y/E on $\delta/\sqrt{F_T/Y}$ under different head deflection angles. (**b**) Three-dimensional surface relationship between $\sin\beta$, Y/E, and $\delta/\sqrt{F_T/Y}$

$$\frac{\delta}{\sqrt{F_T/Y}} = 1.78 + 9.53\frac{Y}{E} - 1.54\sin\beta + 688.02\frac{Y^2}{E^2} + 1.66\sin^2\beta - 5.81\frac{Y}{E}\sin\beta \tag{5.24}$$

The relationship between dimensionless stress, dimensionless scratch width, and Y/E and $\sin\beta$ obtained by fitting is similar to that of previous studies using cone indenter.The scratch width and stress data are obtained through scratch experiments, and the numerical analytical solutions of the yield stress and elastic modulus of the material can be obtained by combining Eqs. (5.18), (5.20), (5.22), and (5.24).

References

1. Simões MI, Martins AX, Antunes JM, et al. Numerical simulation study of the Knoop indentation test. In: COMPLAS XI: Proceedings of the XI International Conference on Computational Plasticity: fundamentals and applications. CIMNE; 2011. p. 287–94.
2. Dasari A, Yu ZZ, Mai YW. Fundamental aspects and recent progress on wear/scratch damage in polymer nanocomposites. Mater Sci Eng R Rep. 2009;63(2):31–80.
3. Zhang J, Li Y, Zheng X, et al. Determination of plastic properties of surface modification layer of metallic materials from scratch tests. Eng Fail Anal. 2022;142:106754.
4. Hodzic A, Kalyanasundaram S, Kim JK, et al. Application of nano-indentation, nano-scratch and single fibre tests in investigation of interphases in composite materials. Micron. 2001;32(8):765–75.
5. Dash R, Bhattacharyya K, Bhattacharyya AS. Film failure at earlier and later stages of nanoindentation in static and sliding modes. Eng Fail Anal. 2023;150:107353.

Chapter 6
Scratch Mechanics After CO_2-Rock Interactions

Conventional mechanical methods are difficult to evaluate the differentiated characteristics of grain layer mechanics in supercritical CO_2-water softening mechanism. This chapter introduces the changing law of the mechanical properties of grain layer shale scratches under fluid action. Based on the nano- and centimeter-scratch techniques and microscopic electron microscopy, the mechanism of change in the microscopic pore structure, mineral composition, and mechanical properties of the grain layer shale under CO_2-water action is highlighted, and the differentiated changing law of the grain layer with different mechanical properties under fluid action is explored.

6.1 CO_2-Shale Interaction Experiments

6.1.1 Centimeter-Scale Experiments

Different laminae are usually distinguished by various mineral compositions and different thicknesses ranging from tens of micrometers to several millimeters, which significantly affects shale deformation and fracture extension during shale gas/oil exploitation. Figure 6.1 depicts a shale sample. Lamina I is primarily composed of clay-rich material inside the yellow-bordered area pictured below. It has a thickness of 4 mm. Lamina II is mostly calcite-rich and has a thickness of 2.5 mm. Lamina III is a quartz/calcite lamina with a thickness of 2.5 mm.

The experiment is divided into four sections: (1) exposure of shale samples to $ScCO_2$-water; (2) large-area elemental scanning analysis using X-ray fluorescence spectroscopy; (3) scratch tests; and (4) microscope testing. The process is illustrated in Fig. 6.2. All shale samples were shipped in small sealed helium bags to minimize changes during storage and were soaked in $ScCO_2$-water under the same conditions.

To assure accuracy, the scratch test used shale cores from the same block and stratum to prevent influence from rock layers and anisotropy. The shale was

L. Yang, *Multiscale Scratch Mechanics of Laminated Shale*,
https://doi.org/10.1007/978-981-95-7041-6_6

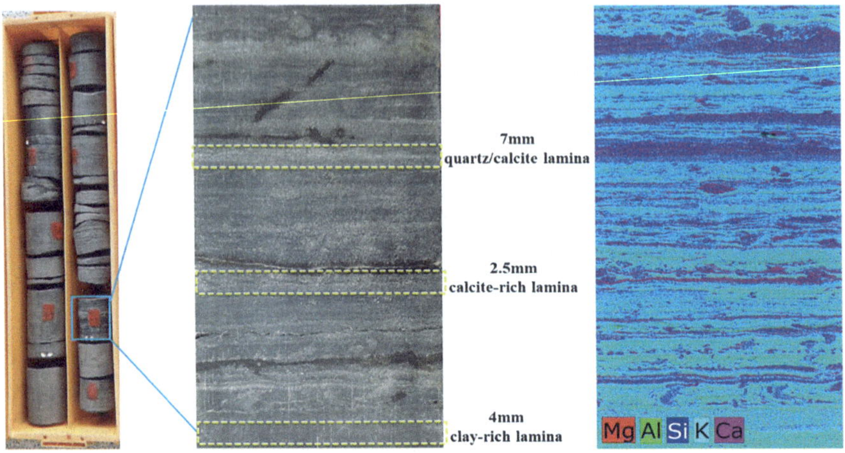

Fig. 6.1 Shale sample

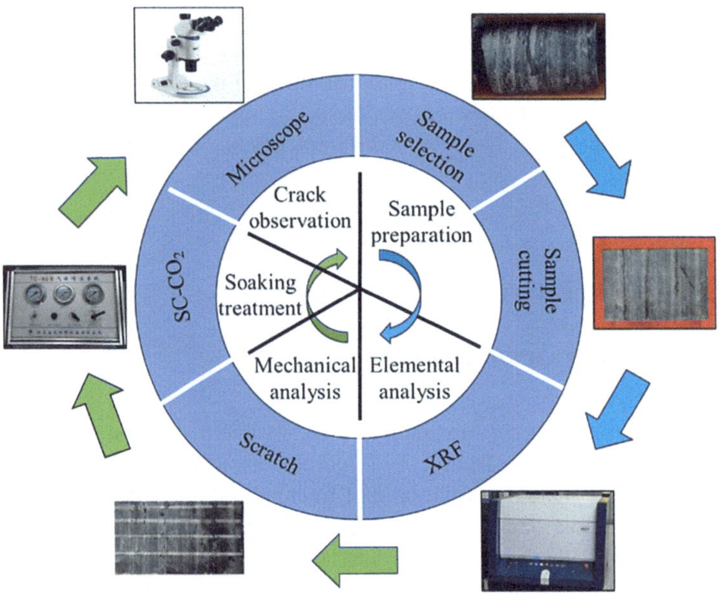

Fig. 6.2 Experimental flow chart

machined into cubic specimens (Fig. 6.1) before being polished to meet scratch test standards. An X-ray fluorescence spectrometer was used to study the shale, which revealed that it comprised predominantly five elements: Mg, Al, Si, K, and Ca. This showed that the shale contains a substantial amount of quartz and calcite, as well as clay minerals and feldspar. This research focuses on the clay-rich lamina, calcite-rich lamina, and quartz/calcite lamina.

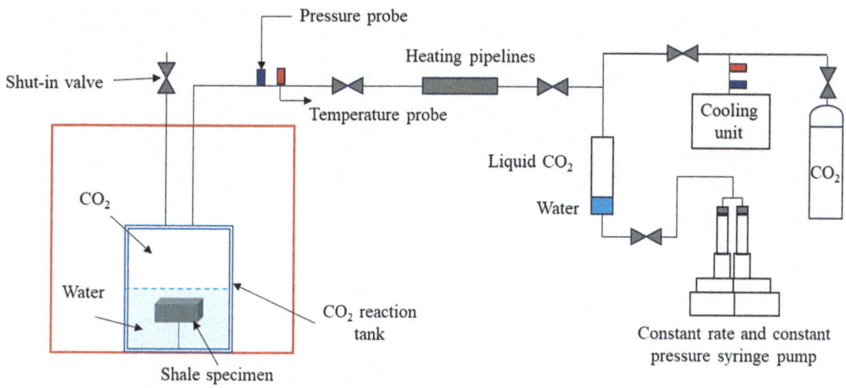

Fig. 6.3 Schematic of the setup used for the static soaking experiments

The $ScCO_2$-water processing test setup is shown in Fig. 6.3. The CO_2 pressurization system is a self-developed experimental device that converts CO_2 to a supercritical state by controlling the temperature and pressure. The immersion temperature is 65 °C, and the pressure is 15 MPa.

The specific practical steps are as follows:

(1) Before $ScCO_2$ treatment, the equipment is evacuated. Then the CO_2 is injected to a certain pressure and the temperature is controlled to bring the CO_2 into a supercritical state. (2) The typical shale specimen is then placed in the $ScCO_2$ treatment chamber. (3) The immersion tank must be adequately sealed and maintain high temperature and pressure conditions for 1d, 2d, and 3d. (4) After treatment, the specimens with different treatment times were subjected to scratch tests, microscopic tests, and X-ray fluorescence spectroscopy.

To continuously and quantitatively analyze the mechanical properties of shale with soaked $ScCO_2$-water treatment, we utilized a self-devised HADHJ-25/100-IV experimental setup for shale scratch measurement. This methodology allowed us to determine the longitudinal continuous distribution characteristics of the mechanical properties, demonstrated in Fig. 6.4. The test blade had a width range of 2–8 mm, while the measurement range for depth was between 0 and 1 mm. Horizontal movement ranged from 0 to 500 mm. Before the scratch test, a leveling procedure was required to make the surface flat. The scratch test was performed along the core, perpendicular to the laminated line, with a constant depth of cut and speed. The cutting depth utilized for this study was within the range of 0.5 millimeters. To obtain trustworthy data, the length of the scratch should be several centimeters long. The width of the tool used in the experiment was 2 mm.

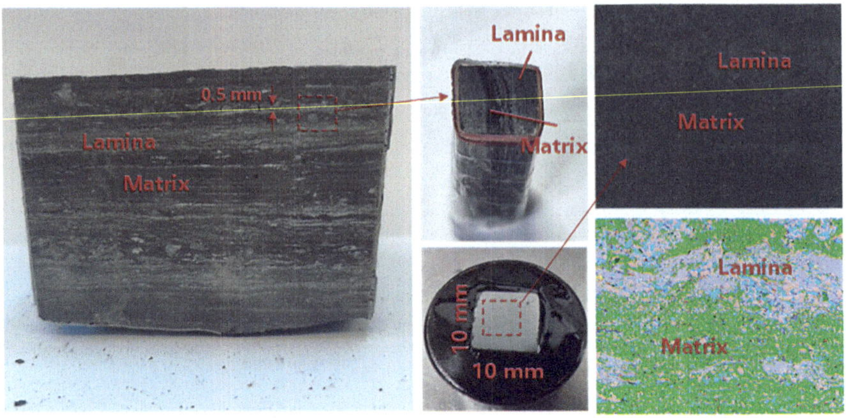

Fig. 6.4 Schematic diagram of laminated samples

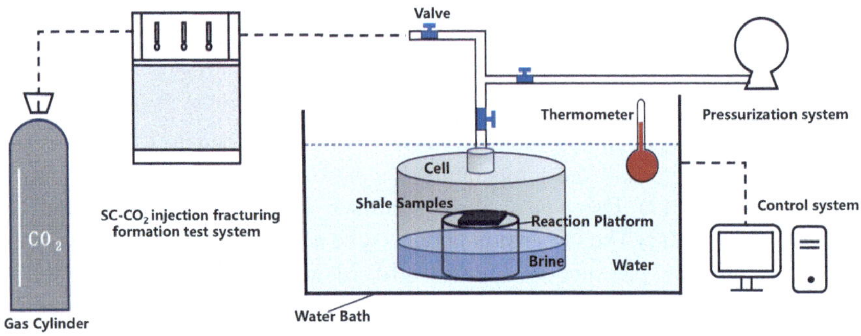

Fig. 6.5 Carbon dioxide immersion process

6.1.2 Nanoscale Experiments

The terrestrial lake basins in China exhibit intricate geological characteristics and have significant variations in depositional settings and diagenesis. As a consequence, the sedimentary rocks in these basins are heterogeneous, and their laminated structure plays an important role in fracture propagation. The samples shown in Fig. 6.4 were cut from a reservoir core of laminated shale in the northern Songliao Basin, China. Specifically, it belongs to the Lower Cretaceous Qingshankou Formation. Thickness of the lamina is less than 0.5 mm. Sample sizes were 10 mm × 10 mm × 5 mm sections. Furthermore, to satisfy the surface smoothness and flatness for the nanoscratch tests, the sample surface was roughly polished using sandpaper, followed by a further precision polishing process using an argon ion polisher. The shale samples were carefully managed to maintain a root-mean-square (RMS) roughness ranging from 120 to 150 nm.

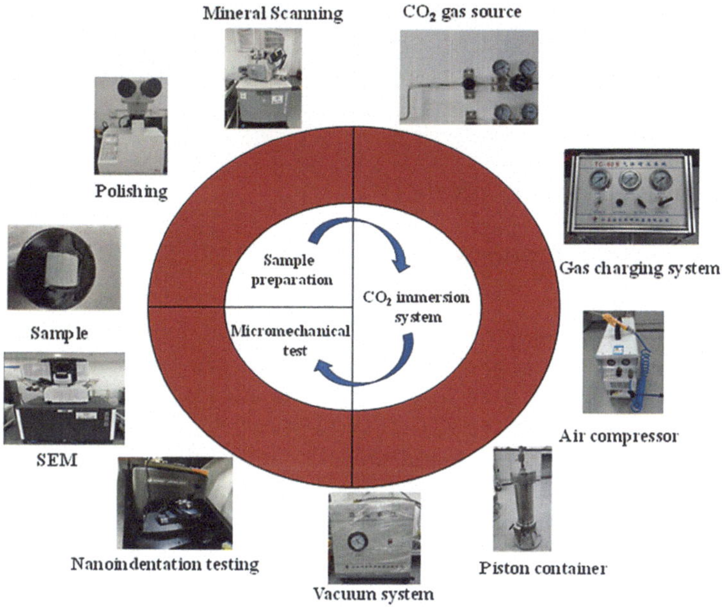

Fig. 6.6 Experimental flow chart

The CO_2-brine saturation procedure is shown in Fig. 6.5. To avoid clay swelling and simulate the in situ groundwater conditions, a brine solution was made by including 2 w % NaCl and 1 w % KCl28. As shown in Fig. 6.2, the shale sample was placed on a reaction table within a vacuum vessel. Brine was subsequently pumped into the vessel, and then CO_2 was injected into the vessel until the pressure raised to 15 MPa. The temperature of the entire CO_2-brine treatment system was maintained at 70 °C using a water bath system. The present experimental conditions exceed the critical temperature and pressure of supercritical carbon dioxide (SC-CO_2), keeping CO_2 in the supercritical phase. The pressure in the reaction vessel was monitored by a pressure sensor. To study the influence of time on the interactions between CO_2-brine and shale, the saturation time was designed to 2, 4, 6, and 8 days, then conducted micro-scratch tests. To avoid the impact of ambient air saturation on the samples, the nanoscratch tests were conducted within 6 h after the samples removed from the vessel.

The experimental flow chart is shown in Fig. 6.6. The spectroscopic probe of the QEMSCAN instrument detected characteristic X-ray signals, allowing the identification of elemental information at specific points. QEMSCAN provided the information of mineral composition and distribution within the laminae and matrix. Subsequently, scanning electron microscopy (SEM) detected the microstructure of the shale laminae and matrix. Then, micro-scratch technique was used to selectively scratch the laminae and matrix at various saturation time. In addition, SEM was used to observe the location and morphology of the scratches. After each saturation time, SEM was employed to document the microstructural alterations occurring in the shale laminae and matrix specifically at the site of the scratches. After the completion of the final scrape, the QEMSCAN-SEM-EDS technique was employed once more to examine the influence of SC-CO_2 on the shale laminae and matrix.

6.2 Centimeter-Scale Changes

6.2.1 Elemental Changes

Figure 6.7 displayed the alterations in the shale samples' elemental composition before and after exposure to ScCO_2-water. Figure 6.7a presented the variations in the elements after soaking. Figure 6.7b showed the modifications in the elemental content before and after soaking with ScCO_2-water. As shown in Fig. 6.7a, the hue of elemental Ca in the shale considerably diminished after soaking with ScCO_2-water, implying that elemental Ca was dissolved following ScCO_2-water contact. Figure 6.7b illustrated that exposure to ScCO_2-water leads to a decrease in Ca, Al, K, and Mg elements in the shale. However, there was an increase in the Si element. As ScCO_2-water did not react with quartz, the overall quality of the shale decreases as some minerals dissolve, resulting in a gradual increase in quartz proportion. Simultaneously, clay minerals reacted with CO_2, resulting in the production of quartz. As a result of CO_2 treatment under dual effect, the percentage of quartz increased significantly. (The content of quartz was significantly increased after a double influence of factors.)

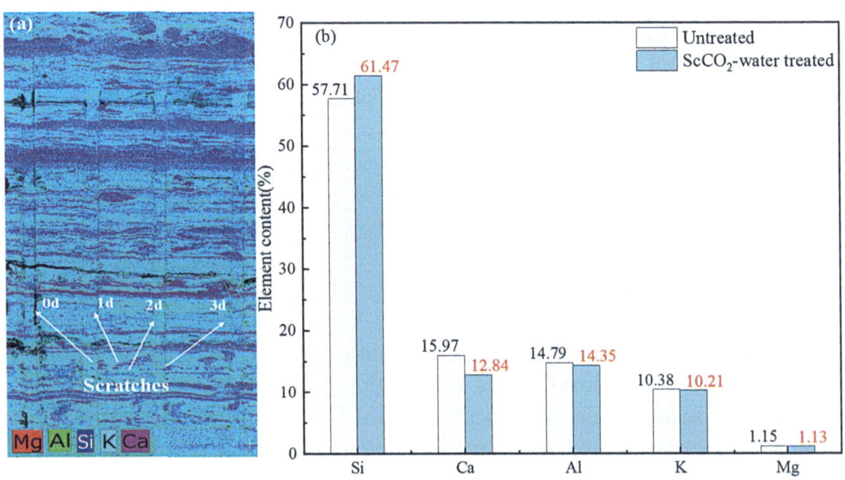

Fig. 6.7 Elements changes in shale surface after soaking with ScCO_2-water: (**a**) XRF element results and (**b**) content variation

Dissimilarities in mineral component dissolution capacity correspond to $ScCO_2$-water-shale interactions. When carbon dioxide dissolves in water to create carbonic acid, the said solution reacts with shale minerals as follows [1]:

$$CO_2 + H_2O \leftrightarrow H_2CO_3 \leftrightarrow H^+ + HCO_3^- \quad (6.1)$$

$$\text{calcite}(CaCO_3) + 2H^+ \leftrightarrow Ca^{2+} + CO_2 + H_2O \quad (6.2)$$

$$\text{quartz}(SiO_2) + 4H^+ \leftrightarrow Si^{4+} + 2H_2O \quad (6.3)$$

$$\begin{aligned} &\text{illite}\left(KAl_2(Si_3, Al)O_{10}(OH)_2\right) + 8H^+ \\ &\leftrightarrow 0.6K^+ + 2.3Al^{3+} + 0.25Mg^+ + 3.5SiO_2 + 5H_2O \end{aligned} \quad (6.4)$$

$$\begin{aligned} &K - \text{feldspar}(2KAlSiO_8) + 2H^+ + 9H_2O \\ &\leftrightarrow 2K^+ + 4H_4SiO_4 + Al_2SiO_3(OH)_4 \end{aligned} \quad (6.5)$$

Equations (6.1)–(6.5) show that CO_2 and water react with inorganic minerals, mainly calcite and clay minerals. This causes the minerals in the shale to dissolve and ions to be released into the solution. The dissolution of calcite in shale results in Ca^{2+} movement, as demonstrated in Eq. (6.4), while the dissolution of specific clay minerals is associated with the release of some Al^{3+} in solution, as shown in Eq. (6.4). Equation (6.5) demonstrates that the migration of K^+ can be ascribed to the dissolution of K-Feldspar. Furthermore, mineral hydrolysis and ion exchange have an impact on mobilizing the elements of shale in acidic conditions [2, 3]. During mineral hydrolysis, hydrogen and hydroxide ions from water ionization enter the mineral lattice and react with their respective positive and negative ions to produce the above process.

Elemental mobility can reflect the rate of $ScCO_2$-water dissolution of different minerals in shale. The equation to quantify the elemental mobility is [4]

$$\eta = \frac{|C_b - C_a|}{C_b} \times 100\% \quad (6.6)$$

where C_b is the elemental content of $ScCO_2$-water before exposure and C_a is the corresponding elemental content after exposure.

As shown in Fig. 6.8, the highest dissolution rate of Ca^{2+} was observed after soaking with $ScCO_2$-water, which indicated that calcite was significantly dissolved. The calcite content decreased, which led to a rapid decrease in the mechanical strength. In addition, carbonation, hydrolysis, and elemental exchange were associated with the reduction of carbonate and clay minerals.

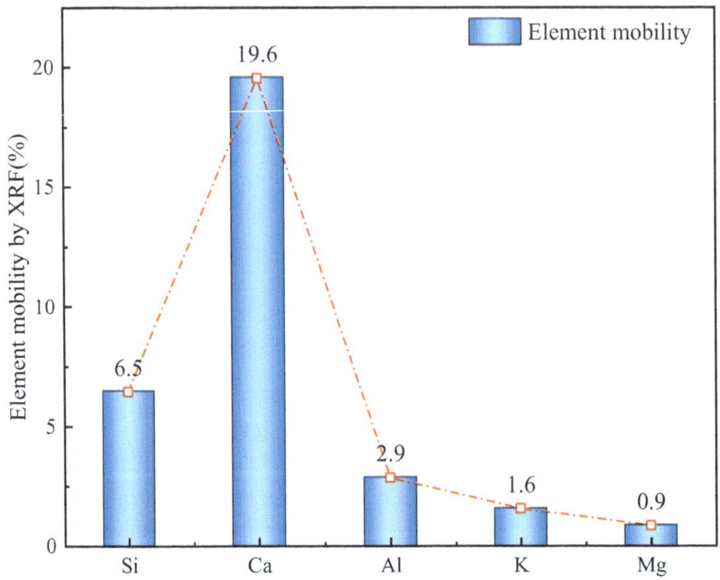

Fig. 6.8 Mobility of major elements in shale induced by $ScCO_2$-water soaking

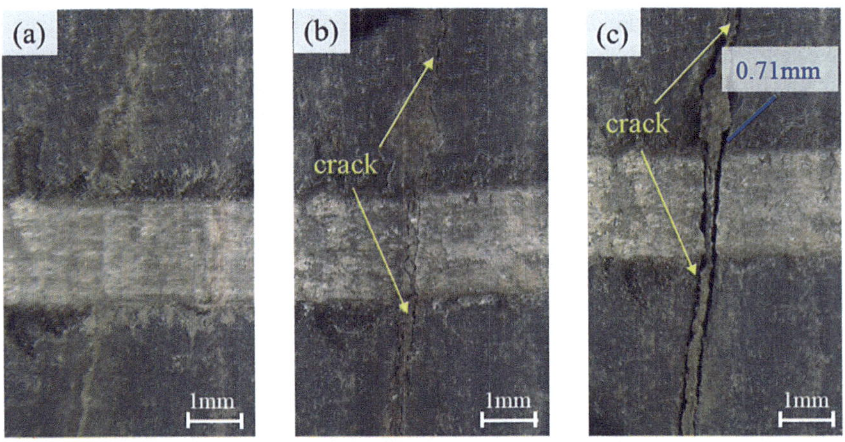

Fig. 6.9 Changes of in situ fractures after soaking with $ScCO_2$-water: (**a**) 0d, (**b**) 1d, and (**c**) 2d

6.2.2 Surface Damage Characteristics

First, the shale surface at the same location was observed microscopically as shown in Fig. 6.9. By comparing the microscopic images of the shale surface before and after soaking with $ScCO_2$-water, it became evident that the gradual widening of the crack was related to the dissolution of carbonate minerals. After soaking for 1d, the calcite within the crack underwent dissolution, which led to a gradual opening. After

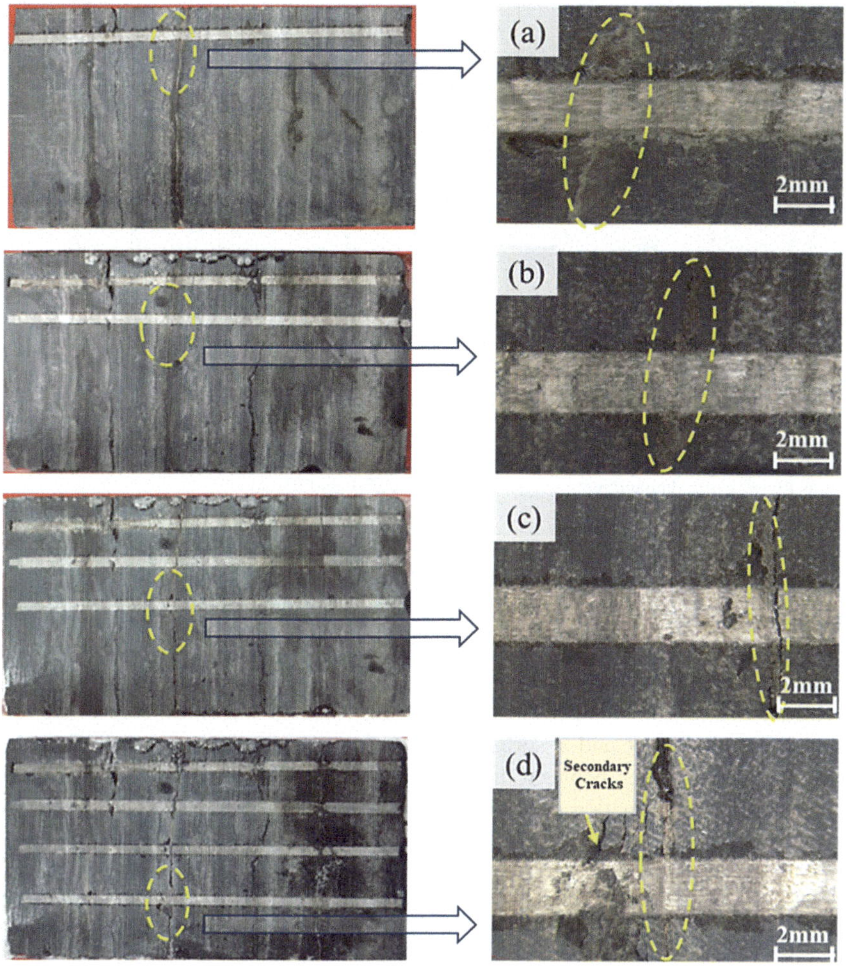

Fig. 6.10 Changes of cracks after soaking with $ScCO_2$-water: (**a**) 0d, (**b**) 1d, (**c**) 2d, and (**d**) 3d

soaking for 2d, the largest point of crack aperture in the microscope image had increased to 0.71 mm.

Figure 6.10 showed the microscopic observation of shale surfaces at different positions with different soaking times. By comparing microscope images of the shale surface following $ScCO_2$-water exposure, it became apparent that as the soaking period increases, $ScCO_2$ adsorption into the broadest point of the aperture proliferates gradually from the opening of the crack. After soaking for 2d, the aperture of the cracks on the microscope images widened, while diffusion of narrow secondary cracks and mineral dissolution occurred. After soaking for 3d, the crack opening was evident with secondary cracks and mineral-increased dissolution.

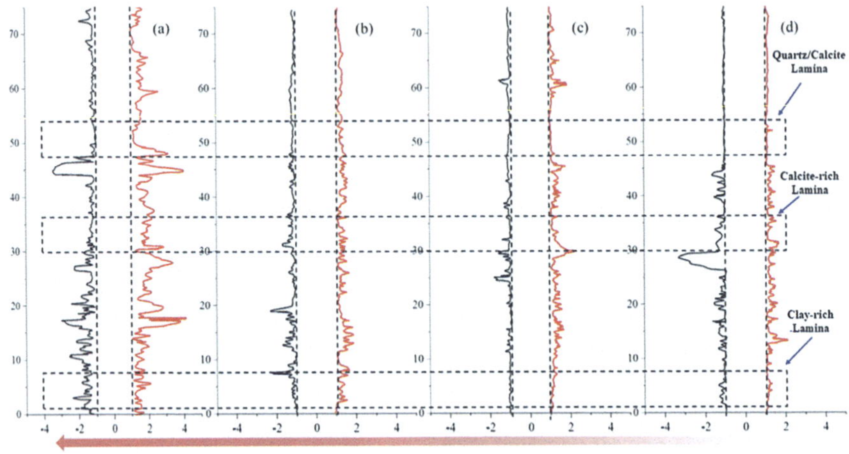

Fig. 6.11 Fracture damage pattern: (**a**) 0d, (**b**) 1d, (**c**) 2d, and (**d**) 3d

Under supercritical CO_2-water interaction, illite–smectite mixed-layer clay minerals in shale tend to undergo hydration and swelling, whereas illite is more susceptible to structural collapse. As a result, obvious particle detachment, clay mineral crystal layer stripping, and induced crack opening are visible in the interior. Supercritical CO_2 adsorption in shale can lead to shale swelling and deformation. During the immersion experiment, CO_2 mainly enters the shale core along the permeable grain layer and adsorbs on the surface of clay mineral particles under the pressure difference between the inside and outside of the core. Therefore, the expansion and deformation caused by adsorption mainly occurred inside the grain layer structure. When the swelling stress is higher than the mechanical strength of the grain layer, induced cracks are formed, and the width of the induced cracks induced by CO_2 adsorption gradually increases with the increase in soaking time. In addition, due to the dissolving effect of CO_2, the original pores on the shale surface were enlarged or new micropores were formed.

To reveal the degree of softening after soaking with $ScCO_2$-water, we observed the surface after the scratch tests. The scratch damage characteristics were quantified in Fig. 6.11. Before soaking with $ScCO_2$-water, distinct scratch damage features were observed at the sample surfaces. The damage was particularly notable in the clay-rich lamina, where the sample surfaces displayed obvious "fishbone" cracks. In harder laminas, like those calcite-rich or quartz/calcite lamina, scratch damage is less noticeable, resulting in samples with smoother surfaces. When subjected to soaking with $ScCO_2$-water, the sample surface remained essentially smooth, and cracks caused by scratches were significantly reduced.

Because of the distinctive scratch damage characteristics and large crack widths near the clay and quartz/calcite lamina, we used a microscope to observe the shale surface near the quartz/calcite lamina. The results showed that the shale surface became softer and the crack width decreased significantly with the increase in soaking time. As shown in Fig. 6.12a, the crack extended at the interface of different

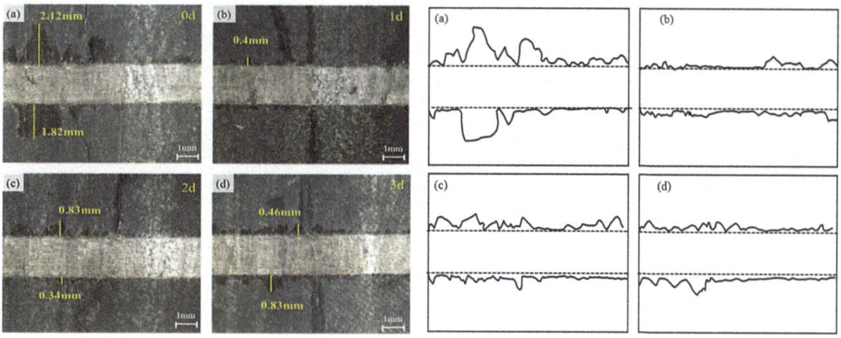

Fig. 6.12 Typical laminar fracture failure pattern: (**a**) 0d, (**b**) 1d, (**c**) 2d, and (**d**) 3d

laminas with a width of 2.12 mm. In contrast, after soaking with $ScCO_2$-water for 1d, the crack extension dropped markedly to 0.4 mm. Following an increase in soaking time from 2 to 3d, the crack width subsequently increased from 0.4 to 0.83 mm, while the overall size decreased by approximately 60.8%.

In summary, the results showed that the degree of scratch damage characterization on the shale surface was closely related to soaking with $ScCO_2$-water. Different scratch damage characteristics were exhibited in different laminas, which were closely related to shale fracture extension. This study helps to understand the crack extension pattern of shale.

6.2.3 Mechanical Alteration

The scratch test is capable of continuously measuring and characterizing the macroscopic mechanical properties of the laminated shale. It has been extensively applied in previous research. Scratch test allows continuous measurement of rock mechanical strength in longitudinal sections of shale samples, quantitative assessment of non-uniform distribution characteristics of rock mechanical properties, and identification of mechanically weak surfaces. To examine the impact of $ScCO_2$-water on the macroscopic mechanical properties of various laminas during soaking, we analyzed the variation modes of the three laminas. Figure 6.13 illustrates the variation of force signal with displacement after soaking for different times.

As depicted in the left side of Fig. 6.13, the force signal went through a down phase followed by a small increase. Before soaking with $ScCO_2$-water, scratch force signals averaged around 102 N. After soaking with $ScCO_2$-water, scratch force remained below the average value. When traversing through the quartz/calcite lamina (as shown in the red dashed box in the left side of Fig. 6.12), the force peak shifts toward the lower left. This is related to the dissolution of calcite on the shale surface after soaking with $ScCO_2$-water and possible mineral migration.

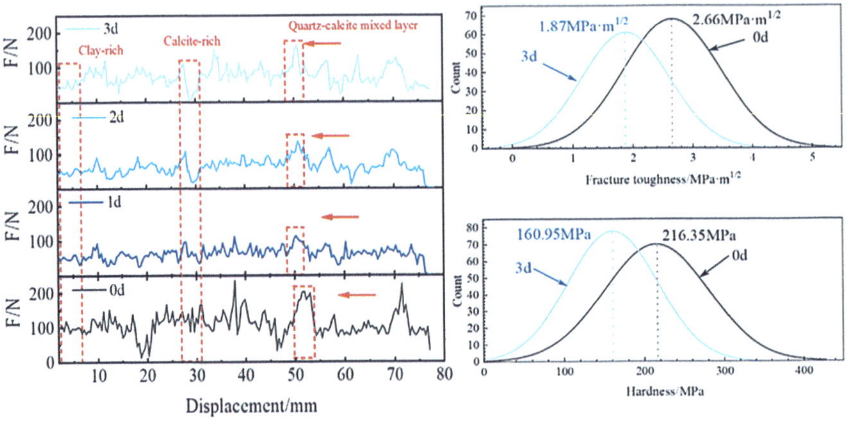

Fig. 6.13 Changes of shale mechanical properties after soaking with $ScCO_2$-water

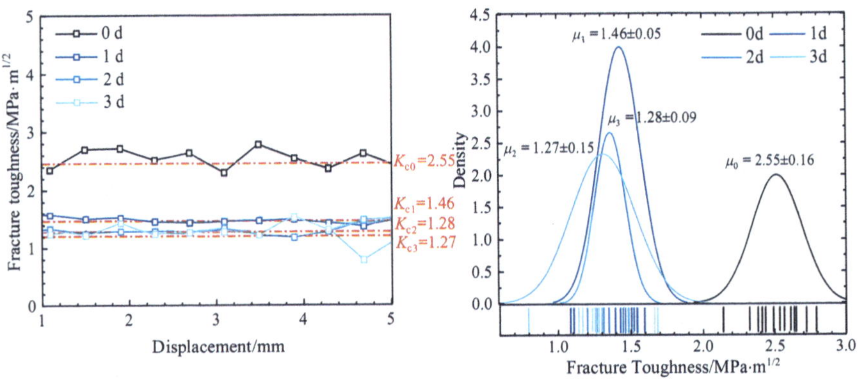

Fig. 6.14 Change in fracture toughness of clay-rich lamina after soaking with $ScCO_2$-water

To further discuss the influence of shale softening behavior on mechanical property, the modes of mechanical change at different laminas were characterized by scratch tests. Then three typical laminas were summarized by combining the scratch curves and the evolution modes of mechanical parameters.

Scratches Located in Clay-Rich Lamina

The left panels of Figs. 6.14 and 6.15 indicated that the fracture toughness and hardness of the clay-rich lamina decreased significantly after exposure to $ScCO_2$-water. The drop was the largest after soaking for 1 d. After soaking for 3 d, these values decreased by 45.8% and 47.1%, respectively. The value remained essentially unchanged with the soaking time from 2 to 3d. The fracture toughness and hardness of the clay-rich lamina were significantly less than the other two laminas.

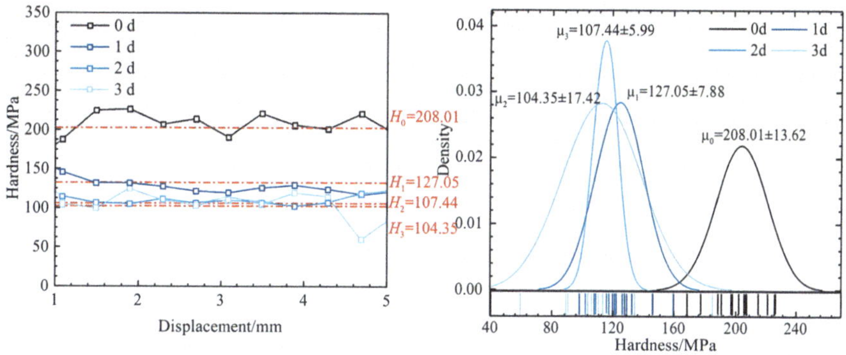

Fig. 6.15 Change in hardness of clay-rich lamina after soaking with $ScCO_2$-water

The fracture toughness of the original shale specimens was primarily distributed between 2.0 and 3.0 MPa $m^{1/2}$, with a peak value of 2.55 MPa $m^{1/2}$ observed in the normal distribution curve. Additionally, the hardness of the original shale specimens was mainly distributed between 160 and 250 MPa, with a peak value of approximately 208 MPa. These results were shown in the right panels of Figs. 6.14 and 6.15. The distribution curves of the clay-rich lamina exhibited a leftward shift with increasing soaking time in $ScCO_2$-water. After soaking for 1 d, the distribution of fracture toughness was concentrated within the 1.0–1.8 MPa $m^{1/2}$ range, with the peak at 1.46 MPa $m^{1/2}$. After soaking for 2d, the normal distribution curve peak of fracture toughness shifted to the left. The values were 1.28 MPa $m^{1/2}$. From the right panel of Fig. 6.15, it is evident that the hardness follows an identical pattern. After soaking for 1d, the hardness is predominantly distributed within the range of 80–180 MPa, with the peak value of the normal distribution curve being 127.05 MPa. After soaking for 2 d and 3 d, the corresponding hardness values were 107.44 and 104.33 MPa, respectively.

Scratches Located in the Calcite-Rich Lamina

The results presented in Figs. 6.15 and 6.16 indicate that the fracture toughness and hardness of the calcite-rich lamina decreased over time during $ScCO_2$-water exposure. The greatest decrease occurred after soaking for 2 d. The fracture toughness and hardness decreased by 40.9% and 34%, respectively, after soaking for 1 d. Finally, after soaking for 3d, the value decreased by 80% and 83.6%, respectively. This could be due to the calcite mineral dissolution by $ScCO_2$-water. As displayed in Figs. 6.16 and 6.17, the fracture toughness of the initial shale specimens was predominantly distributed within 1.5–4.0 MPa $m^{1/2}$ range and reached a peak of 2.81 MPa $m^{1/2}$. Moreover, the hardness of the initial was primarily distributed within the 120–300 MPa range, and its peak value was 216.04 MPa. Based on Figs. 6.15 and 6.16, the fracture toughness and hardness of calcite-rich lamina were notably

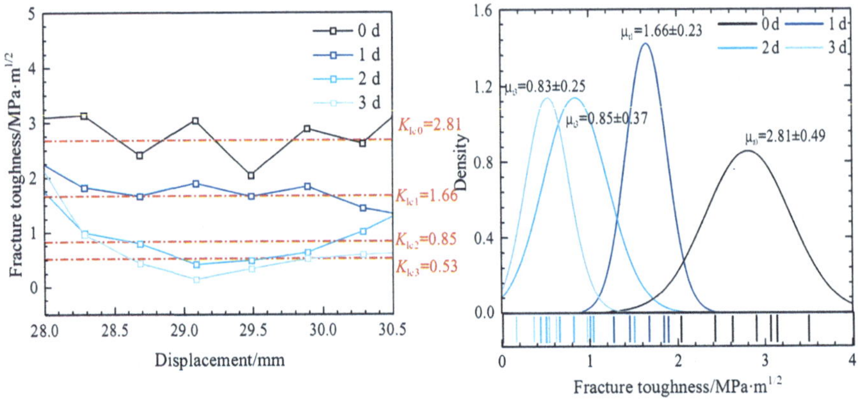

Fig. 6.16 Change in fracture toughness of calcite-rich lamina after soaking with $ScCO_2$-water

higher than those of clay-rich lamina. Upon $ScCO_2$ soaking, the peaks of the

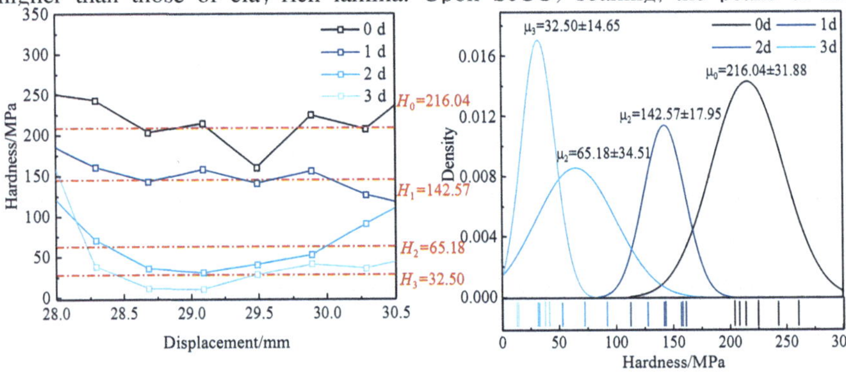

Fig. 6.17 Change in hardness of calcite-rich lamina after soaking with $ScCO_2$-water

distribution curves for fracture toughness and hardness also shift to the left as soaking time increases.

The fracture toughness was primarily distributed between 0.8 and 2.5 MPa $m^{1/2}$, with a peak value of 1.66 MPa $m^{1/2}$ after soaking for 1d. The peak value then shifted to the left after soaking for 2d, with the peak value at 0.85 MPa $m^{1/2}$: After soaking for 3d, the peak value of the normal distribution curve of fracture toughness shifted to the left. Similarly, after soaking for 1d, the hardness was primarily distributed from 75 to 200 MPa. The peak was 142.57 MPa. After soaking for 2d, the hardness peak decreased and mainly ranged from 0 to 160 MPa with a peak value of 65.18 MPa. The hardness peak shifted leftward after soaking for 3d, but with a significantly smaller amplitude. Hardness was predominantly distributed within the 0–75 MPa range. The highest point of the normal distribution curve was 32.50 MPa.

Scratches Located in the Quartz/Calcite Lamina

The fracture toughness and hardness of the quartz/calcite lamina exhibited an initial decrease and a slight subsequent increase with an increase in soaking time. But overall, the trend was downward. These results were illustrated in the left panels of Figs. 6.18 and 6.19. The peak of the fracture toughness was 3.69 MPa $m^{1/2}$, 2.25 MPa $m^{1/2}$, 2.58 MPa $m^{1/2}$, and 2.74 MPa $m^{1/2}$ with the soaking time increasing from 0 to 3 d, while the hardness was 291.03 MPa, 191.57 MPa, 220 MPa, and 232.21 MPa, respectively. After soaking for 1 d, the fracture toughness and hardness decreased by 40.3% and 37.4%, respectively. After soaking for 2 d and 3 d, the scratch curves showed a slight rise, possibly due to the reaction of $ScCO_2$-water with calcite and subsequent precipitation. The fracture toughness and hardness ultimately decreased by 27.3% and 24.7%, respectively. This type of lamina had the highest fracture toughness and hardness values of the three laminas.

The fracture toughness of the original shale specimens was distributed within the range of 0–6 MPa $m^{1/2}$, as shown in the right panels of Figs. 6.18 and 6.19. After

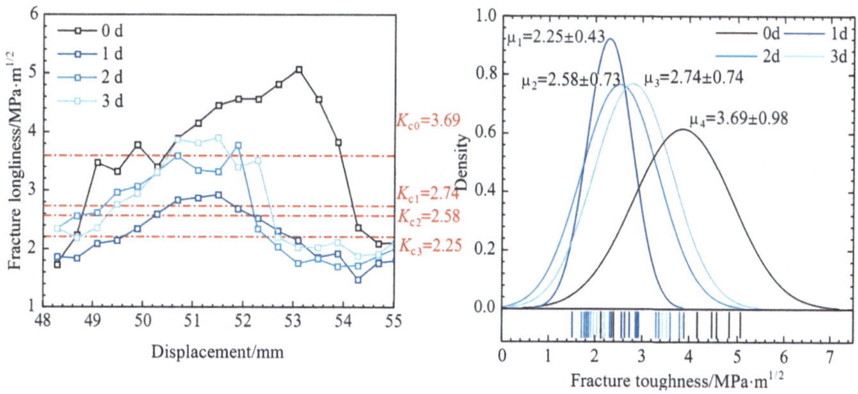

Fig. 6.18 Change in fracture toughness of quartz/calcite lamina after soaking with $ScCO_2$-water

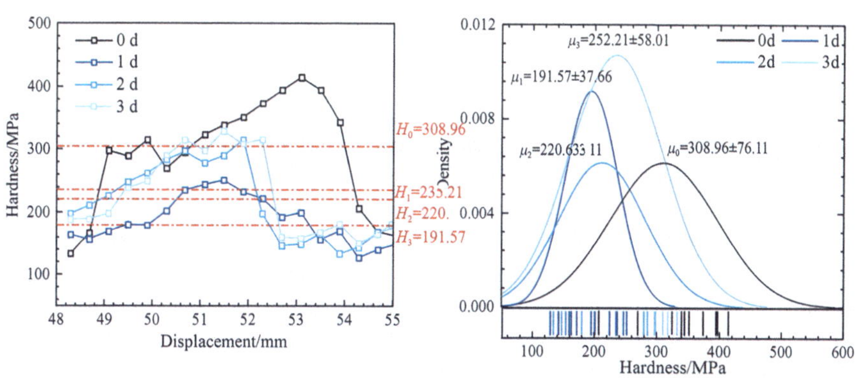

Fig. 6.19 Change in hardness of quartz/calcite lamina after soaking with $ScCO_2$-water

soaking for 1d, the fracture toughness was primarily distributed in the 0–4 MPa $m^{1/2}$ range. After soaking for 2d, the peak begins to shift to the right, and the fracture toughness was mainly distributed in the range of 0–5 MPa $m^{1/2}$. After soaking for 3d, the peak continued to shift to the right, but the magnitude was relatively smaller. Fracture toughness was mainly distributed in the range of 0–6 MPa·$m^{1/2}$. Similarly, the hardness of the original shale specimens was distributed mainly between 0 and 500 MPa. After soaking for 1d, the hardness was mainly distributed between 0 and 325 MPa as shown in the right panel of Fig. 6.18. After soaking for 2d, the peak of the normal distribution curve for hardness shifted to the right. Hardness was primarily distributed within the range of 0–450 MPa. After soaking for 3d, the peak value continued to shift to the right. It may be the dissolution of minerals and softening of the shale surface, and then the appearance of new precipitates, which led to the appearance of a decline followed by a small increase. (It was probably the dissolution of minerals and the softening of the shale surface, followed by new precipitation, that led to the appearance of a decline and then a small increase.)

6.2.4 Heterogeneity

Shale laminas are highly developed and influenced by paleo sediment sources, climate, and water environment. The main mineral compositions are quartz, carbonate, and clay. Differences in mineral components between different laminas lead to extreme inhomogeneity, which in turn leads to different mechanical properties. During the exploration and development of shale gas, much attention is being given to the inhomogeneity of shale reservoirs. This study has become vital in promoting shale theory. The centimeter-scratch experiment possesses an advantage in that it continuously measures mechanical parameters, thereby characterizing shale inhomogeneity. Therefore, a proposed parameter δ aims to quantify the inhomogeneity in shale by utilizing the scratch fracture toughness curve.

$$\delta = \sqrt{\frac{1}{N}\sum_{i=1}^{N}(x_i - \mu)^2 \times a} \tag{6.7}$$

where "a" is the time when the peak-to-trough (or trough-to-peak) distance in the curve is greater than or equal to the standard deviation (derived from Fig. 6.20). "N" represents the total number of samples of the scratch surface fracture toughness. "x_i" represents the fracture toughness value at each point where the scratch crosses the surface of the sample, and "μ" represents the average value of the fracture toughness.

As shown in Fig. 6.21, the inhomogeneity of the shale surface initially decreases and then slightly increases with prolonged soaking time, eventually leading to an overall decrease in inhomogeneity. This phenomenon is attributed to variations in shale surface minerals. After soaking for 1d, $ScCO_2$-water dissolves calcite found on

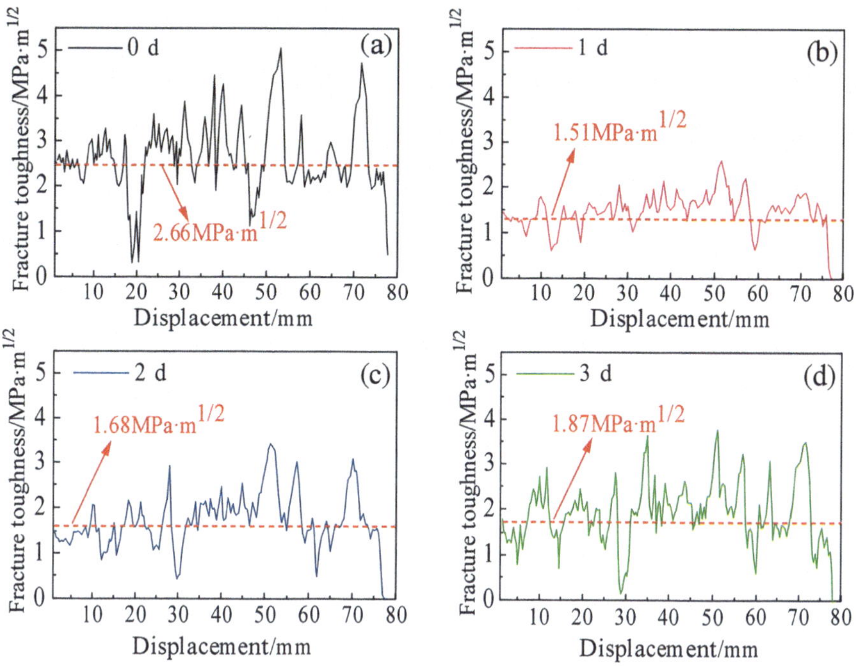

Fig. 6.20 Scratch fracture toughness curve of different soaking times: (**a**) 0d, (**b**) 1d, (**c**) 2d, and (**d**) 3d

the shale surface. This causes a significant reduction in the inhomogeneity of the shale. After soaking for 2 d and 3 d, most of the calcite in the shale samples dissolves, which leads to the formation of new precipitates. This process causes a slight increase in the inhomogeneity. After soaking, the total inhomogeneity decreased by 33.47%.

6.3 Nanoscale Alteration

6.3.1 Micro-Mineral Distribution

The rock-forming minerals were analyzed based on the QEMSCAN results (Fig. 6.22), and the distribution of mineral composition is shown in Fig. 6.23. The striated shale is composed of illite, quartz, calcite, sodium feldspar, dolomite, muscovite, apatite, kaolinite, and pyrite. The striated shale matrix consists mainly of 43.94% illite, 25.2% quartz, 12.36% natrium feldspar, and 6.26% calcite. The striated shale grains are mainly composed of 59.93% calcite, 16.58% quartz, 6.27% illite, and 8.83% Na feldspar. The overall mineral percentage of dolomite,

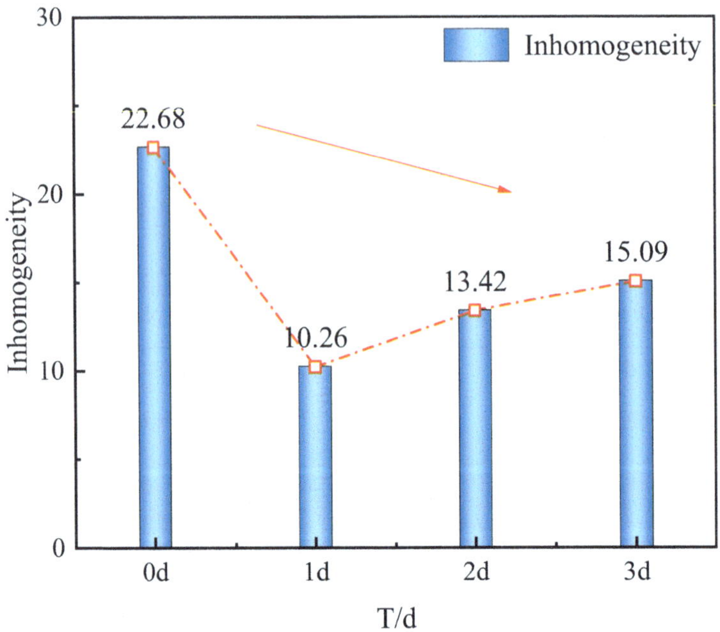

Fig. 6.21 Inhomogeneity changes for different soaking times

muscovite, kaolinite, apatite, and pyrite is less than 5%, ignoring the changes produced by SC-CO_2-brine on these minerals.

The change in mineral composition after 8 days of SC-CO_2-brine soaking is shown in Fig. 6.22, where the calcite, natrium feldspar, and clay content of the matrix decreased by 0.25%, 0.20%, and 0.04%, respectively, and the quartz content increased by 2.32%. Weak carbonic acid (H_2CO_3) was formed when SC-CO_2 came in contact with water, dissolving clay and carbonate minerals. Illite dissolves in the presence of H+ ions in solution and forms small amounts of quartz. The resulting quartz particles are stable in an acidic environment. This is the reason for the increase in quartz content and decrease in clay minerals in the matrix.

Calcite decreased by 15% in the grain layer, and clay content increased by 0.74% and natrium feldspar content increased by 0.69%. Mineral carbonation of calcite occurs over long periods of time and can dissolve in water and precipitate as a carbonate solid. The calcite content decreased due to the dissolution of calcite by SC-CO_2 at the grain layer, and the content of other minerals increased at the grain layer.

In summary, it can be seen that the mineral reaction effect produced by SC-CO_2 with the matrix is weak, the distribution of matrix minerals does not change dramatically, and SC-CO_2 produces limited mineral reaction effect with the grain layer. This is due to the fact that the amount of carbonic acid produced by the reaction between water vapor and $SCCO_2$ is much smaller than the amount of carbonic acid produced by the direct contact between water and $SCCO_2$. The

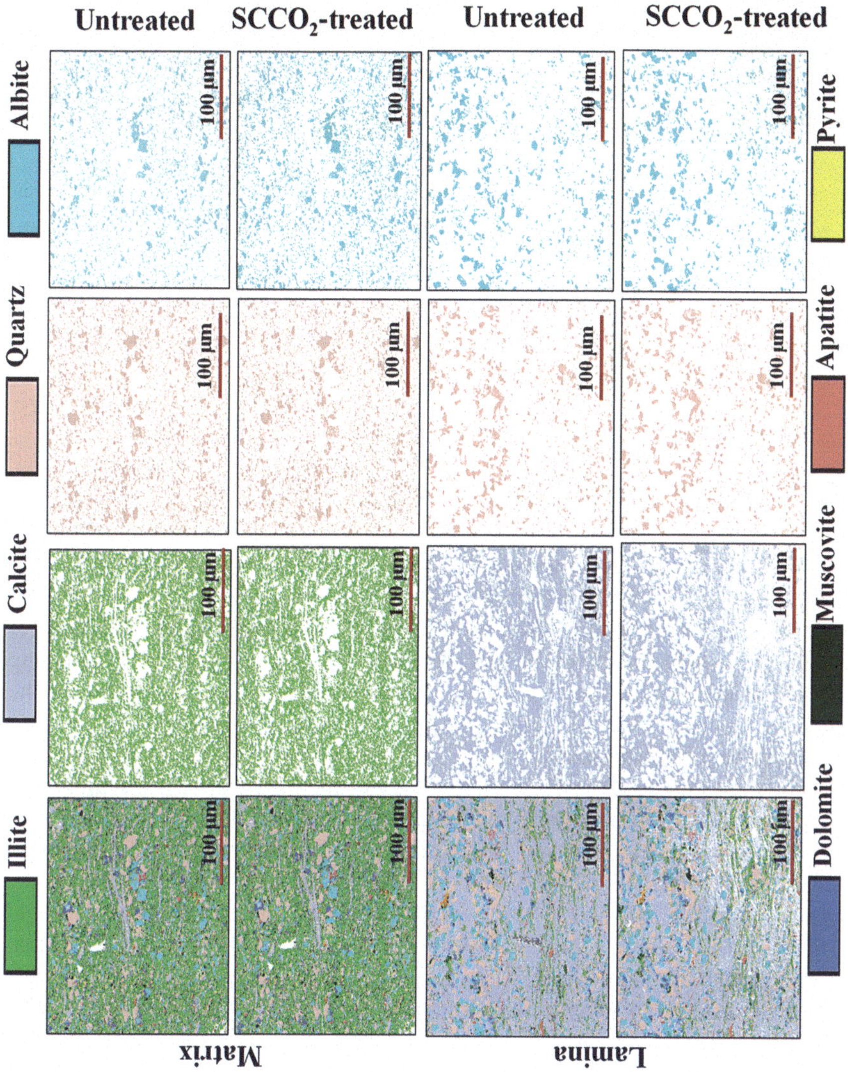

Fig. 6.22 QEMSCAN map of shale mineral distribution

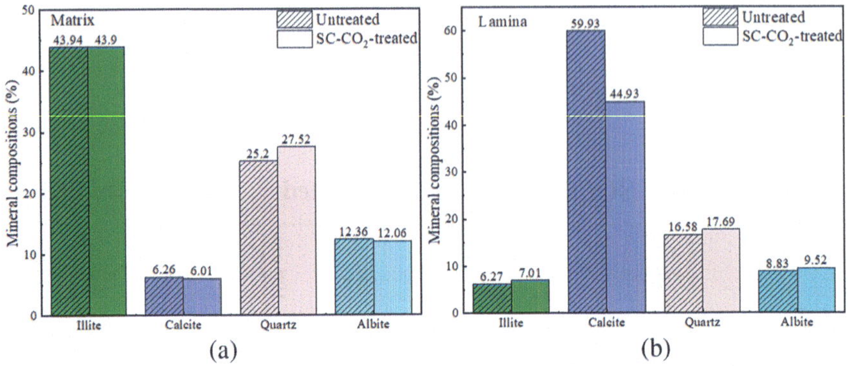

Fig. 6.23 Changes in mineral composition of shale before and after saturation: (**a**) Matrix and (**b**) laminae

mineralization process of shale with SC-CO_2-brine needs to be further studied to understand the long-term mineral trapping mechanism.

6.3.2 Microstructural Changes

The scratch results in the alteration of the shale's initial micro-morphology, leading to the formation of scratch grooves accompanied by changes in pore structure and the development of microfractures. This study aimed to investigate the impact of SC-CO_2-brine on the microstructure of shale. This was accomplished by conducting SEM observations of the static evolution of the same scratch groove position and the undamaged surface location after various treatments involving saturation in SC-CO_2-brine for varied time.

Figure 6.24a shows the surface morphology of the matrix after being immersed for a period of 2 days. In comparison to the initial matrix surface, certain minerals within the scratched grooves were observed to have been displaced and transported as a result of the interaction between SC-CO_2 and shale. This phenomenon led to the formation of additional pores. There is an absence of notable alterations observed on the intact surface. Figure 6.24b illustrates the surface morphology of the matrix that was immersed in the solution for a period of 4 days. The presence of scratch grooves induced a minor chemical reaction, resulting in the formation of flocculated crystals and the concurrent development of additional pores. The surface that remained intact did not experience significant alterations in its chemical reactions. Figure 6.24c shows the surface morphology of the matrix that underwent a saturation period of 6 days, specifically focusing on the scratched surface. The chemical reaction occurring in the scratched grooves exhibited heightened intensity, resulting in the formation of granular crystals and the concurrent development of novel holes. Additionally, the recently formed minerals served to obstruct the pores inside the

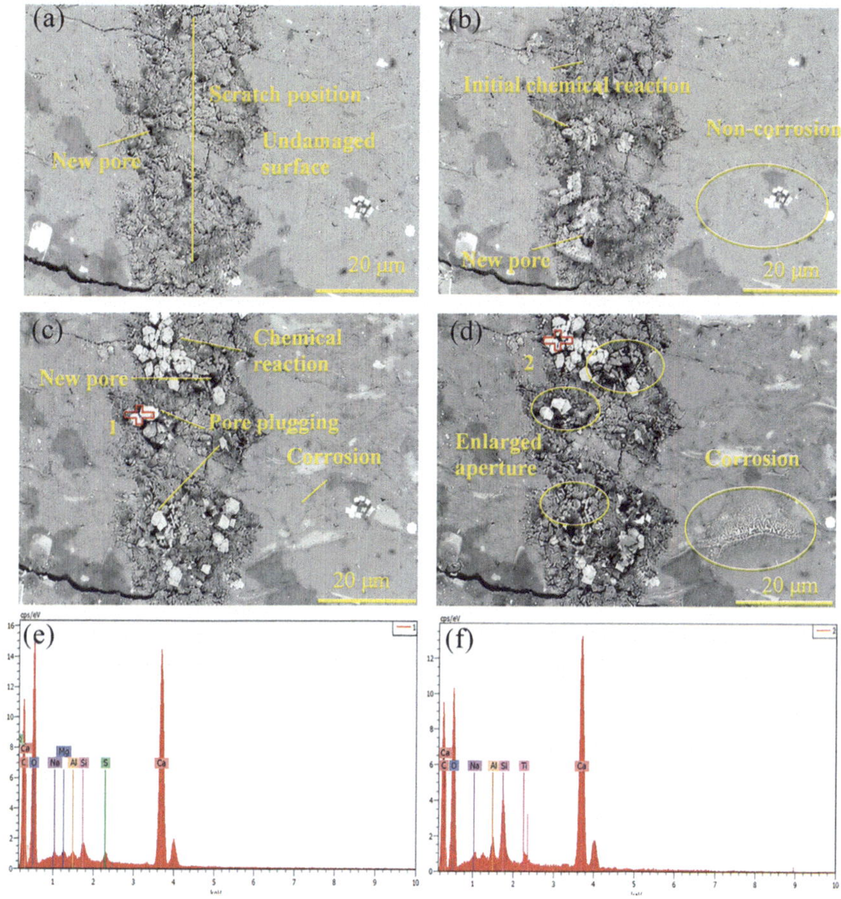

Fig. 6.24 SEM images of scratching at the same position of the matrix for different saturation time: (**a**) 2 d, (**b**) 4 d, (**c**) 6 d, (**d**) 8 d, (**e**) EDS results of point 1, and (**f**) EDS results of point 2

scratched region of the matrix. The unaltered surface experiences significant alterations in response patterns, leading to the emergence of novel minerals on its surface. The occurrence of these newly discovered minerals can perhaps be attributed to the dissolving of shallow minerals at the surface, leading to the emergence of deeper minerals. Conversely, it is possible that the minerals were generated through the mineral transportation facilitated by supercritical carbon dioxide (SC-CO_2) and brine. Figure 6.24d shows the surface morphology of the matrix that was immersed in the solution for a period of 8 days. It is evident that the number of pores and the size of the pores at the sites of the scratches saw a significant increase. In Fig. 6.24c, the minerals with a triangular shape present at the dissolution site underwent dissolution and decomposition, resulting in the formation of fragmented mineral grains, as a consequence of the interaction between supercritical carbon dioxide and

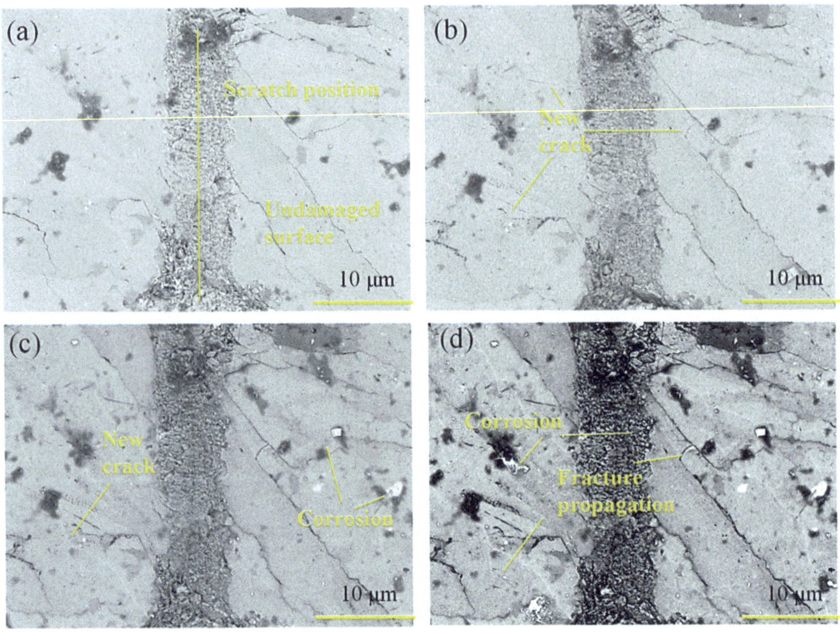

Fig. 6.25 SEM images of the same scratch at the grain layer for different immersion times: (**a**) 2 d, (**b**) 4 d, (**c**) 6 d, and (**d**) 8 d

brine. Furthermore, Figs. 6.24e, f present the EDS spectra of the granular crystals at the scratch places after 6 and 8 days of saturation. It is observed that points 1 and 2 consist primarily of a combination of calcite minerals, which are predominantly composed of calcium carbonate. This observation suggests that supercritical carbon dioxide (SC-CO_2) had a reaction with the minerals present in the matrix, resulting in the formation of a limited quantity of calcium carbonate minerals.

Figure 6.25a shows the surface morphology of the laminae which were saturated for 2 days. In contrast to the laminated surface, a portion of the minerals within the scratched grooves were found to have undergone disappearance due to the transport or dissolution facilitated by SC-CO_2-brine. Conversely, no discernible alterations were found on the undamaged surface. Figure 6.25b shows the morphology of the scratched surface of the laminae after being saturated for 4 days. Some chemical reactions took place at the place of the scratch, resulting in the formation of many microfractures initiating from the unaffected surface. Figure 6.25c shows the surface morphology of the scratches after the saturation for 6 days in the laminae. The unaltered surface experiences a significant alteration in response, leading to the manifestation of mineral dissolution phenomena. This exacerbates the formation and propagation of microfractures on the surface. Figure 6.25d illustrates the surface morphology of the scratches after a saturation period of 8 days. The previously intact surface experiences significant changes, resulting in the dissolution of minerals and the emergence of numerous microfractures. Numerous microfractures present at the site of the scratch undergo dissolution, resulting in the formation of granular bands.

In summary, the ability of SC-CO_2 to penetrate into the shale micropores and thin layers is facilitated by its low interfacial tension and viscosity. SC-CO_2 tends to adsorb in equilibrium at the fracture and pore development. Therefore, SC-CO_2-brine has a larger effect on the matrix and laminae at the pore and fracture development. Since the adsorption capacity and pore development of the matrix are larger than that of the laminae, the number in laminae microfractures is more than that in matrix. Therefore, after SC-CO_2-brine saturation, the number of pores and pore size on the matrix surface increased significantly, and the number and size of microfractures in the laminae increased significantly too. In addition, SC-CO_2-brine induces a dissolving effect on the matrix and the surface of the laminae, and the degree of this dissolving effect is affected by time.

6.3.3 *Initial Scratch Damage Morphology*

Scratches induce the formation of grooves on the surface of the rock, penetrating into the rock with a specific depth. The fracture, occurring throughout the entire thickness of the rock, is commonly referred to "through-thickness fracturing." The mode of failure that arises between the area of the groove and the adjacent minerals is known as the interfacial failure mode. The aforementioned phenomena pertain to the observed patterns of damage in shale when subjected to the influence of scratches. This study employed scanning electron microscopy (SEM) technique to examine the residual scratch morphology at both matrix and laminae sites. The study also provided a comprehensive summary of the typical scratch morphology and scratch curve characteristics observed in the matrix and laminae.

As shown in Fig. 6.26a, the scratches are located in the matrix area and along the direction from bottom to top. The yellow-striped areas are the grooves formed by the scratches penetrating into the matrix at specific depths. The fracture damage pattern observed in the matrix grooves can be attributed to the relatively low bond strength among the clay particles. Thickness-through fractures in the matrix are ductile fractures with a high density, mainly in the form of fine flexural fractures. The red bands in the figure show the damage zones at the interface between the matrix grooves and the surrounding minerals. Plastic deformation and shear damage can be observed in the damage zone. The scratched interface shows morphological features of wavy edges and mineral flaking. The observed phenomena can be attributed to the stress concentration resulting from the scratch-induced interaction between the grooves and the adjacent minerals.

The scratches are located at the laminar region, as shown in Fig. 6.26c, and the direction of the scratches is from bottom to top. Through-thickness fractures in the laminae are curved fractures with directional arrangement. The angles between the fractures and grooves are about 45°. Those are the yield fractures resulting from shear damage. After yield damage, the fracture extends to the edge of the scratch

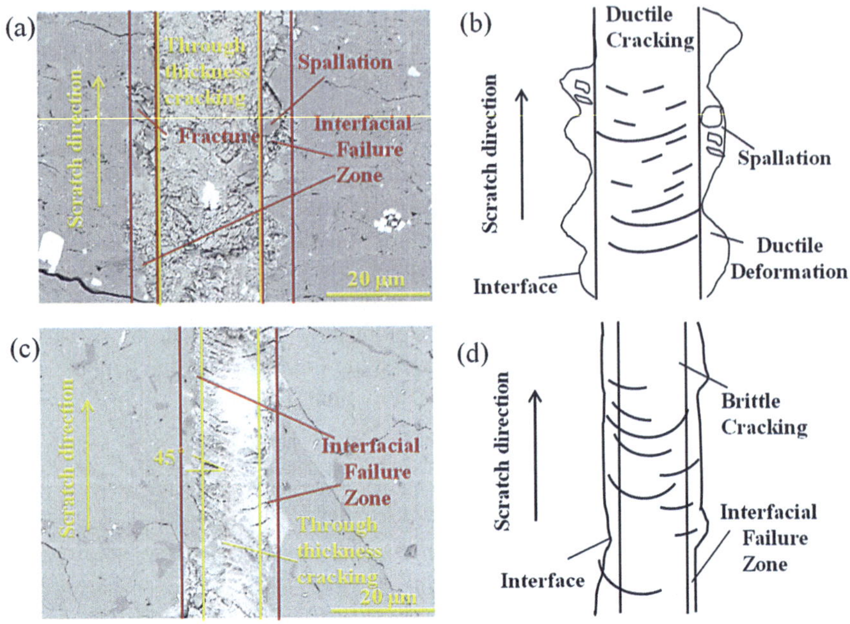

Fig. 6.26 Typical morphological characteristics of scratches: (**a**) Matrix SEM surface, (**b**) matrix damage pattern, (**c**) laminate SEM surface, and (**d**) laminate damage pattern

interface. In addition, the scratch interface is smooth and straight, and there is no mineral stripping.

The scratch depth curve and scratch horizontal force curve of the matrix are shown in Fig. 6.27a, b, respectively, and both curves are sawtooth distribution characteristics. The maximum horizontal force of the scratch is 43.01 mN, the minimum is 31.13 mN, and the average horizontal force is 39.24 mN. The maximum scratch depth is 4759.89 nm, the minimum scratch depth is 3563.04 nm, and the average depth is 4138.32 nm. The horizontal force curve and depth curve of laminar scratch are shown in Fig. 6.27c, d, respectively, and both curves are sawtooth distribution characteristics. The maximum horizontal force of scratches is 42.5mN, the minimum horizontal force is 32.8mN, and the average horizontal force is 37.8mN. The average horizontal force of layer 37.8 mN is lower than that of matrix 39.24 mN, indicating that matrix has better shear resistance. The maximum scratch depth is 2501.61 nm, the minimum scratch depth is 1569.46 nm, and the average depth is 1899.57 nm. The average scratch depth at the grain layer is 1899.57 nm less than the average scratch depth of the substrate 3897.97 nm. Smaller scratch depth means that the harder the sample is, the hardness of the grain layer is greater than that of the substrate.

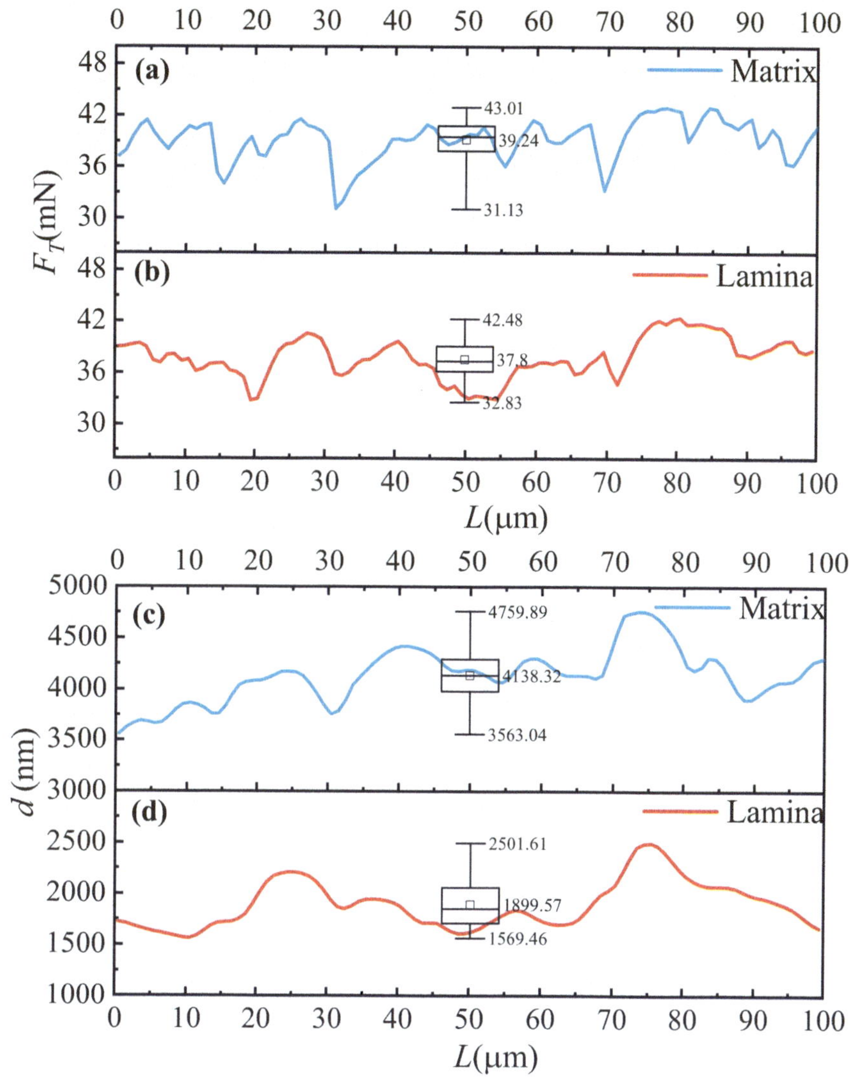

Fig. 6.27 Characteristics of typical matrix scratch curves: (**a**) Horizontal force-displacement curve of matrix scratch, (**b**) horizontal force-displacement curve of laminar scratch, (**c**) matrix depth-displacement curve, and (**d**) matrix scratch horizontal force-displacement curve

6.3.4 Microdestruction Model

After the nanoscratch experiment, the horizontal force-displacement curves of shale before and after soaking are obtained. Brittleness and toughness failure have obvious horizontal force signals, combined with the characteristics of horizontal force-scratch curve at different soaking times; the scratch horizontal force-displacement

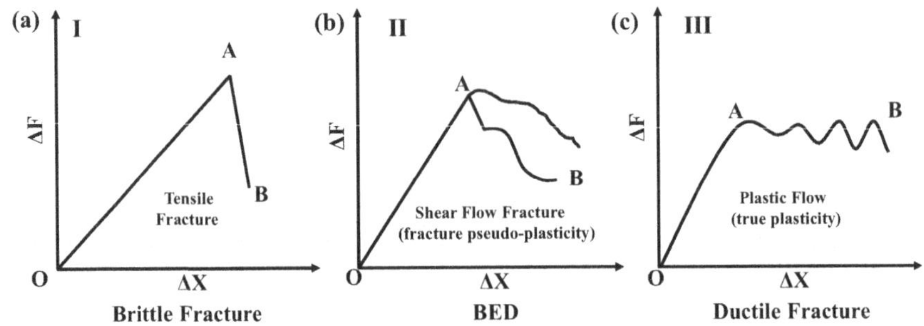

Fig. 6.28 Morphological characteristics of horizontal force-displacement curves of typical scratches in shale: (**a**) Brittle failure, (**b**) BED, and (**c**) plastic failure

curve can be roughly divided into three categories. Class I is shown in Fig. 6.28a. The OA section is basically a straight line, during which the initial microcracks in the rock are closed under pressure, which belongs to the category of elasticity. After reaching the peak point A (section AB), the curve rapidly and linearly declines. Acoustic emission monitoring has shown that stress-induced damage events usually correspond to a sudden drop in horizontal force when a scratch is created in a brittle material. The failure mechanism of this brittle failure is the "tensile" fracture.

As shown in Fig. 6.28c, the OA section is basically straight and belongs to the category of elasticity. After reaching the peak point A (section AB), the curve presents an irregular slope decline. The failure mechanism is shear flow fracture. The deformation of rock mainly shows the coexistence of plastic deformation and elastic deformation, and the stress-strain curves of rock show complex characteristics.

As shown in Fig. 6.28c, the OA section is basically straight and belongs to the category of elasticity. After reaching the peak point A (section AB), the horizontal force does not increase, and the displacement continues to increase. The failure mechanism of this ductile failure is plastic flow fracture. The plastic flow does not show the phenomenon of fast strain energy release, which is caused by the poor cementation of the particles in the rock.

Brittle and ductile failure have obvious transverse force signal. Brittle failure has a pronounced "zigzag" lateral force response with large fluctuations, while ductile failure has a less volatile profile[34]. The horizontal force and displacement results of the substrate under different soaking times are shown in Fig. 6.29a, where the red dashed line frame represents class I, the blue dashed line frame represents class II, and the black dashed line frame represents class III. The damage surface without any fluid treatment is shown in Fig. 6.29b, and the average initial scratch width is 26.83 μm. In the damage mechanism of scratch process, class I accounted for 27.4%, class II accounted for 47.4%, and class III accounted for 25.2%. The scratch depth of SC-CO_2-brine immersion for 2 days is shown in Fig. 6.29c. The average scratch width is 27.76 μm. In the damage mechanism of scratch process, class I accounted for 30.9%, class II accounted for 20%, and class III accounted for 30.9%.

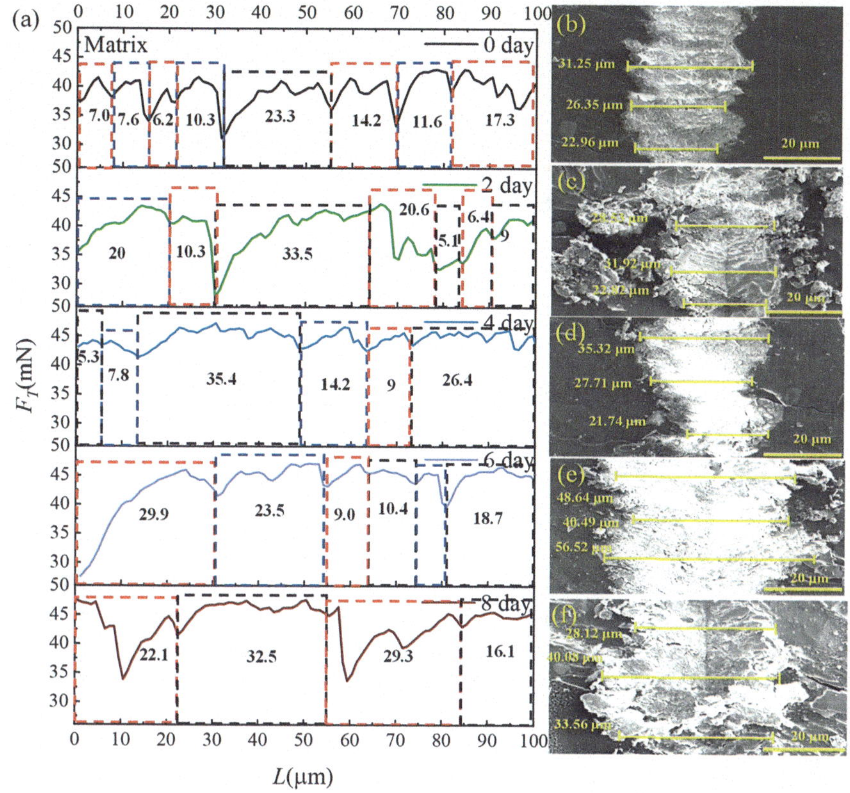

Fig. 6.29 Failure characteristics of matrix under different soaking times: (**a**) Matrix horizontal force-displacement curve, (**b**) matrix 0d, (**c**) matrix 2d, (**d**) matrix 4d, (**e**) matrix 6d, and (**f**) matrix 8d

The scratch width of SC-CO_2-brine after 4 days of immersion is shown in Fig. 6.29d. The average scratch width is 28.26 μm. In the damage mechanism of scratch process, class I accounted for 9%, class II accounted for 23.9%, and class III accounted for 67.1%. The scratch width of SC-CO_2-brine after soaking for 6 days is shown in Fig. 6.29e. The average scratch width is 48.55 μm. In the damage mechanism of scratch process, class I accounted for 38.9%, class II accounted for 29.8%, and class III accounted for 38.9%. The scratch width of SC-CO_2-brine after 8 days of immersion is shown in Fig. 6.29f. The average scratch width is 33.92 μm, which increased by 26.42% compared with that before treatment. In the damage mechanism of scratch process, the proportion of class I accounted for 51.2%, which increased by 86.86% compared with that before treatment. Class II accounted for 11.5%, which was 75.74% lower than before treatment. The proportion of class III was 37.3%, which was 48.01% higher than that before treatment. Statistical results of matrix failure mechanisms under different soaking times are shown in Fig. 6.30a. SC-CO_2-brine immersion improves the brittleness and plasticity of the matrix,

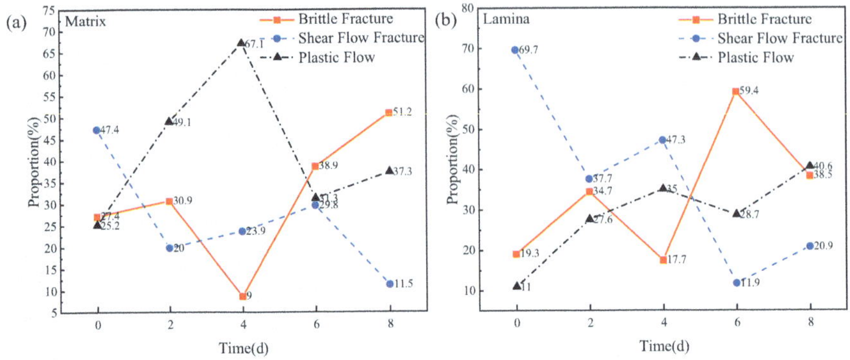

Fig. 6.30 Statistical results of matrix and laminar failure mechanism under different soaking times: (**a**) Matrix and (**b**) lamination

because SC-CO_2-brine immersion significantly weakens the BED zone. The weakened part improves the shale plasticity, and the other part improves the brittleness of the shale (Fig. 6.30a).

The results of horizontal force and displacement of substrate under different soaking times are shown in Fig. 6.31a. The damage surface without any fluid treatment is shown in Fig. 6.31b, with an initial average scratch width of 13.08 μm. The damage mechanism of the scratch process accounted for 19.3% of class I, 68.7% of class II, and 11% of class III. The scratch depth of SC-CO_2-brine immersion for 2 days is shown in Fig. 6.31c. The average scratch width is 11.61 μm. In the damage mechanism of scratch process, class I accounted for 34.7%, class II accounted for 37.7%, and class III accounted for 27.6%. The scratch width of SC-CO_2-brine after 4 days of immersion is shown in Fig. 6.31d. The average scratch width is 11.23 μm. In the damage mechanism of scratch process, class I accounted for 17.7%, class II accounted for 47.3%, and class III accounted for 35%. The scratch width of SC-CO_2-brine after 6 days of immersion is shown in Fig. 6.31e. The average scratch width is 12.54 μm. In the damage mechanism of scratch process, class I accounted for 59.4%, class II accounted for 11.9%, and class III accounted for 28.7%. The scratch width of SC-CO_2-brine after 8 days of immersion is shown in Fig. 6.31f. The average scratch width is 13.62 μm, which increased by 4.12% compared with that before treatment. In the damage mechanism of scratch process, class I accounted for 38.5%, which increased by 2.5 times compared with that before untreated. The proportion of class II was 20.9%, which was 70.01% lower than that before treatment. The proportion of class III was 40%, which increased by 2.69 times compared with that before treatment. The statistical results of lamination failure mechanism under different soaking times are shown in Fig. 6.30b. SC-CO_2-brine immersion significantly improved the brittleness and plasticity of lamination and significantly reduced the proportion of BED zone.

It can be observed that the scratch width of the matrix is larger than the scratch width of the striate. However, under the same exposure time, the surface morphology of the substrate and the scratch width of the scratch are different at each stage. The

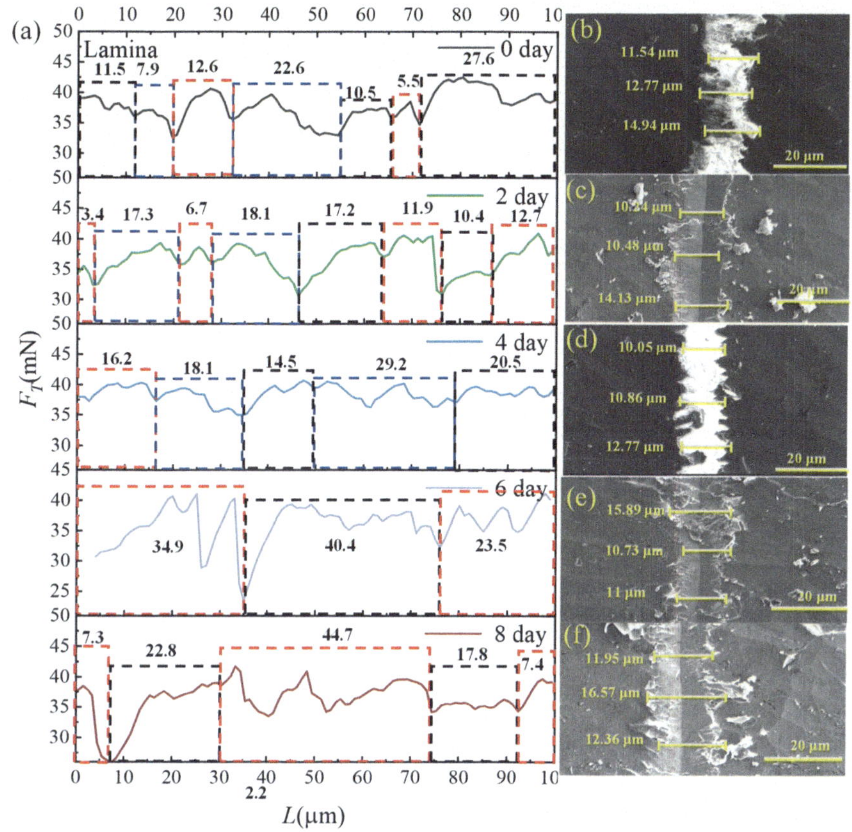

Fig. 6.31 Failure characteristics of laminate under different soaking times: (**a**) Laminate horizontal force-displacement curve, (**b**) laminae 0d, (**c**) laminar 2d, (**d**) lamellar 4d, (**e**) laminae 6d, and (**f**) laminae 8d

initial scratch width of the matrix was 24.7 μm, which was larger than that of the striatum 12 μm. This indicates that the matrix is soft and has high plasticity. The lamination plasticity is low. Substrate scratch width increased by 41.9% and laminates scratch width increased by 2.1% after SC-CO_2-brine immersion, indicating that SC-CO_2-brine improved the plasticity of substrate and laminatus.

6.3.5 Scratch Displacement Changes

Both point plots and box plots are used to visualize changes in parameter values. In a box diagram, the line below the box represents the upper limit, the line above the box represents the lower limit, and the black dot represents the average. The scratch depth of the substrate after immersion in SC-CO_2-brine is shown in Fig. 6.32. In the

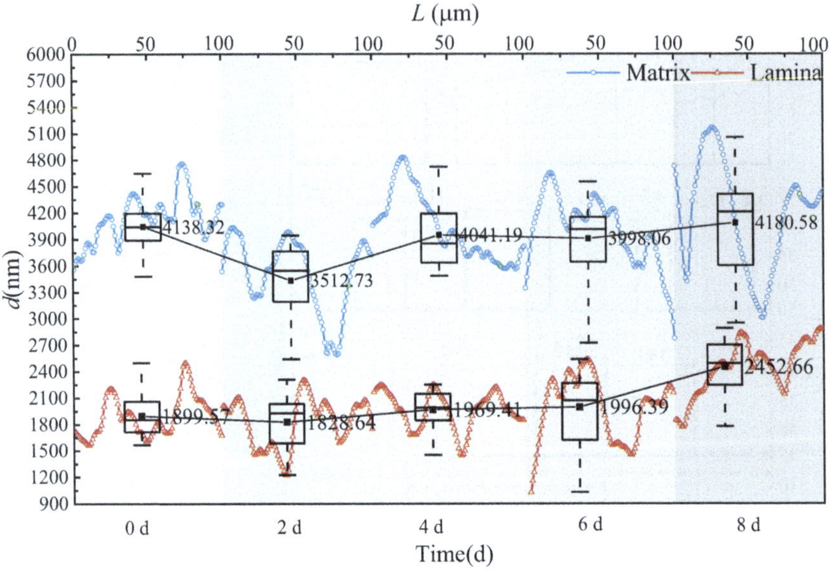

Fig. 6.32 Evolution of striation and substrate scratch depth under different soaking times

absence of any fluid treatment, the initial average scratch depth was 4138.32 nm. The scratch depth of SC-CO_2-brine immersion for 2 days is 3512.73 nm, and the scratch depth of SC-CO_2-brine for 4 days is 4041.19 nm. The scratch depth after 6 days of reaction was 3988.06 nm, and the scratch depth after 8 days of final reaction was 4180.58 nm, which increased the average scratch depth by 1.02%. After SC-CO_2-brine immersion, the scratch depth of the substrate showed a trend of decreasing at first and then increasing.

The scratch depth of the striate layer after immersion in SC-CO_2-brine is shown in Fig. 6.34. In the absence of any fluid treatment, the initial average scratch depth was 1899.57 nm. The scratch depth of SC-CO_2-brine immersion for 2 days is 1828.24 nm, and the scratch depth of SC-CO_2-brine for 4 days is 1969.41 nm. The scratch depth after 6 days of reaction was 1996.39 nm, and the scratch depth after 8 days of final reaction was 2452.66 nm, which increased the average scratch depth by 29.12%. After SC-CO_2-brine immersion, the scratch depth of the striated layer also showed a trend of decreasing first and then increasing.

Further, the surface softening behavior of SC-CO_2-brine immersion was described by height difference (Hd), maximum depth (Md), and peak area (Pa) of the nanoscratch depth-displacement curve. The peak area (Pa) is a scratch displacement-depth curve integral that takes into account the length and depth of the scratch and is used to assess the overall damage of the scratch to the material. The larger the Hd, Md, and Pa, the smaller the surface strength of the sample. In addition, the number of contour peaks (Pc) and full width at half maximum (FWHM) were used to analyze the changes of matrix and laminar heterogeneity and adsorption after

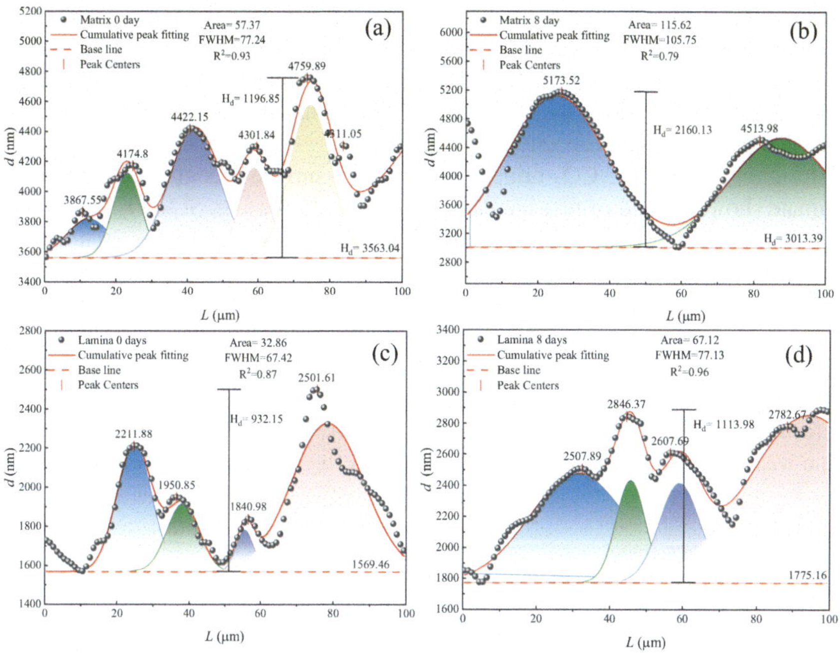

Fig. 6.33 Morphological characteristics of scratch depth-displacement curves before and after laminar and substrate immersion: (**a**) Substrate 0 d, (**b**) matrix 8 d, (**c**) laminae 0 d, and (**d**) laminae 8 d

SC-CO_2-brine effect. The greater the number of peaks, the stronger the heterogeneity. The larger the FWHM, the stronger the adsorption on the shale surface.

The initial scratch curve of the substrate is shown in Fig. 6.33a. Hd, Md, and Pa are 1196.85 nm, 4759.89 nm, and 57.37 nm^2, respectively; Pc is 5 and FWHM is 77.24 μm. The scratch depth-displacement curve of the substrate after soaking for 8 days is shown in Fig. 6.33b, and Hd, Md, and Pa are 2160.13 nm, 5173.52 nm, and 115.62 nm^2, respectively. Hd increased by 80.48%, Md increased by 24.37%, and Pa increased by 101.53%. This indicates that SC-CO_2-brine has a weakening effect on the surface strength of the substrate. In addition, the number of PCS has been reduced from 5 to 2. This suggests that SC-CO_2-brine weakens the heterogeneity of the substrate. FWHM increased to 105.75 μm, an increase of 36.89%. This indicates that SC-CO_2-brine enhances the adsorption capacity of the substrate.

The initial scratch curve of the grain layer is shown in Fig. 6.33c, and the number of PCS remains variable. This indicates that SC-CO_2-brine has little effect on lamellar heterogeneity. Hd, Md, and Pa are 932.15 nm, 2501.61 nm, and 32.86 nm^2, respectively; Pc is 4 and FWHM is 67.42 μm. The scratch depth-displacement curve of the substrate after soaking for 8 days is shown in Fig. 6.33d, and Hd, Md, and Pa are 1113.98 nm, 2846.87 nm, and 67.12 nm^2, respectively. Hd increased to 19.5 percent, Md increased to 13.8 percent, and Pa increased to 104.26 percent. This indicates that SC-CO_2-brine has a weakening effect on the laminate surface strength. From the fact that the increase of Pa in lamellar (104.26%) is greater than that in matrix (101.53%), it can be found that the overall damage degree of SC-CO_2-brine lamellar is greater than that in matrix. In addition, FWHM increased by 14.4% to 77.13 μm. This indicates that SC-CO_2-brine enhances the adsorption capacity of laminatus.

In summary, after SC-CO_2-brine immersion, the surface strength of substrate and laminate shows a trend of first increasing and then decreasing. SC-CO_2-brine has a weakening effect on the laminae and substrate surface strength, resulting in increased surface damage after scratching. The different weakening effects of SC-CO_2-brine on matrix and laminate surface are related to mineral composition and mineral structure arrangement. Because the clay matrix has high adsorption capacity, weak structure, and low density, the height difference and maximum depth of the clay matrix are relatively large. The calcite lamellar structure is relatively more uniform and stable, so the height difference is relatively small. In addition, SC-CO_2-brine weakens matrix heterogeneity. It has little effect on laminar heterogeneity.

6.3.6 Fracture Toughness Changes

During the micro-scratching process of the shale surface, horizontal forces result in surface fractures. The alteration in fracture toughness of the matrix after SC-CO_2-brine saturation is denoted by the blue curve in Fig. 6.34. The general trend of the scratch fracture toughness of the matrix after SC-CO_2-brine saturation indicates a slight increase. After 0 days of SC-CO_2-brine saturation, the initial average fracture toughness measured 0.61 MPa/m$^{1/2}$. Subsequently, the fracture toughness after 2,4, 6, and 8 days of saturation are 0.8, 0.73, 0.73, and 0.69 MPa/m$^{1/2}$. This final increase in fracture toughness amounted to 13.11%. The weak increase in the fracture toughness of the matrix after saturation with SC-CO_2-brine may, on the one hand, be caused by the change in the location of scratches on the shale surface. On the other hand, it is possible that the SC-CO_2-brine solution may have facilitated the penetration and filling of the internal pores, leading to an increase in internal densification and thus a slight increase in matrix fracture toughness.

The red curve in Fig. 6.34 shows the fracture toughness of the laminae after SC-CO_2-brine saturation. The scratch fracture toughness of the laminae exhibits a pattern of initial increase followed by a subsequent decrease. Laminae in the absence of SC-CO_2-brine immersion, the initial average fracture toughness measures 1.92 MPa/m$^{1/2}$. After 2 days of saturation, the fracture toughness increases to 2.03 MPa/m$^{1/2}$, but then decreases to 1.86 MPa/m$^{1/2}$ after 4 days and further to 1.79 MPa/m$^{1/2}$ after 6 days. The fracture toughness decreases to 1.26 MPa/m$^{1/2}$ after 8 days, resulting in a 34.38% reduction in the average fracture toughness. This reduction in fracture toughness of the laminae due to SC-CO_2-brine saturation is attributed to the high permeability and solubility of SC-CO_2-brine. The strong penetration of SC-CO_2-brine enlarges the microfractures in laminae, leading to the reduction in fracture toughness. Additionally, the chemical reactions between

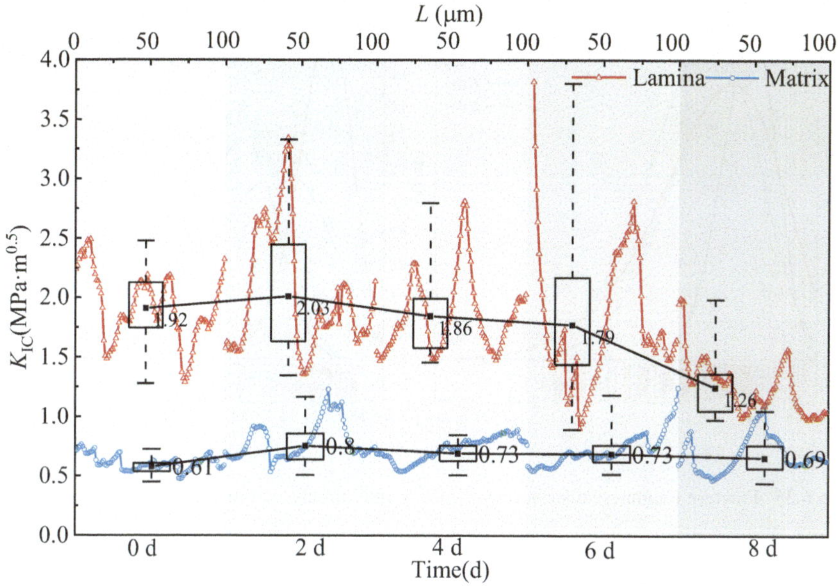

Fig. 6.34 Evolution of laminae and matrix fracture toughness at different immersion times

SC-CO_2-brine and carbonate minerals contribute to the decrease in fracture toughness. Furthermore, the initial fracture toughness of the laminae, measuring 1.92 MPa/m$^{1/2}$, exceeds that of the matrix, which is 0.61 MPa/m$^{1/2}$, indicating a higher fracture toughness of the laminae compared to the matrix.

The variation in shale fracture toughness is associated with the presence of different minerals and the extent of SC-CO_2-brine reaction. As indicated by the mineral analyses, illite constitutes the highest percentage in the matrix, while calcite is predominant in the laminae. The distribution of fracture toughness in the matrix before and after SC-CO_2-brine saturation is showed in Fig. 6.32 and Fig. 6.34, revealing a normal distribution of fracture toughness, with the highest density value in the curve corresponding to the illite mineral in the matrix. The maximum density value of the matrix after 0 days of SC-CO_2-brine saturation is 6.45. The fracture toughness at this point is 0.60 MPa/m$^{1/2}$. The difference in fracture toughness at the maximum density value of 0.60 MPa/m$^{1/2}$ and the average fracture toughness value of 0.61 MPa/m$^{1/2}$ before saturation are only 1.67%, indicating that the illite mineral significantly influences the fracture toughness of the matrix. The maximum density value of the matrix after 8 days of SC-CO_2-brine saturation is 5.25. The fracture toughness at this point is 0.64 MPa/m$^{1/2}$. This represents a 6.67% increase in fracture toughness at the maximum density value after saturation, which is consistent with the 13.11% increase in the average fracture toughness value after saturation. These findings suggest that the interaction of SC-CO_2-brine with illite in the matrix leads to the changes in the overall fracture toughness of the matrix.

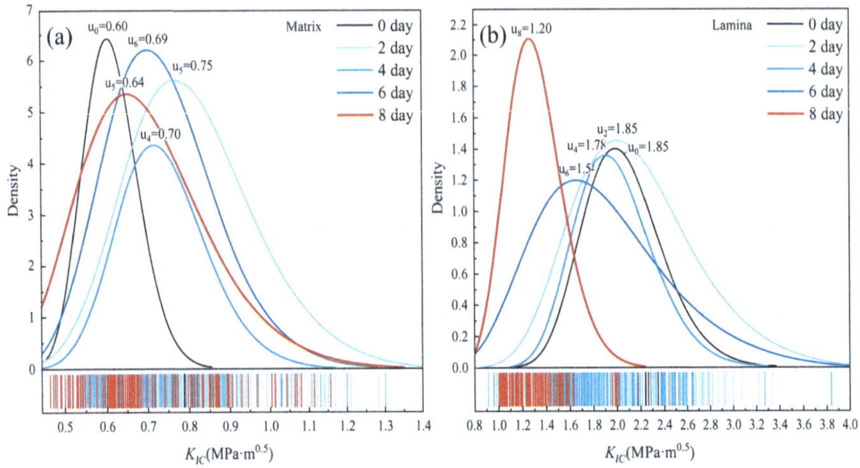

Fig. 6.35 Fracture toughness distribution of matrix and laminae: (**a**) matrix and (**b**) lamina

The fracture toughness distribution of laminae before and after SC-CO_2-brine saturation is showed in Fig. 6.35b. The fracture toughness values of the laminae exhibit a normal distribution, with the highest density value in the distribution curve corresponding to the calcite mineral, which has the largest percentage in the laminae. The maximum density value of the laminae after 0 days of SC-CO_2-brine saturation is 1.40. The fracture toughness at this point is 1.85 MPa/m$^{1/2}$. The difference in fracture toughness at the maximum density value of 1.85 MPa/m$^{1/2}$ and the average fracture toughness value of 1.92 MPa/m$^{1/2}$ before saturation are only 3.78%, indicating that the calcite mineral significantly influences the fracture toughness of the laminae. The maximum density value of the laminae after 8 days of SC-CO_2-brine saturation is 2.1. The fracture toughness at this point is 1.20 MPa/m$^{1/2}$. This represents a 35.13% decrease from the initial fracture toughness, which is consistent with the average reduction of 34.38% in the fracture toughness of the laminae after saturation. This suggests that the interaction of SC-CO2-brine with calcite in the laminae led to an overall reduction in fracture toughness.

References

1. Akono AT. Energetic size effect law at the microscopic scale: application to progressive-load scratch testing. J Nanomech Micromech. 2016;6(2):04016001.
2. Liu JH, Zeng Q, Xu SL. The state-of-art in characterizing the micro/nano-structure and mechanical properties of cement-based materials via scratch test. Constr Build Mater. 2020;254:119255.
3. Liu A, Liu S, Liu Y, et al. Characterizing mechanical heterogeneity of coal at nano-to-microscale using combined nanoindentation and FESEM-EDS. Int J Coal Geol. 2022;261
4. Hernandez-Uribe LA, Aman M, Espinoza DN. Assessment of mudrock brittleness with micro-scratch testing. Rock Mech Rock Eng. 2017;50:2849–60.

Chapter 7
Scratch Brittleness Index

Set the input vector of the engineering sweet-spot evaluation model to include grain-layer structural characteristics, in addition to elastic parameters and mineral composition. This is necessary because methods based only on elastic parameters and mineral composition cannot capture the influence of grain-layer structure on field engineering operations. Therefore, such methods are not suitable for engineering sweet-spot classification in laminated shale. In this chapter, we quantitatively characterize the striated shale striation development index based on the scratch curve and propose to construct an engineering sweet spot evaluation model based on the two indexes of compressive strength and striation development index. Using the continuous scratch curve to compare and correlate the logging data, the results of core scratching are generalized to the field, the accuracy of the traditional logging curve is improved, and the engineering sweet spot evaluation method applicable to striated shale is established.

7.1 Scratch Characteristics of Full-Well Section

7.1.1 Geologic Background

The Lower Cretaceous Qingshankou Formation in the northern part of the Songliao Basin is the largest lakeshore deposit, and it develops widely distributed organic-rich, dark-colored mud shales with relatively high maturity. The formation of the Qingshankou Formation is the most dominant hydrocarbon source rock in the Songliao Basin, and its stratigraphic thickness ranges from about 260 to 500 m. Controlled by the depositional conditions, three different types of shale oils are developed: interbedded, intercalated, and pure shale types. In addition, the content of brittle minerals is relatively high, the content of clay minerals is between 30 and 60%, and micro- and nanopores and microfractures are highly developed. The Heroic Ridge tectonic zone is situated in the core area of the Chaixi depression in

L. Yang, *Multiscale Scratch Mechanics of Laminated Shale*,
https://doi.org/10.1007/978-981-95-7041-6_7

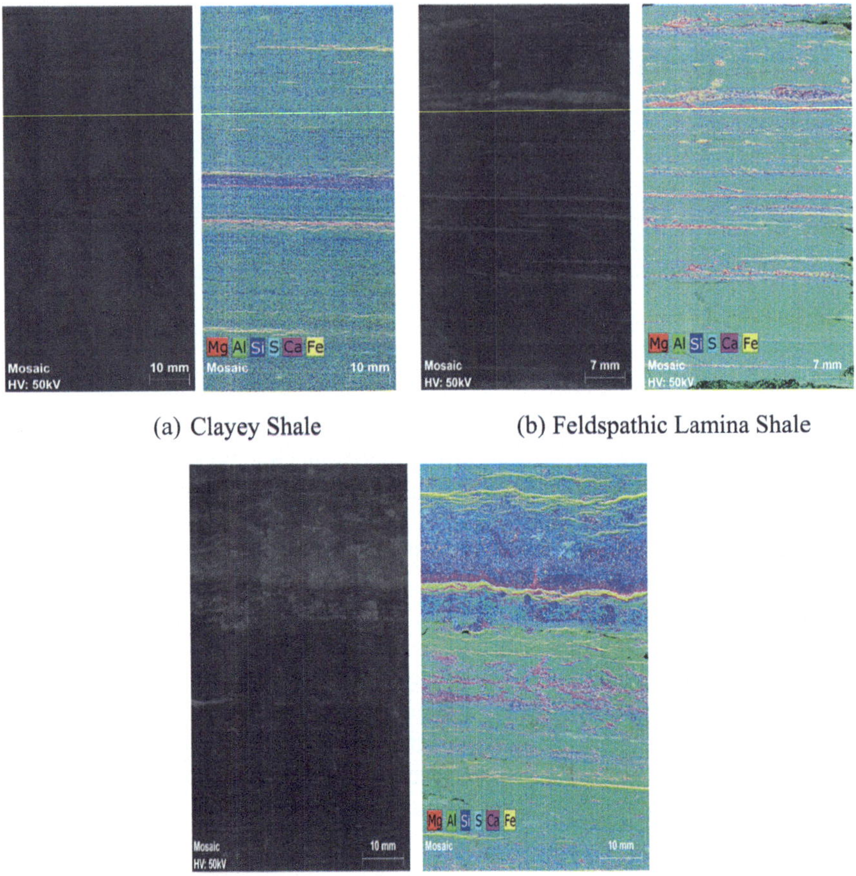

(a) Clayey Shale (b) Feldspathic Lamina Shale

(c) Mixed-lithology Lamina Shal

Fig. 7.1 Results of core observation and XRF analysis of shale samples

the Qaidam Basin and is part of the Paleoproterozoic to Neoproterozoic oil and gas-bearing system in Chaixi. It is estimated that the oil resources in the area are as high as 1.9 billion tons, and the proven oil geological reserves are 500 million tons, showing great potential for resource development. The thickness of the shale reaches 1000–2000 m, forming a large area of mixed-accumulation type lake-phase carbonate rocks and striated mudstone high-frequency stacked assemblage, and the sediments are dominated by fine-grained sedimentary rocks, with a high degree of mixing of the source reserves.

Core samples from wells C41-70 and TY-2 in the Songliao Basin and well C2-4 in the Hero's Ridge area of the Qaidam Basin were selected to carry out experimental tests of centimeter scratches in the whole well section. The results show that the core samples from the Songliao Basin are mainly classified as shale, siltstone, sandstone interbedded, chert, and as containing dolomite nodules. They can be further

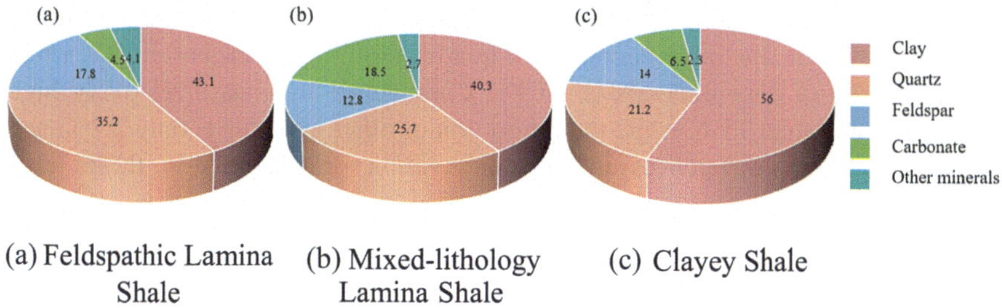

Fig. 7.2 XRD mineral composition analysis of different lithologic shales. (**a**) Feldspathic lamina shale. (**b**) Mixed-lithology lamina shale. (**c**) Clayey shale

classified into two main categories: clayey shales and striated shales (e.g., Fig. 7.1). The number of siltstones, sandstone interlayers, cherts, and dolomitic interlayers is relatively small and the thickness is relatively thin. Clayey shales and striated shales account for more than 95% of the thickness of the Qingshankou Formation stratigraphy, whereas other types of lithology account for less than 5% of the thickness. The lithofacies types of the Lower Dry Chaigou Formation of the Hero's Ridge are most commonly characterized by grain-laminated clayey shales, laminated mudstones, grain-laminated cloudy tuffs, laminated gray dolomites, and sandstones, while the other lithofacies are negligibly less abundant. Further the grained shales can be classified into four main categories: long quartz grained, dolomitic grained, calcite grained, and clay grained.

Based on the description of the cores and the elemental face-scanning technique of XRF, the distribution of elements in the shale was characterized in detail. As shown in the elemental scanning results of the striated shale samples demonstrated in Fig. 7.1, it can be observed that the distribution of elements in the clayey shale is relatively uniform, whereas the striated shale can be further categorized into two types, long quartz striated and mélange striated, based on the compositions of the different elements. Clayey shales are usually black in color, with very well-developed shale texture and relatively uniform distribution of elements, with little variation among the grain layers. Long quartz shale is usually gray-black, the grain layer is mainly composed of silicon and iron elements, the surface has obvious different grain layer interlayer phenomenon, and the grain layer is also developed in the pyrite bands. Most of the mafic shales are grayish-black, and the thickness of the grain layer is usually more than 1 mm; the grain layer is mainly composed of silicon, iron, and calcium, while the high-organic shale layer alternates with the calcium-containing mafic grain layer.

In order to analyze the mineral composition of the main lithological samples in a targeted manner, the shale samples were sampled for mineral composition analysis based on XRF elemental scanning analysis. Figure 7.2 demonstrates the mineral composition of the three different lithological phases, in which the clayey shale contains 56% of clay minerals, while the quartz and feldspar minerals account for 21.2% and 14% of the mineral content. The mineral content of carbonate rocks is only 6.5%. In the long quartz grained shale, the mineral content of quartz and feldspar amounted to 53%, while the percentage of clay minerals was 43.1%. In the mixed quartz shale, the mineral content of quartz and feldspar reaches 38.5%, while the percentage of clay minerals is 40.3%. Carbonatite also has a relatively high content of more than 15%, while the proportion of other minerals is only 2.7%, and there are obvious differences in the composition of minerals in these three lithologies.

7.1.2 *Comparison and Verification*

The strength calculated by traditional uniaxial experiments was used to verify the reasonableness of the scratch curve of the whole well section. Conventional uniaxial experiments were carried out on the striated shale at different depths in well C2-4 in the Hero's Ridge area (Fig. 7.3), and the results of the uniaxial experiments show that the shale presents obvious brittle cleavage damage characteristics, with laminar shear misalignment of the striated layer in the loading process. The overall uniaxial compressive strength of the shale of the Dry ChaiGou Formation in the Qaidam Basin is large, with most of the uniaxial compressive strengths around 160 MPa; additionally, the maximum value of the compressive strength is 174.15 MPa, and the modulus of elasticity is relatively large, with most of them around 28 GPa, and the maximum value of the modulus of elasticity is 34.61 GPa.

The results obtained from the uniaxial tests were compared with those from the scratch tests for further analysis. Figure 7.4 shows that the results of scratch compressive strength and traditional uniaxial compressive strength experiments are close to each other with smaller errors, and the results of scratch experiments are reliable and continuously varying relative to traditional uniaxial experiments, which are better applicable to grain shale. The scratch test can reveal the non-homogeneity of the mechanical properties of shale samples. In addition, the scratch test can not only quantitatively characterize the strong non-homogeneity of shale but also make up for the shortcomings of the traditional experimental methods, which are unable to obtain the rock strength profile and difficult to universally test the strength of shale and other complex formations.

The scratch experiment was carried out in well C41-70, well TY2, and well QY1 in Songliao Basin, in which wells TY2 and QY1 were selected as typical shale samples for the scratch test, and well C41-70 was a full-well section for the scratch test and was divided into four horizons, namely, Q1 (2398–2410 m), Q2 (2384–2398 m), Q3 (2370–2384 m), Q4 (2370–2384 m), Q4 (2384 m), and Q4 (2370–2384 m). In total, more than 150 shale samples were scratched. Some of the results of scratch experiments in each well are as follows. And the results of scratch experiments of the grain layer shale in well C2-4 in Hero's Ridge area are compared.

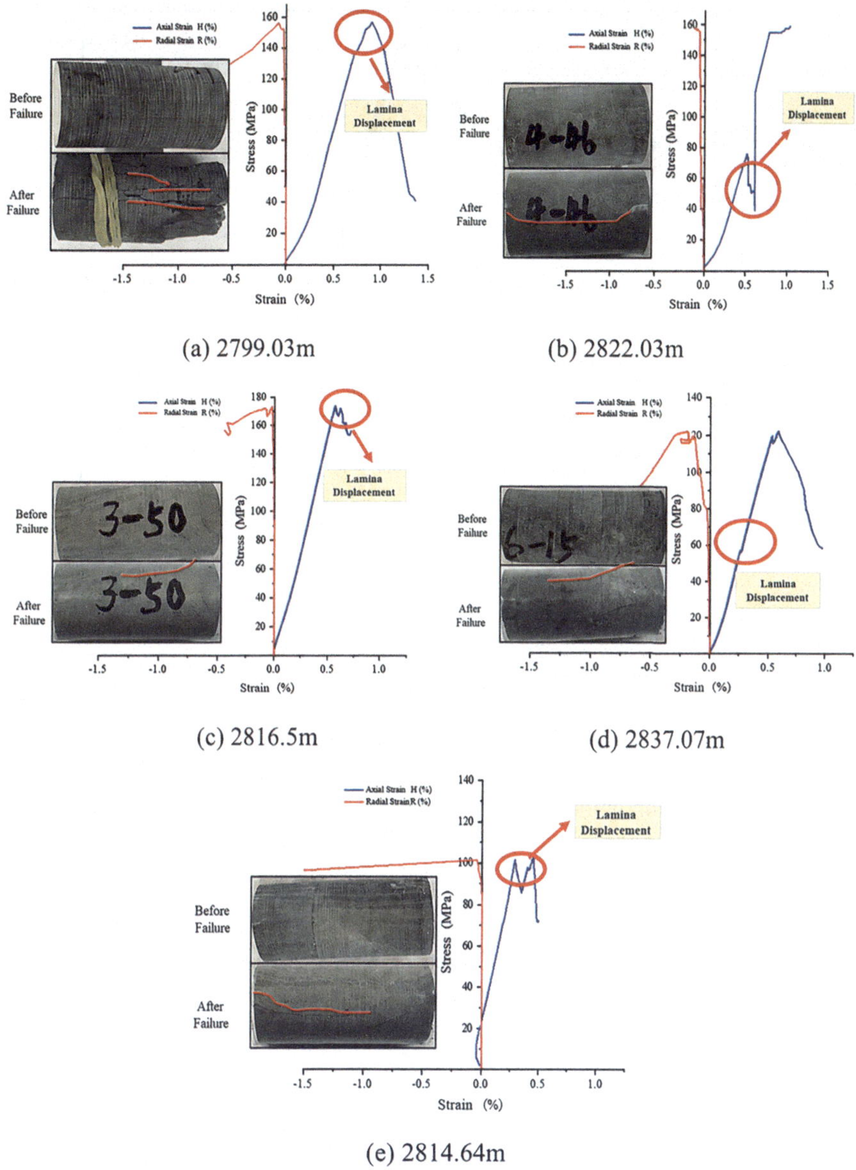

Fig. 7.3 Uniaxial experiment results

Scratch of the Whole Well Section of C41-70 Well in Songliao Basin

Scratch of the Whole Well Section of Layer Q1

Scratch results of some samples from Q1 layer of C41-70 well are shown in Fig. 7.5. Shale sample (a) is a long quartz shale, the overall compressive strength is large, the average value is 72.68 MPa in the 10–35 mm test section of the mechanical

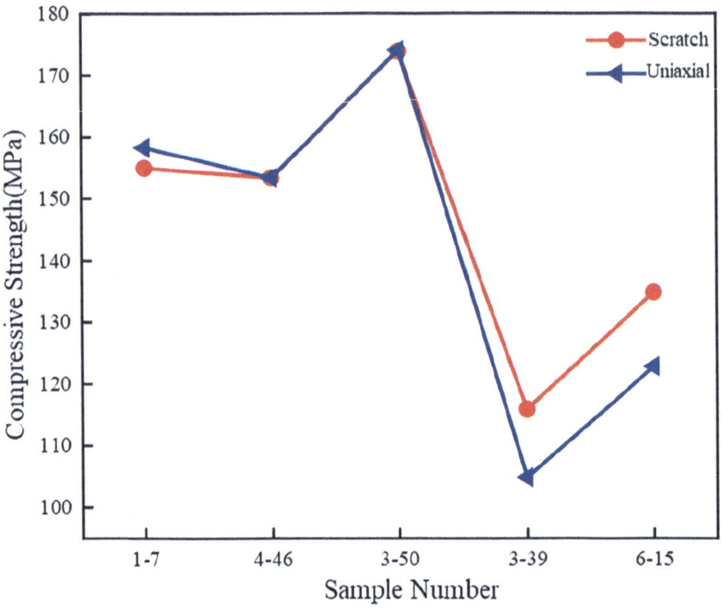

Fig. 7.4 Comparison of scratch results and conventional uniaxial results

properties of high-frequency cyclic fluctuations, the scratch surface has obvious damage, the upper and lower surfaces are smoother, and the curve fluctuation number is significantly reduced. For shale sample (b), the 30–60 mm interval corresponds to amphibole-bearing shale, whereas the 0–30 mm interval corresponds to clay shale. The average compressive strength of the upper interval is clearly higher than that of the lower interval. In the upper interval, the sample surface is relatively smooth, with no obvious scratch-induced damage. The corresponding curve also shows fewer fluctuation cycles. In contrast, the lower interval exhibits evident scratches and damage features on the sample surface. The number of fluctuation cycles in the curve also increases, indicating a more pronounced heterogeneous mechanical response. Shale sample (c) is quartz-rich laminated shale, with an average compressive strength of 104.41 MPa. In the 20–40 mm test interval, the interaction between adjacent laminae is evident. A distinct mechanical weak plane develops in this interval. As a result, the mechanical response shows high-frequency cyclic fluctuations, and obvious scratch-induced damage appears on the sample surface. In contrast, the upper and lower intervals are relatively smooth. The curve fluctuations are also significantly reduced in these intervals. The shale sample (d) is a muddy mesomorphic tuff, the overall compressive strength is larger, the average value is 72.64 MPa, the whole sample surface is relatively smooth, there is no obvious scratch damage, and the curve fluctuation of the number of rotations is obviously reduced. The scratch curve fluctuates within the mean value. Shale sample (e) consists of mixed shale in the 52–90 mm test interval and clay shale in the 0–52 mm test interval. The compressive strength of the upper interval is clearly higher

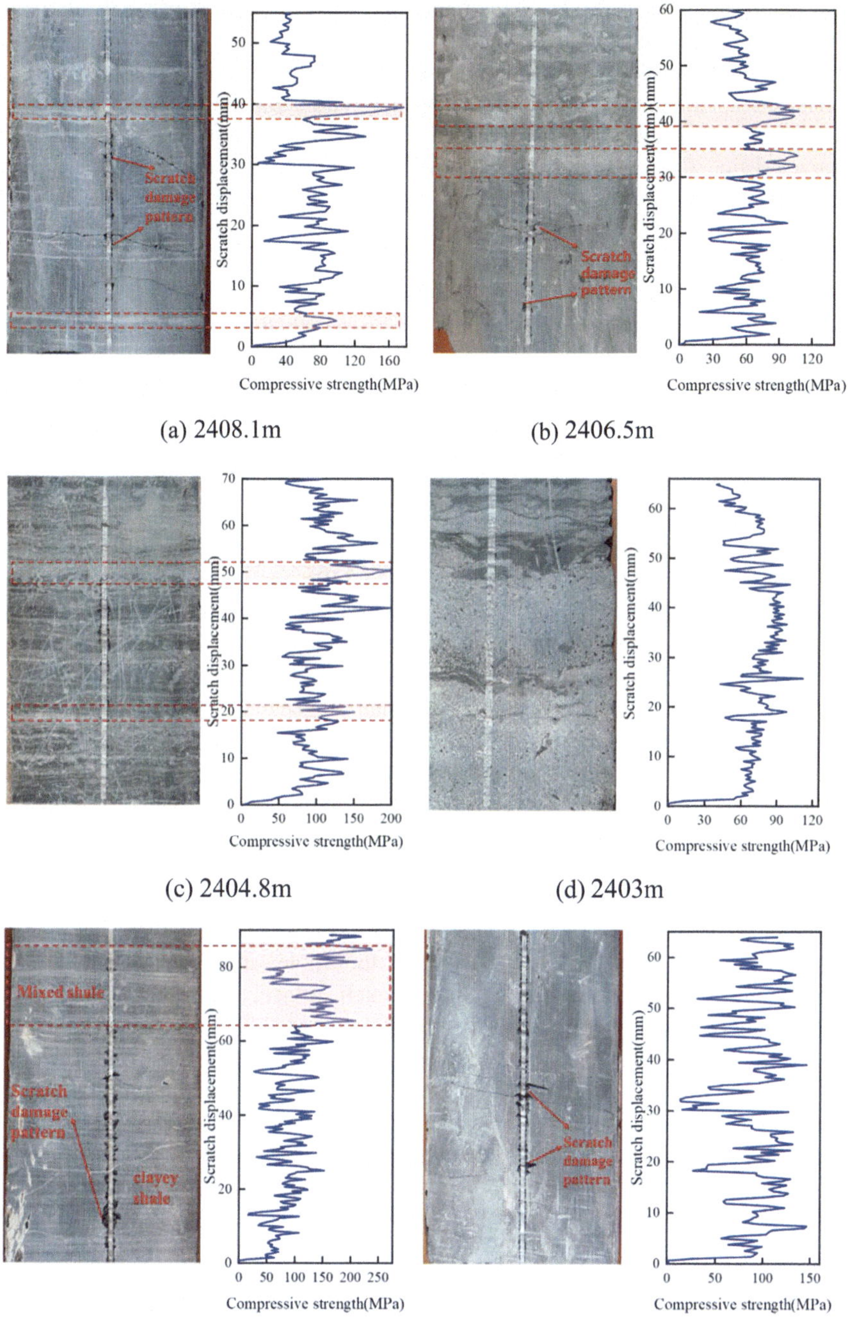

(a) 2408.1m (b) 2406.5m

(c) 2404.8m (d) 2403m

(e) 2401.6m (f) 2399.8m

Fig. 7.5 Results of partial scratches in Q1 formation of well C41-70

than that of the lower interval. In the upper interval, the sample surface is relatively smooth and shows no obvious scratch-induced damage. The corresponding curve also exhibits fewer fluctuation cycles. In contrast, the lower interval shows evident scratch-induced damage on the sample surface. The damaged area develops into a Feng-shaped structure. The lower surface of the sample has obvious scratch damage characteristics, forming a "Feng" structure. Shale sample (f) is quartz-rich laminated shale. The interaction between adjacent laminae is evident, and a distinct mechanical weak plane develops within the sample. The scratch curve shows pronounced fluctuations, indicating high-frequency cyclic variations in mechanical properties. This response reflects strong heterogeneity in the compressive strength of the shale. Along the scratch direction, the compressive strength fluctuates sharply between 0 and 150 MPa.

Scratch of the Whole Well Section of the Q2 Layer

Scratch results of some samples from Q2 layer of C41-70 well are shown in Fig. 7.6 The shale sample (a) is a clayey shale with a long quartz grain layer in the middle, the average value of compressive strength is 72.11 MPa, the mechanical properties of the 10–45 mm test section have high-frequency cyclic fluctuations, and the surface of the scratches has obvious damage. In shale samples (b) for mixed shale, compressive strength is larger, the average value is 90.12 MPa, the sample surface is relatively smooth without obvious scratch damage characteristics, and the curve fluctuations of the number of rotations are small. Shale sample (c) is felsic shale, with a relatively high overall compressive strength and an average value of 101.42 MPa. In the 40–60 mm test interval, a distinct mechanical weak plane is present. The mechanical properties show high-frequency cyclic fluctuations in this interval, and obvious scratch-induced damage appears on the sample surface. In contrast, the upper and lower intervals are relatively smooth, and the curve fluctuations are significantly reduced. The shale sample (d) is a silty sandy interlayer, the average value of compressive strength is 103.45 MPa, the whole sample surface is relatively smooth, there is no obvious damage to the scratch surface, and the curve fluctuation of the number of cyclic fluctuations is obviously reduced. The scratch curve fluctuates within the mean value. The shale sample (e) and shale sample (f) are clayey shale, it can be clearly seen that the mechanical properties of high-frequency cyclotron fluctuations found that the whole section of the scratch surface can be seen on the surface of a clear high-frequency thin interlayer structural features, scratches on both sides of a large number of damage surfaces, the formation of the "Feng" zigzag structure, the overall existence of the obvious mechanics of weak surfaces, the mechanical properties of high-frequency cyclotron fluctuations, the scratch curve fluctuation is obvious, and the scratch curve fluctuates significantly. The scratch curve exhibits pronounced fluctuations. The mechanical properties show high-frequency cyclic variations, indicating strong heterogeneity in the compressive strength of the rock. Along the scratch direction, the compressive strength fluctuates

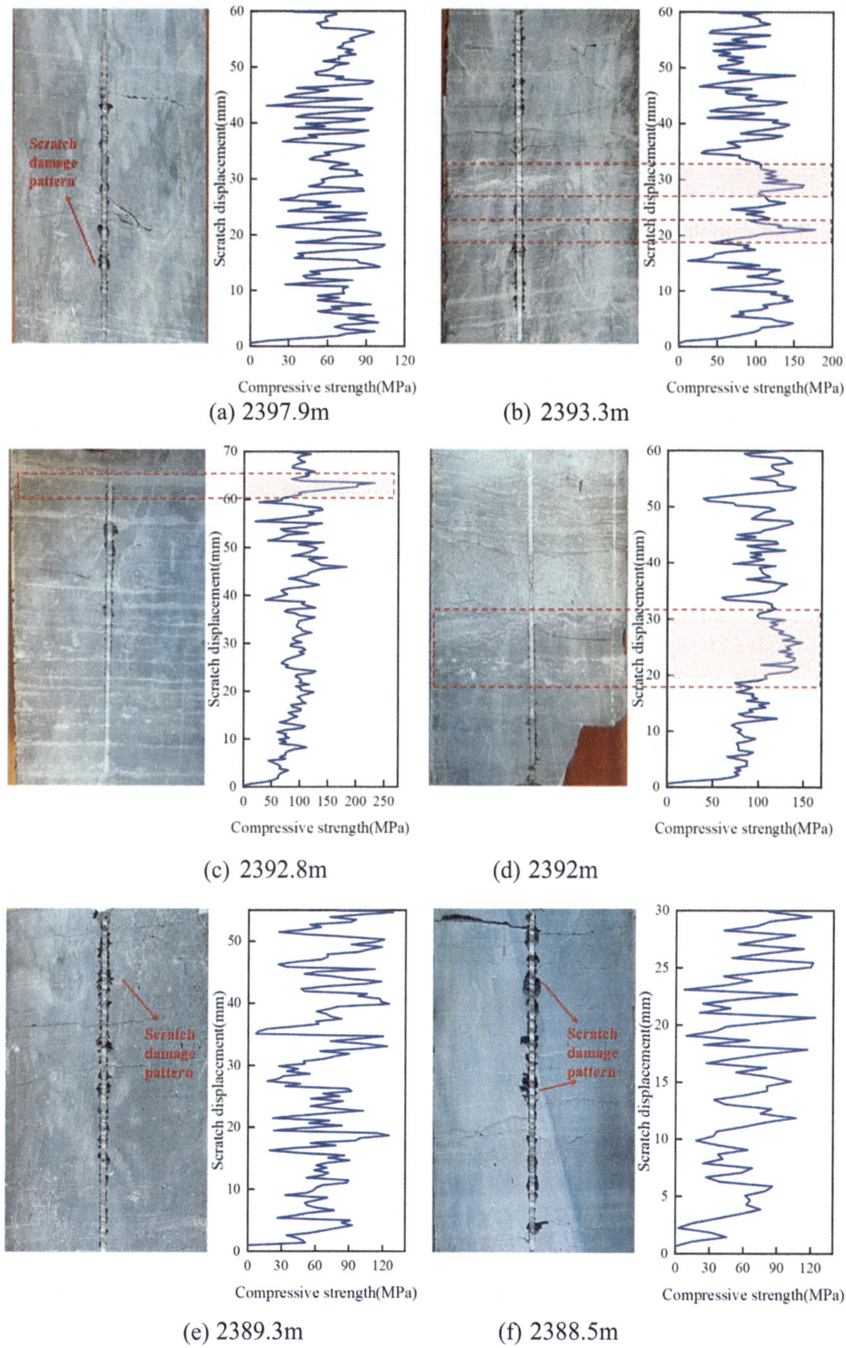

Fig. 7.6 Results of partial scratches in Q2 formation of well C41-70

sharply. The average compressive strength is relatively low, with values of 69.70 and 63.02 MPa, respectively.

Scratches of the Whole Q3 Formation

The scratch results of some samples of Q3 layer in C41-70 well are shown in Fig. 7.7 The shale sample (a) is a clayey shale with small average compressive strength of 45.23 MPa, and a large number of damaged surfaces appear on both sides of the scratches, forming a "Feng" shaped structure. The upper part of shale sample (b) is mafic shale, and the lower part is clayey shale, it can be clearly seen that the compressive strength of the upper part of the sample is greater than that of the lower part, and the surface of the upper part of the sample is smooth, and the surface of the lower part of the sample is characterized by obvious damage of scratches. Shale sample (c) is siltstone, the compressive strength is larger, the average value is 127.43 MPa, the sample surface is smooth and there is no scratch damage on the surface, and the curve fluctuation is less. Shale sample (d) 0–25 mm test section is mafic shale, 25–55 mm test section is clayey shale, the whole sample surface is smooth, no obvious scratch damage, and curve fluctuations significantly reduce the number of rotations. Shale sample (e) is clayey shale, with a relatively low average compressive strength of 54.72 MPa. Numerous damaged surfaces develop on both sides of the scratch, forming a Feng-shaped damage pattern. The scratch curve shows pronounced fluctuations, and the mechanical properties exhibit high-frequency cyclic variations. These features indicate strong heterogeneity in the compressive strength of the shale. Along the scratch direction, the compressive strength fluctuates within the range of 0–90 MPa. The shale sample (f) 0–30 mm test section is a clayey shale, 30–45 mm test section is amphibolite shale, and the whole surface of the sample is relatively smooth, only in the middle part of the obvious scratch damage.

Scratches in the Whole Q4 Layer

The scratches results of some samples of Q4 layer in well C41-70 are shown in Fig. 7.8. The shale samples (a), (b), and (f) are clayey shale, and the average values of compressive strength are small, respectively, 61.21, 60.67, and 47.7 MPa. The fluctuation of the scratch curve is obvious, and the mechanical properties of the rocks fluctuate with high-frequency cyclic fluctuation, and the non-homogeneity of the rock's compressive strength is very intense, and the surface of the whole section of the scratch can be seen with clear and high-frequency thin and interlayer structural features, and a large number of damaged surfaces appear to form a "Feng" shaped structure. It is found that the surface of the whole section is characterized by a clear high-frequency thin interlayer structure, and a large number of damaged surfaces appear on both sides of the scratch, forming a "Feng" structure. Shale sample (c) 0–10 mm test section is mafic shale, 10–40 mm test section is clayey shale, it

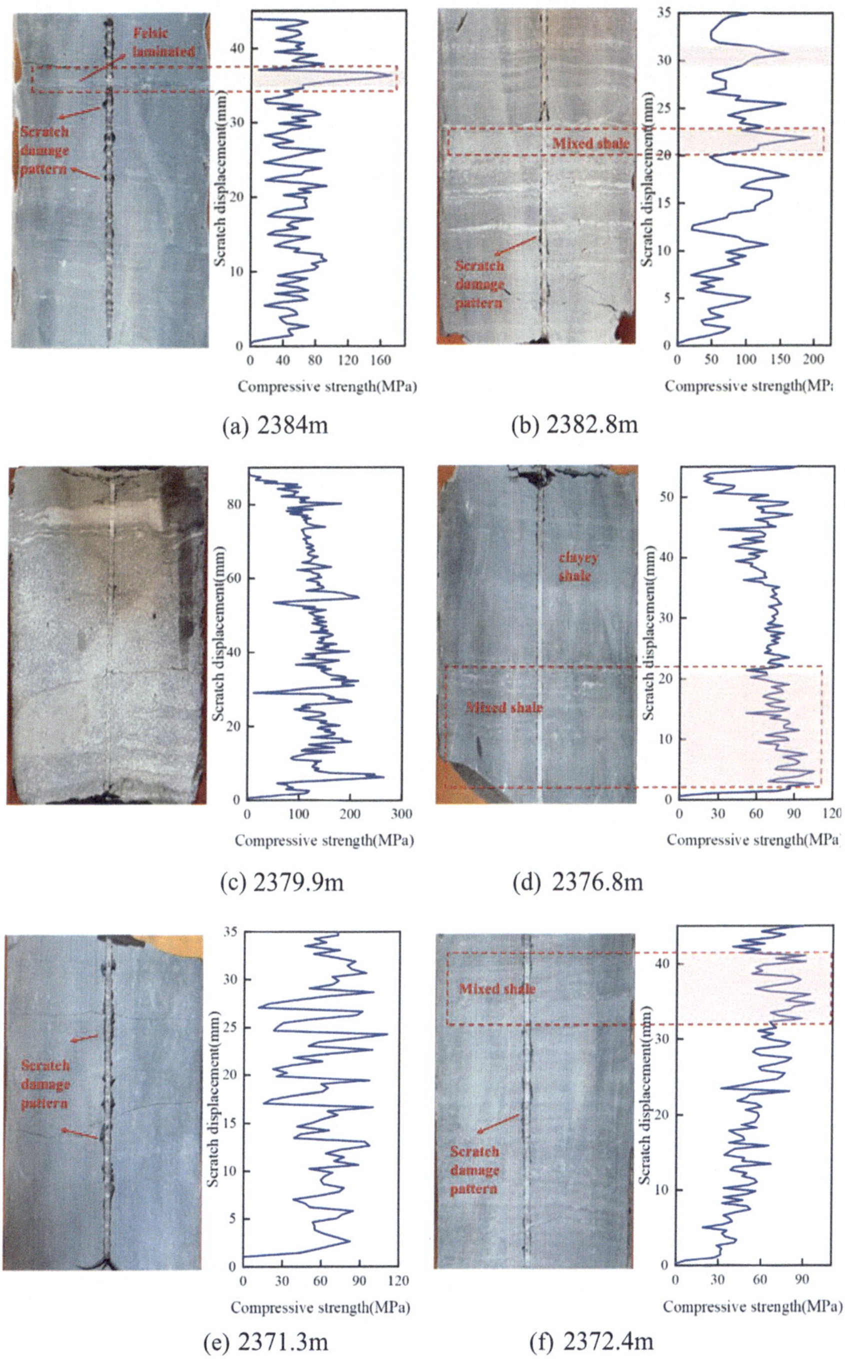

Fig. 7.7 Results of partial scratches in Q3 formation of well C41-70

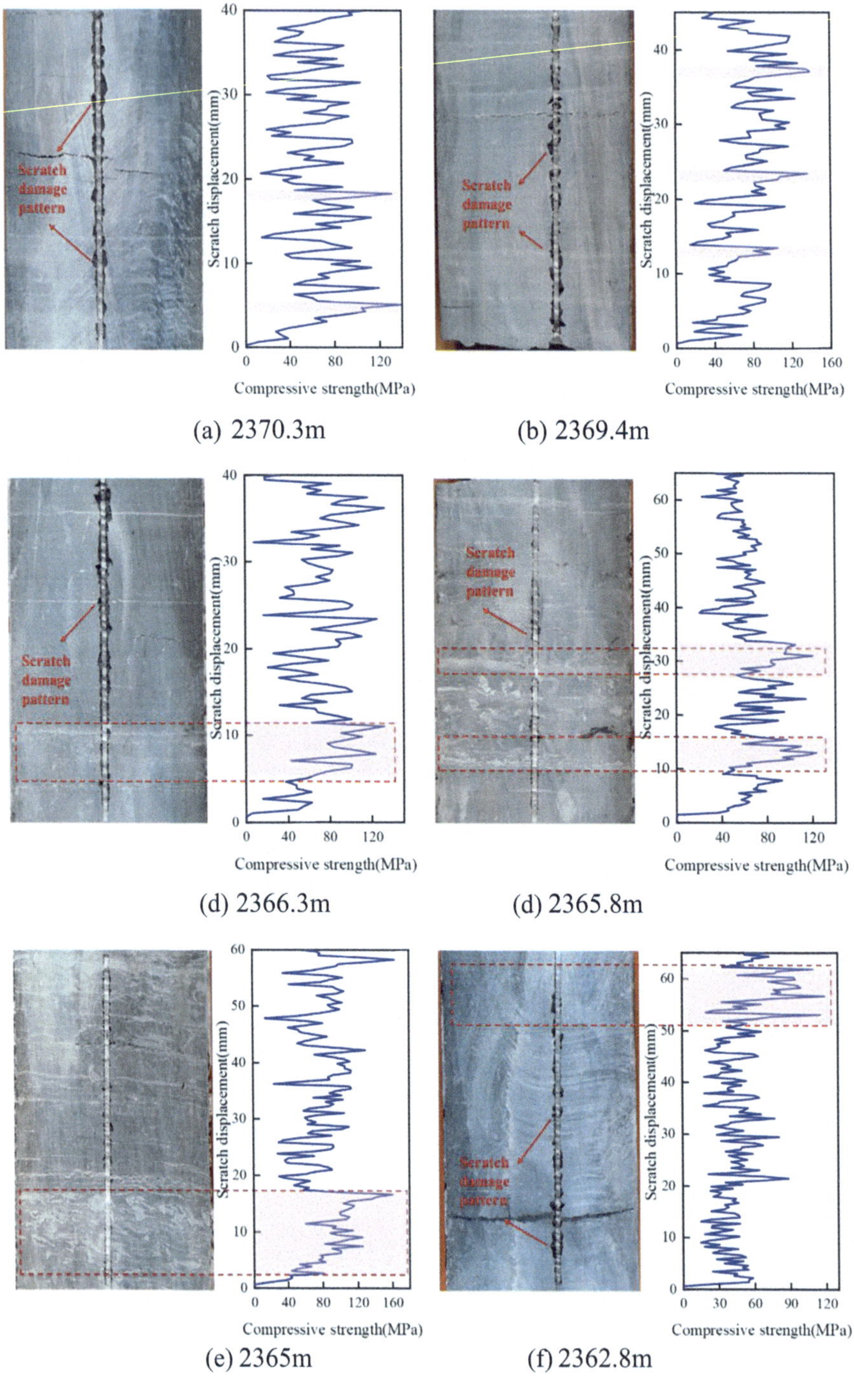

Fig. 7.8 Results of partial scratches in Q4 formation of well C41-70

can be clearly seen that the upper part of the sample compressive strength is greater than the lower part of the sample, and the sample in the 10–40 mm test section curve fluctuations increase the number of rotations. The surface of the sample has obvious scratch damage characteristics. Shale sample (d) is mixed cumulus, the overall compressive strength is larger, the average value is 74.23 MPa, the whole sample surface is relatively smooth, there is no obvious scratch damage, and the curve fluctuation cyclotron number is small. Shale sample (e) 0–20 mm test section is mafic shale, 20–60 mm test section is clay shale, the upper part of the compressive strength is obviously smaller than the lower part, the sample surface is smooth without obvious scratches damage characteristics, and curve fluctuation cyclotron number is small.

Scratches of the Whole Well Section of QY1 Well in Songliao Basin

The scratches results of some samples of QY1 well are shown in Fig. 7.9. Shale sample (a) is a mafic shale, the average value of compressive strength is 120.84 MPa, the whole sample surface is relatively smooth, there is no obvious scratch damage, and the curve fluctuates with a small number of rotations. Shale sample (b) is silty shale, with an average compressive strength of 92.29 MPa. Only weakly developed scratch-induced damage is observed on the laminated sample surface. A small number of horizontal damage surfaces appear on both sides of the scratch. The scratch curve shows limited fluctuations, with relatively few fluctuation cycles. Shale sample (c) is clayey shale, the compressive strength is small, the average value is 61.18 MPa, and it is found that clear high-frequency thin interlayer structural features can be seen on the scratch surface at the top of the sample, and a large number of damage surfaces appear on both sides of the scratch, forming a "Feng" shaped structure. The shale sample (d) is a long quartz shale, with an average compressive strength of 79.93 MPa, mainly developing long quartz silt grain layer, with obvious layering features on the scratch surface and a few horizontal damage surfaces on both sides of the scratch.

Scratches of the Whole Well Section of TY2 Well in Songliao Basin

The scratch results of some samples from the TY2 well are shown in Fig. 7.10. The shale sample (a) is mainly long quartz shale, mainly developing long quartz silt grain layer, the scratch surface has obvious stratification characteristics, a small number of horizontal damage surfaces on both sides of the scratch, and the average value of compressive strength is larger, which is 73.92 MPa. The top of the shale sample (b) is mainly clay shale, the scratch surface has obvious stratification characteristics, and the scratch surface has a large number of horizontal damage surfaces on both sides of the scratch, which forms a higher density "Feng" zigzag. The 0–25 mm test section is mixed shale, mainly developing mesomorphic grain layer, the scratch stratification is not obvious, the average compressive strength of 25–70 mm test

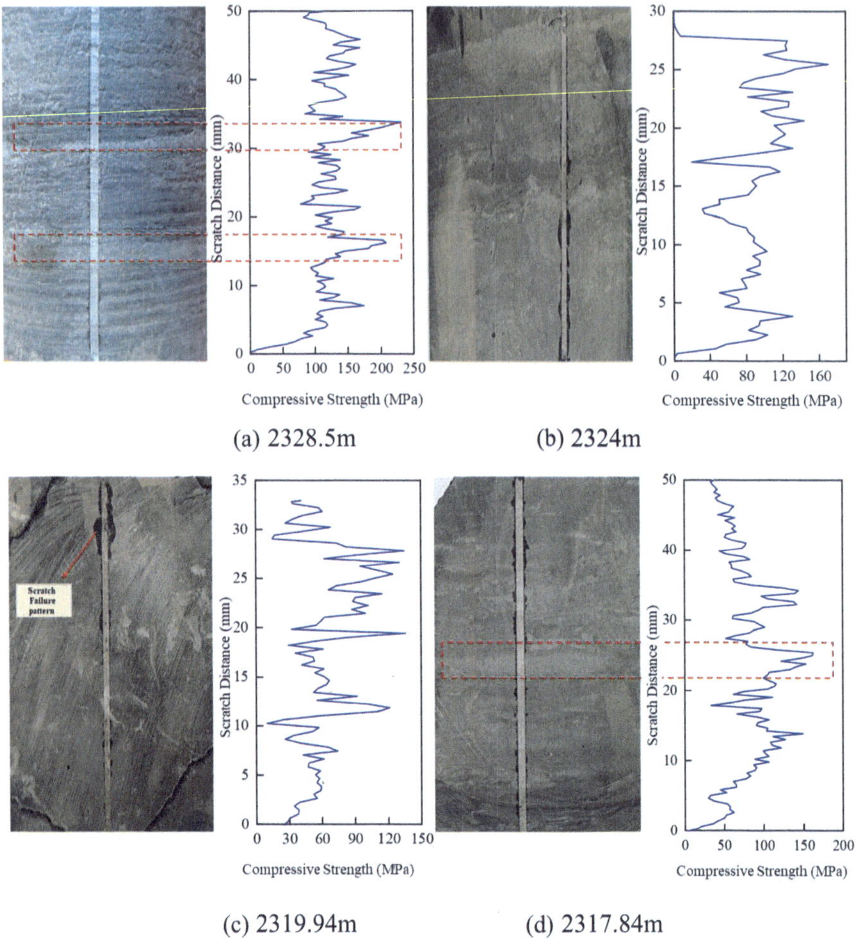

Fig. 7.9 QY1 well partial scratch results

section is 63.58 MPa, and the average compressive strength of 0–25 mm test section is 71.63 MPa. The lower part of the sample has smooth surface, no obvious scratch damage, and fewer fluctuating rotations in curve. The number of fluctuating rotations is small. Shale sample (c) is clayey shale, with a relatively low average compressive strength of 61.28 MPa. High-frequency thin interlayer structures can be clearly observed across the scratched section. Numerous damage surfaces develop on both sides of the scratch, forming a Feng-shaped damage pattern. Shale sample (d) clayey shale is dominated by clay mineral layer scratches on both sides of a large number of horizontal damage surfaces, forming a high level of damage, and the average compressive strength of the sample (d) clay shale is dominated by clay mineral layer scratches. A large number of horizontal damage surfaces develop on both sides of the scratch, forming a high-density Feng-shaped damage pattern. High-frequency

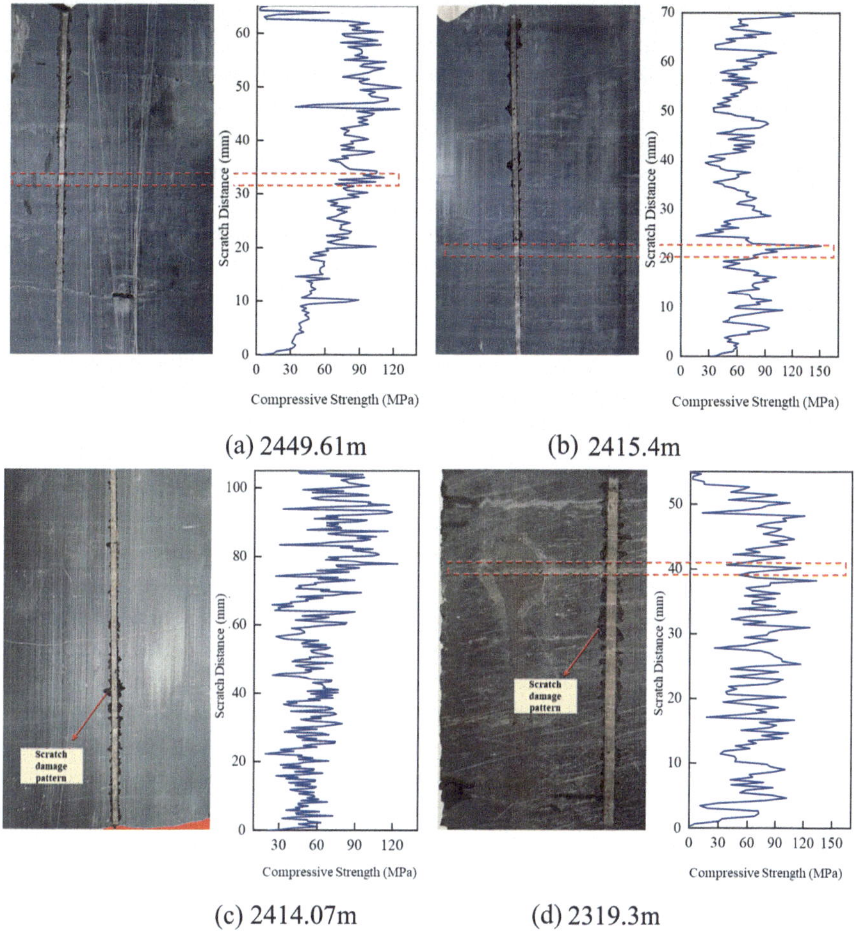

Fig. 7.10 TY2 well partial scratch results

thin interlayer structures can be clearly observed across the scratched section. The middle interval is interspersed with elongated quartz-rich laminae, which appear as white stripes. The scratch-induced lamination features are therefore distinct. The interaction between adjacent laminae is evident, and a distinct mechanical weak plane develops within the sample. The scratch curve shows pronounced fluctuations, indicating high-frequency cyclic variations in mechanical properties. This response reflects strong heterogeneity in the compressive strength of the rock.

Scratches in the Whole Well Section of the C2-4 Well at Hero Ling

In addition, scratch tests were conducted on shale cores from well C2-4 in the Hero's Ridge area of the Qaidam Basin, with rock samples originating from well depths of

2798–2851 m, and the total length of the scratches amounting to 15.26 m, totaling more than 160 samples. The petrographic types include laminated cloudy tuff, laminated gray dolomite, laminated clayey shale, and sandy conglomerate. The mechanical properties of different typical lithologies are analyzed here, and some of the samples cover a combination of 2–3 lithologies, which can be more intuitively recognized as the mechanical differences between different lithologies. The volatility of the scratch curve can reflect the development degree of grain layer, and the steep change of the curve can reflect the change of lithological interface.

As shown in Fig. 7.11, shale (a) is a sandy conglomerate, the overall compressive strength is large, the average value is 144.63 MPa, the compressive strength fluctuates within the average value, the sample surface is smooth, and there is no obvious scratch damage characteristics. In shale sample (b), the upper and lower parts of the core consist of relatively homogeneous gray laminated mudstone. The compressive strength of the gray grain-laminated clayey shale is higher than that of the gray laminated mudstone. Layering features on the scratched surface are mainly observed in the middle interval. The damage surfaces on both sides of the scratch are also concentrated in this interval. The crack width in the middle layer is smaller than that on both sides. The laminated gray dolomite shows a relatively high overall compressive strength, with an average value of 180 MPa. However, its fluctuation range is the smallest. In shale sample (c), the upper and lower parts are composed of grain-laminated clayey shale, whereas the middle interval consists of laminated gray dolomite. Clear laminar features are observed on the scratched surfaces of the upper and lower intervals. A large number of horizontal damage surfaces develop on both sides of the scratch, forming a low density Feng-shaped damage pattern. The compressive strength of the grain-laminated clayey shale is significantly lower than that of the laminated gray dolomite. After scratching, the clayey shale shows more obvious scratch-induced damage than the laminated gray dolomite. After the clayey shale is scratched, the debris accumulates, the damage surface of the scratch is obvious, and the width of the cracks is larger, while the surface of the laminated gray dolomite has almost no damage surface, and the width of the cracks is smaller, and the curve fluctuation of the laminated gray dolomite is smaller than that of the textured laminated clayey shale. The top of shale (d) is a layer of gray-black laminated clayey shale, in which the gray-white laminated layer is millimeters in size, relatively flat and straight, generally continuous in the transverse direction, locally intermittent and lenticular, and the thickness of the gray-black clayey laminated layer is larger, and the contact interface between the laminated clayey shale and the lower gray laminated mudstone is relatively flat and straight, and a relatively straight fracture is developed along the contact interface. The lower gray bedded mudstone is generally more homogeneous, with millimeters of grayish-black clay content visible in the internal grain layer, and the grain layer is distributed in a lens-like manner, with a small overall proportion. A horizontal crack that extends along the level and a high-angle crack that extends vertically along the level are developed in the interior; the horizontal crack is unfilled and the high-angle crack is filled with white gypsum. The grainy clayey shale is stronger than the gray laminated mudstone, and cracks may be present at the junction of the gray laminated mudstone and

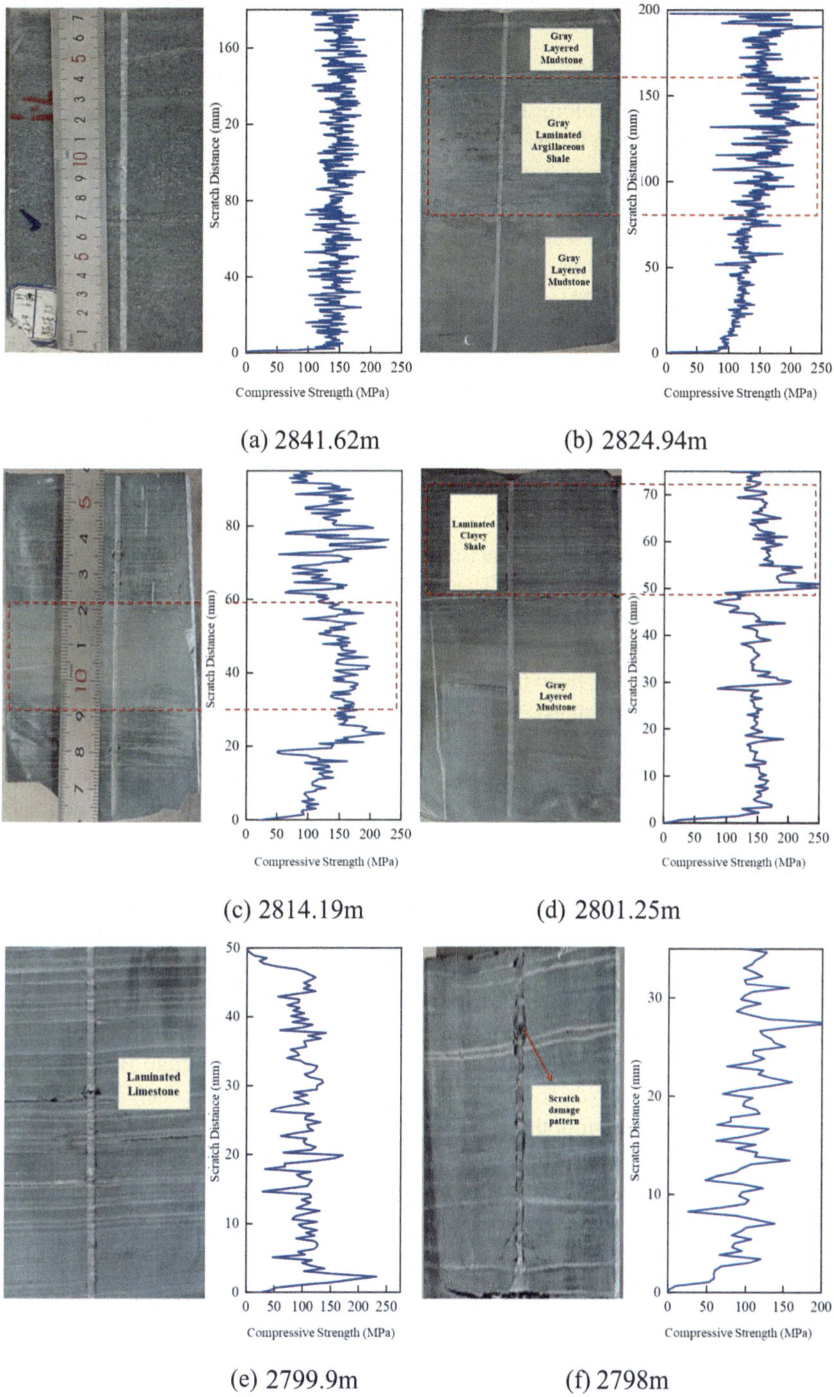

Fig. 7.11 Results of partial scratches in well C2-4

the grainy clayey shale, resulting in a curvilinear bulge. Shale (e) is a grain-stratified cloudy tuff, in which the grayish-black portion is mainly clayey grain, and the off-white, flat, fine-grained grain is carbonate grain, with horizontal cracks developed in the core, mainly along the interface between the two types of grain. Shale (f) clayey shale with grain layer, the surface has obvious clear laminar characteristics after scratching, and a large number of horizontal damage surfaces are formed on both sides of the scratch, the scratch debris is obviously piled up, and the crack extension forms a "Feng"-shaped extension, with a large width of crack, but the overall compressive strength of the grain-laminated clayey shale is relatively small, with an average of 91.83 MPa. The average compressive strength is 91.83 MPa.

Based on the above scratch experiments, the strength profile of well C41-70 in Songliao Basin and the mechanical parameters of some samples from wells QY1 and TY2, as well as the strength profile of well section C2-4 in Hero's Ridge area were constructed to statistically change the laws of change of the mechanical properties in different basins, well sections, and reservoirs. As shown in Fig. 7.12, the Q1 section of well C41-70 in the Qingshankou Formation has the largest mean compressive strength of 88.08 MPa, and the Q3 section has the smallest mean compressive strength and mean hardness and fracture toughness of 68.96 MPa and 138.69 MPa and 1.71 MPa $m^{1/2}$, respectively. The hardness and fracture toughness of the Q2 section have the largest mean hardness and fracture toughness of 157.78 MPa and 2.05 MPa $m^{1/2}$.

As shown in Fig. 7.13, from the results of different regions, the average values of compressive strength, hardness, and fracture toughness of well C2-4 in the Heroes' Ridge region are the largest, followed by well QY1 in the Songliao Basin, and the smallest by well TY2. In terms of the number of fluctuation cycles of the scratch curve and the damage characteristics of the sample surface scratch, the Q4 section of the C41-71 well in Songliao Basin has the highest number of fluctuation cycles and the most obvious damage characteristics of the sample surface scratch, while the Q1 section has the lowest number of fluctuation cycles and the least obvious damage characteristics of the sample surface scratch, and the damage characteristics and the number of curve fluctuations of the sample surface scratch and curve fluctuations of the two sections of the wells of TY2 and C41-70 do not have much difference, and the QY1 well in Songliao Basin has the highest average compressive strength, hardness, and fracture toughness. The QY1 well section in Songliao Basin and the C2-4 well section in Hero's Ridge have the lowest number of scratch curve fluctuation gyrations and the smoothest sample surfaces. The experimental results show that the compressive strength, hardness, and fracture toughness of the shale in Songliao Basin are lower than those in Heroic Ridge, and the damage characteristics of the scratches are obvious, and the number of curve fluctuations is higher than the number of rotations. The fracture extension forms a high-density "Feng"-shaped extension, and the width of the fracture is large.

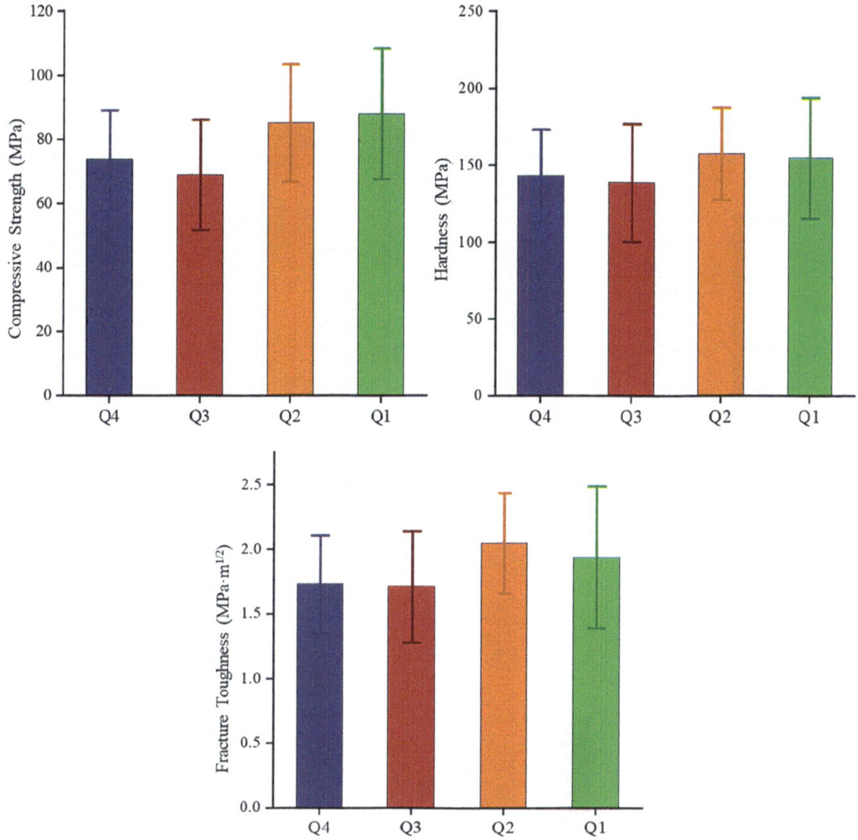

Fig. 7.12 Comparison of mechanical parameters of different reservoir segments in well C41-70

7.1.3 *Effects of Different Lithologies*

Compressive Strength and Modulus of Elasticity

In order to better understand the mechanical distribution characteristics of different lithology samples in the striated shale samples and to study the interrelationship between shale surface minerals and their mechanical properties, several representative shale samples were selected for the scratch experiments. Through the observation of the scratch location, it is noticed that the whole scratch surface shows a clear high-frequency thin interlayer structure, and at the same time, a large number of damage surfaces appear on both sides of the scratch, forming a "Feng" shaped structure. In the test samples, the compressive strength curves show obvious high-frequency fluctuation characteristics.

The results are shown in Fig. 7.14. Group A shale samples show obvious fluctuations in the scratch curve of clayey shale, with high-frequency cyclic

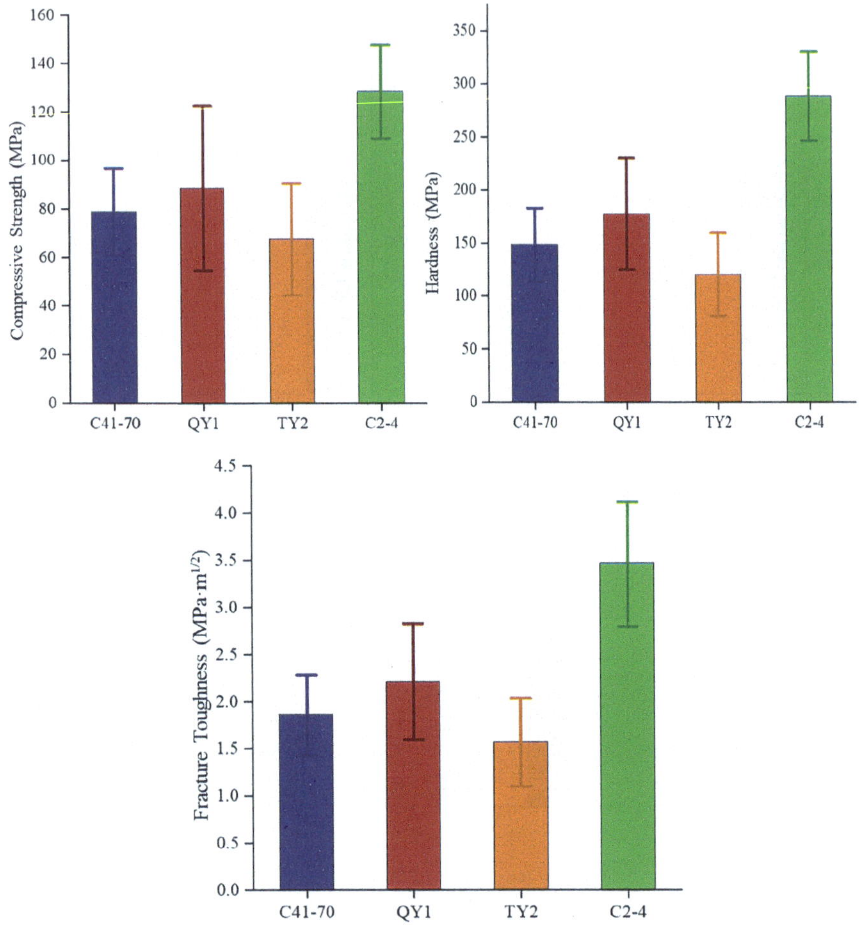

Fig. 7.13 Comparison of mechanical parameters of different locations

fluctuations in mechanical properties, and the number of fluctuations with compressive strength amplitude greater than 40 N in the 85 mm test section is about 77 times, with an average of 9.06 times/cm, nearly 14 cycles per centimeter. The laminar structure is very developed, and the microcracks that are not easily distinguishable by naked eye are also well developed. The compressive strength of the rock is very strongly non-homogeneous, and the compressive strength value along the direction of the scratches varies drastically in the range of 20–150 MPa. In long quartz textured shale, the compressive strength is larger, but the fluctuation amplitude is reduced compared with clayey shale, the number of fluctuations of compressive strength greater than 40 N in the 55 mm test section is about 32 times, with an average of 5.81 times/cm, and nearly 11 rotations per centimeter. Mixed-accumulation type shale compressive strength is larger, but the frequency of fluctuation is more reduced, the number of fluctuations with amplitude greater than

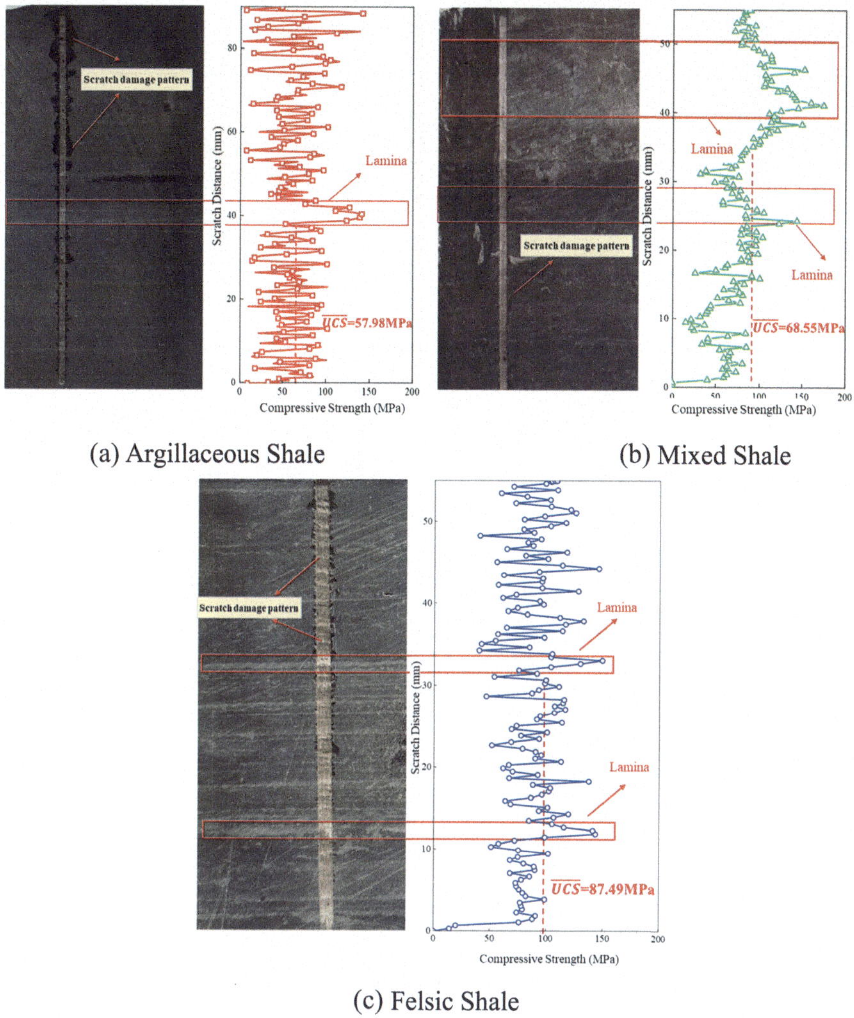

Fig. 7.14 Scratches compressive strength of different lithologic shales of group A

40 MPa in the 55 mm test section is about 18 times, with an average of 3.27 times/cm, and nearly 10 rotations per centimeter, and the clayey layer in the 0–30 mm test section is about 11 times, with amplitude greater than 40 MPa, and the number of fluctuations is significantly greater than that in the 30–55 mm test section, with nearly 12 rotations, but the lower test section compressive strength is obviously smaller than the upper test section. It shows that the scratch experiment can well identify the mechanical weak surface of shale.

The results are shown in Fig. 7.15. Group B shale samples also show obvious fluctuation of scratch curves for clayey shale, high-frequency cyclic fluctuation of mechanical properties, the number of fluctuations of compressive strength amplitude

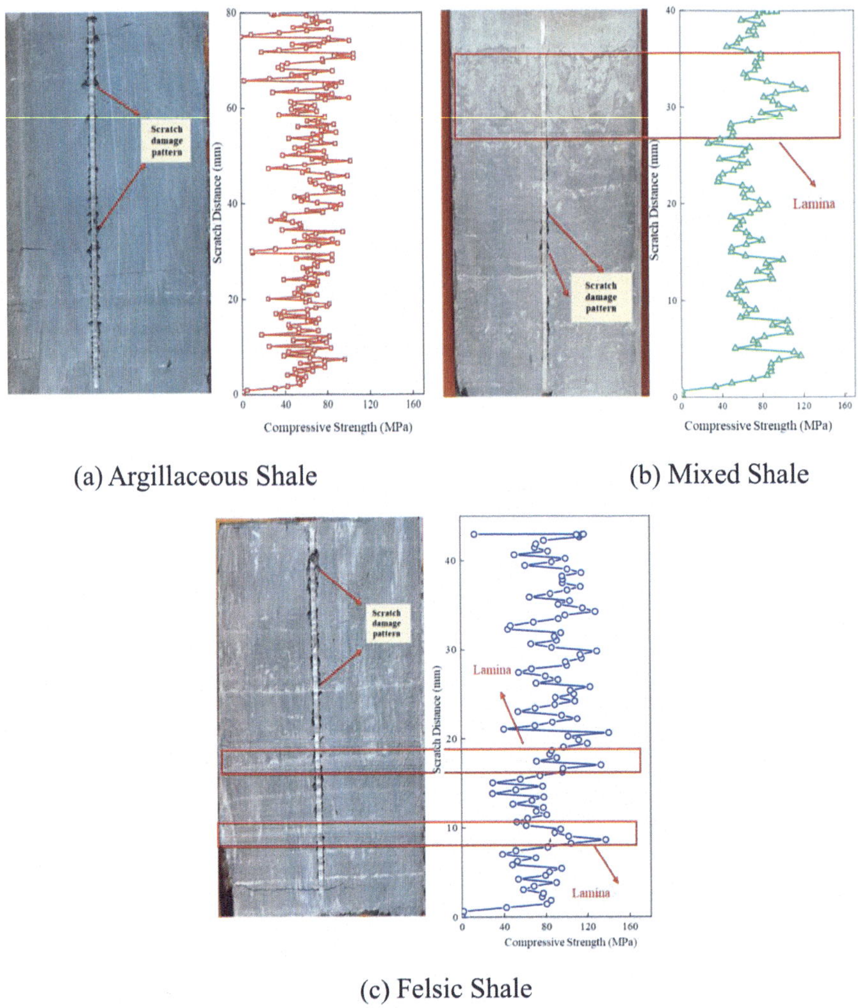

Fig. 7.15 Scratches compressive strength of different lithologic shales of group B

greater than 40 N in the 80 mm test section is about 62, with an average of 7.75 times/cm, and nearly 15 times cyclic per centimeter. In long quartz textured shale, compressive strength is larger, the number of fluctuations of lateral force greater than 40 N in the 45 mm test section is about 35, nearly 12 rotations per centimeter. However, the fluctuation amplitude is reduced compared to the clayey shale of the same group, but increased compared to the long quartz shale of group A. The fluctuation frequency of mafic shale is reduced, the number of fluctuations with amplitude greater than 40 MPa in the 40 mm test section is about 10 times, with nearly 11 rotations per centimeter. 0–26 mm test section is a clayey layer, the number of fluctuations with amplitude greater than 40 MPa is about 7 times, with

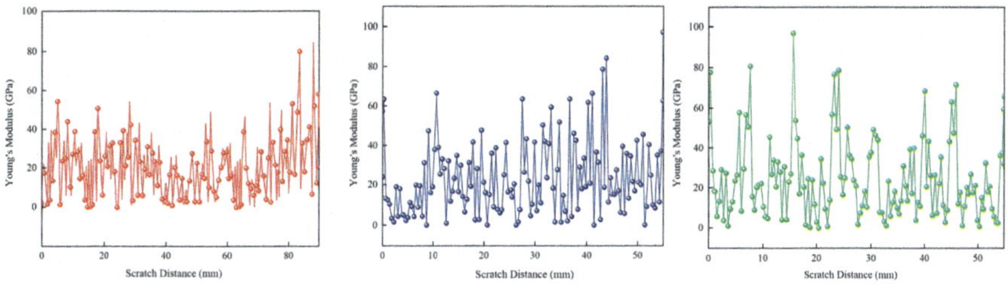

(a) Argillaceous Shale (b) Felsic Shale (c) Mixed Shale

Fig. 7.16 Scratches elastic modulus of different lithologic shales of Group A

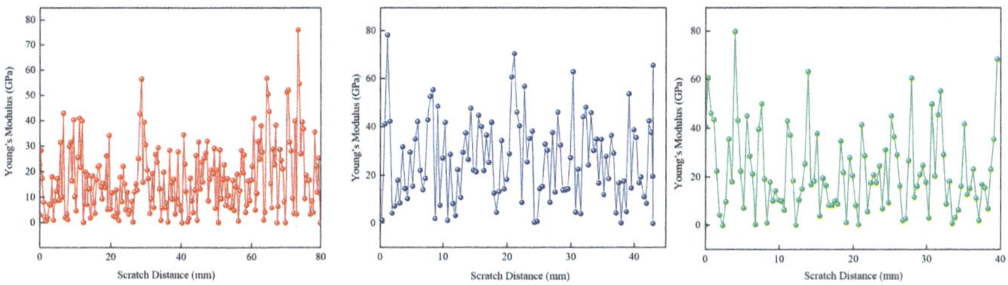

(a) Argillaceous Shale (b) Felsic Shale (c) Mixed Shale

Fig. 7.17 Scratches elastic modulus of different lithologic shales of Group B

fluctuations significantly greater than that of the 30–55 mm test section, with nearly 12 rotations per centimeter, and the value of the compressive strength of the lower test section is significantly smaller than that of the upper test section.

The modulus of elasticity was calculated for three shale samples with different lithologies in Group A and Group B. The modulus of elasticity scratch curves are shown in Figs. 7.16 and 7.17, and the average values of modulus of elasticity in Group A were 21.07, 24.02, and 25.14 GPa, respectively, while the corresponding results of modulus of elasticity in Group B were 17.18, 26.49, and 22.34 GPa, respectively.

For the shale samples of three different lithologies in Group A, one column was processed and the corresponding core scratch locations were recorded to carry out the traditional compressive strength experiments, and the test results were 51.42, 82.69, and 70.21 MPa, respectively, and the average values of the scratch compressive strengths at the corresponding locations were taken, and the corresponding scratch test results were 57.98 and 68.55 MPa The relevant test results and stress-strain curves are shown in Table 7.1 and Fig. 7.18. The uniaxial experiments show that the elastic modulus of the long quartz shale is 22.97 GPa with Poisson's ratio of 0.13, the elastic modulus of the mélange shale is 16.76 GPa with Poisson's ratio of 0.44, and the elastic modulus of the clayey shale is 20.54 GPa with Poisson's ratio of 0.34.

Table 7.1 Comparison of the experimental results of scratch test and traditional compressive strength test

Lithology	Core position	Length (mm)	Diameter (mm)	Density (g/cm^3)	Average compressive strength of scratch (MPa)	Average compressive strength (MPa)	Error %
Argilla-ceous shale	Lower part	50.55	25.28	2.50	57.98	51.42	11.3
Felsic shale	Lower part	50.19	25.11	2.49	87.49	82.69	5.4
Mixed shale	Lower part	50.78	25.28	2.59	68.55	70.21	2.8

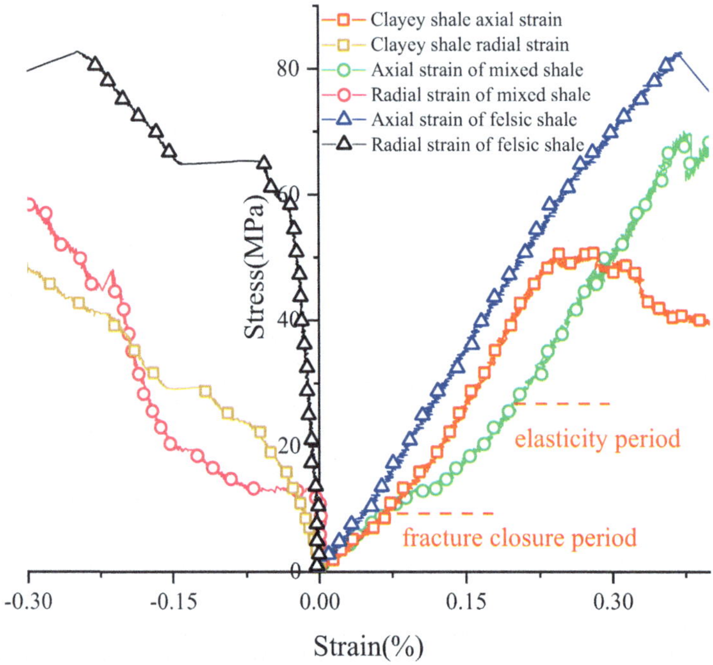

Fig. 7.18 Stress-strain curves of different lithology samples

The comparison revealed that the non-homogeneous shale test results had a large error, but the experimental results for both test methods had an error of 10% (except for the clayey shale) or less, which is within the normal range. The clayey shale test results had the largest error of 11.3%, but the error is real and acceptable. The rock structure and mineral components of unconventional and complex formations with a high degree of non-homogeneity, such as shale, show non-homogeneity, and in addition, the stratigraphy and local features (cracks, fissures, and shear surfaces) are more developed, which leads to non-homogeneity in the expression of the strength of the rock. The results obtained by the traditional testing methods always over- or under-estimate the strength values at the weak or hard layers and other rock segments, which greatly reduces the quality of the results.

Comparison of the two test methods reveals that the conventional test method tends to ignore the non-homogeneity of rock strength and cannot effectively screen out the outliers obtained from the conventional test. According to the average intensity value, the curve can be subdivided into several different regions, which may be caused by various reasons such as changes in microstructure or mineral composition. Scratch tests can reveal the non-homogeneity of the mechanical properties of shale samples. In addition, the scratch test can not only quantitatively characterize the subtle or strong strength non-homogeneity of the core but also make up for the shortcomings of the traditional experimental methods that are unable to obtain rock strength profiles and difficult to universally test the strength of rocks in complex formations such as shale.

Scratch Hardness and Fracture Toughness

From Figs. 7.19 and 7.20, it can be seen that among the three specimens in Group A, the average hardness value of the mafic shale is the largest (185 MPs), while the average hardness value of the clayey shale is the smallest (118 MPa). Whereas among the three specimens in Group B, the average hardness value of the long quartz shale is the largest (164 MPa) and the average hardness value of the clayey shale is the smallest (130 MPa). Comparison by the previous section shows that the compressive strength of the long quartz shale is maximum. It was found that the sample with the largest hardness value does not necessarily have the largest compressive strength, this is because compressive strength and hardness have different physical meanings. Scratch hardness is the hardness of a material when it is scratched along the surface by a tool under a given load. Scratch hardness reveals the resistance of the material to plowing by the tool. Scratch compressive strength is the intrinsic specific energy, which refers to the amount of energy required to cut a piece of rock, whereas hardness refers to the resistance to plowing during the scratching process. Therefore, the amphibolite shale has the greatest resistance to plowing in the Group A samples. As can be seen from the figure, the mean value of fracture toughness of long quartz shale in Group A samples (2.21 MPa $m^{1/2}$) is 1.5 times higher than the mean value of clayey shale (1.48 MPa $m^{1/2}$). The mean value of fracture toughness of the mélange shale is about 2.22 MPa $m^{1/2}$. The mean value of fracture toughness of the long quartz shale is the largest among the samples in group B (2.05 MPa $m^{1/2}$), which is 1.3 times of the mean value of the clayey shale (1.58 MPa $m^{1/2}$). The fracture toughness value of mélange shale is about 1.87 MPa $m^{1/2}$.

The clay minerals of the striated shale are dominated by illite, which is oriented and aligned by diagenetic compaction, which makes the rock exhibit a fissile shale character and is easily stripped at the level to form weak layers. The interaction

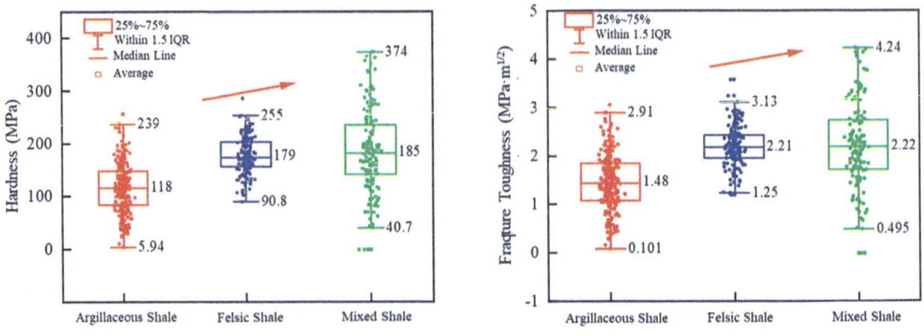

(a) Hardness (b) Fracture Toughness

Fig. 7.19 Mechanical properties of A scratches in different lithologic shales

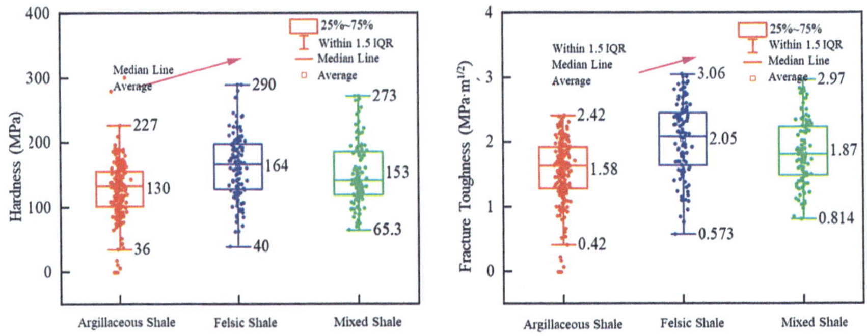

Fig. 7.20 Mechanical properties of B scratches in different lithologic shales

between the composition of the minerals and the weak surfaces of the shale provides a clear basis for progressive damage modeling of the mechanical properties of the striated shale.

The shale scratch load-displacement curves were analyzed to broadly classify the shale into two modes: ductile and brittle damage (Fig. 7.21). Brittle damage is associated with microscopic fracture, where stress-induced damage events tend to correspond to a sudden drop in force at the time of scratching. Ductile damage, however, is the result of interlattice or intergranular slip damage within crystalline grains in rocks, which occurs mainly under shear stresses, and although microfracture and shear expansion can also occur, the important feature of the deformation is plastic flow. Depolymerization of the matrix may produce material, and particles and powders may accumulate in front of the scratch probe and residual scratches on the sides. Plastic flow does not exhibit a signal of rapid release of strain energy. Observe the transverse force signatures of brittle and ductile damage. Brittle damage has a distinct "sawtooth" transverse force response with large fluctuations, while ductile damage has a less fluctuating profile. The existence of distinct transverse force signatures in ductile and brittle damage is confirmed.

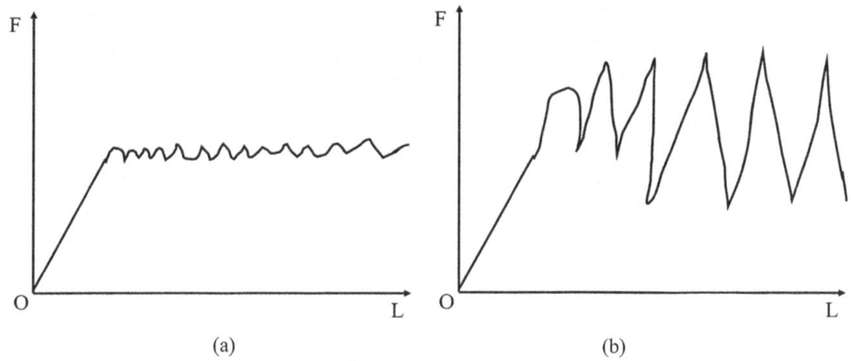

Fig. 7.21 Scratch curve characteristics of different shale samples: (**a**) stable scratch response with smallamplitude fluctuations; (**b**) unstable scratch response with large-amplitude cyclic fluctuations

The surface of the samples was further observed through the scratch test, and the quantification of the scratch damage characteristics is shown in Figs. 7.22 and 7.23. Due to the interaction between the mineral structure and the weak surface of the shale, the mechanical characteristics of the textured shale have different progressive damage patterns, and the two groups of shale samples with different lithologies in Group A and Group B show the same pattern of scratch damage characteristics. For clayey shale, it is found that the whole section of the scratch surface can be seen clearly as high-frequency thin interlayer structural characteristics, and a large number of damage surfaces appear on both sides of the scratch, forming a "Feng" structure, with obvious scratch damage characteristics, obvious "sawtooth" transverse force response, and large fluctuations. There are obvious scratch damage characteristics and obvious "sawtooth" transverse force response with large fluctuation. As for the felsic shale and mafic shale, they belong to the harder grain layer, and the scratch damage is less obvious, resulting in a smoother surface of the samples. Crack extension is not as great as in clayey materials, and the cracks caused by scratches are significantly less than in clayey shales.

7.2 Logging-Scratch Correlation

Rock mechanics data are difficult to obtain and are limited to small cross-sections of the reservoir. In the absence of quality data, rock strength predictions are poor with significantly lower confidence. Rock strength prediction in uncored sections is based on calibration between direct core measurements and logging measurements. This approach has two fundamental limitations: (1) the insufficient number of samples tested and (2) the difference in resolution between test cores and sonic logging (Figs. 7.24 and 7.25).

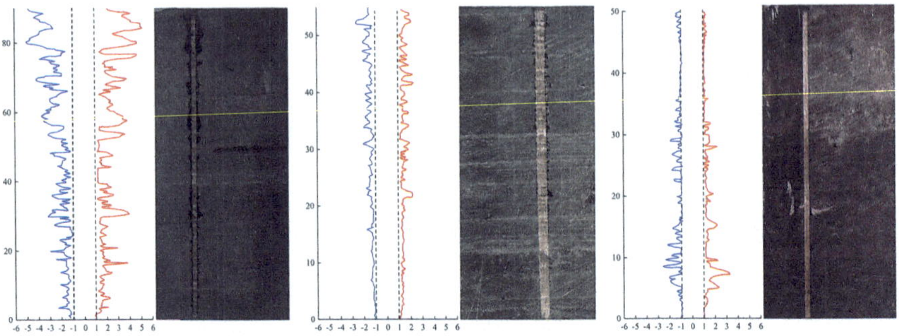

(a) Argillaceous Shale (b) Felsic Shale (c) Mixed

Fig. 7.22 Scratches damage characteristics of Group A shale of different lithology

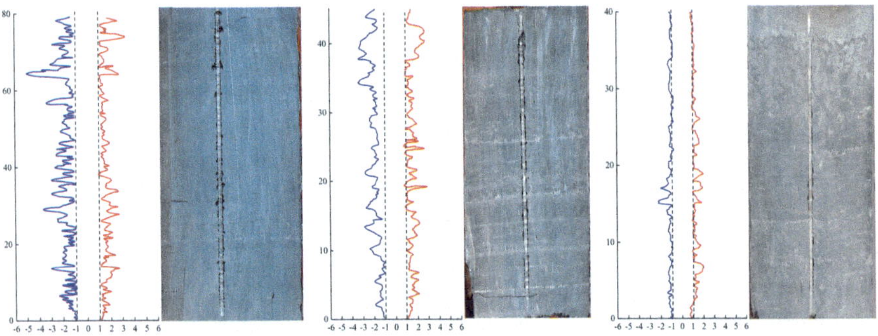

(a) Argillaceous Shale (b) Felsic Shale (c) Mixed

Fig. 7.23 Scratches damage characteristics of Group B shale of different lithology

The results of the distribution of mechanical parameters in the whole well section of the striated shale are shown in Fig. 7.2, which indicate that the compressive strength, hardness, and fracture toughness values of the striated shale of the Qingshankou Formation are smaller than those of the shale of the Dry Chaigou Formation. The minimum value of compressive strength of shale in the Dry ChaiGou Formation is 72.95 MPa, the maximum value is 197.08 MPa, and the average value is 128 MPa; the minimum value of hardness is 160.17 MPa, the maximum value is 490.85 MPa, and the average value is 299 MPa; the minimum value of fracture toughness is 1.77 MPa-m1/2, the maximum value is 5.22 MPa-m1/2, and the average value is 3.29 MPa-m1/2; and the average value is 3.29 MPa-m1/2, and the minimum value is 3.29 MPa-m1/2. 3.29 MPa-m1/2. Conversely, the minimum value of compressive strength of Qingshankou Formation grain shale is 37.02 MPa, the maximum value is 161.46 MPa, and the average value is 86.7 MPa; the minimum value of hardness is 89.2 MPa, the maximum value is 305.47 MPa, and the average value is 169 MPa; and the minimum value of fracture

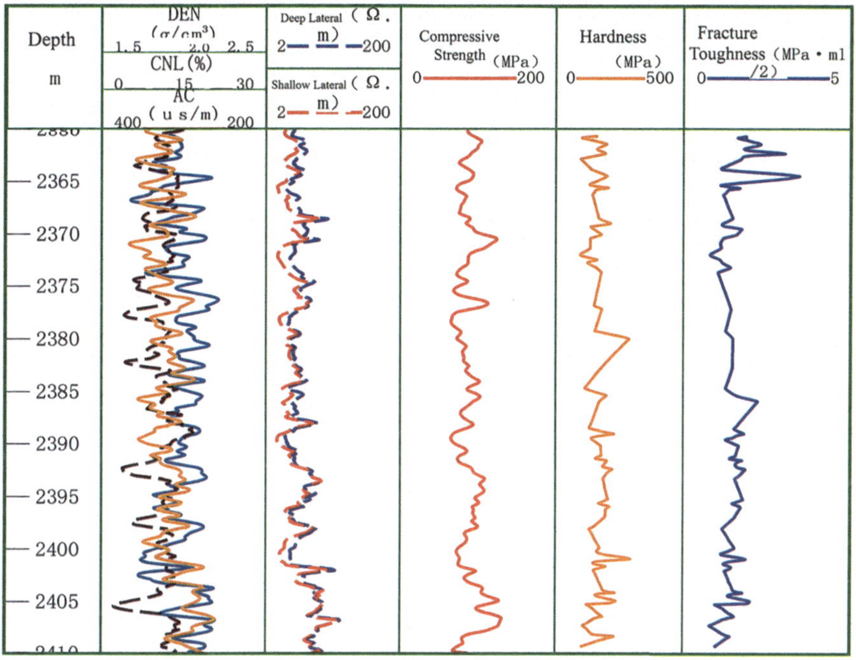

Fig. 7.24 C41-70 scratch results across the whole well

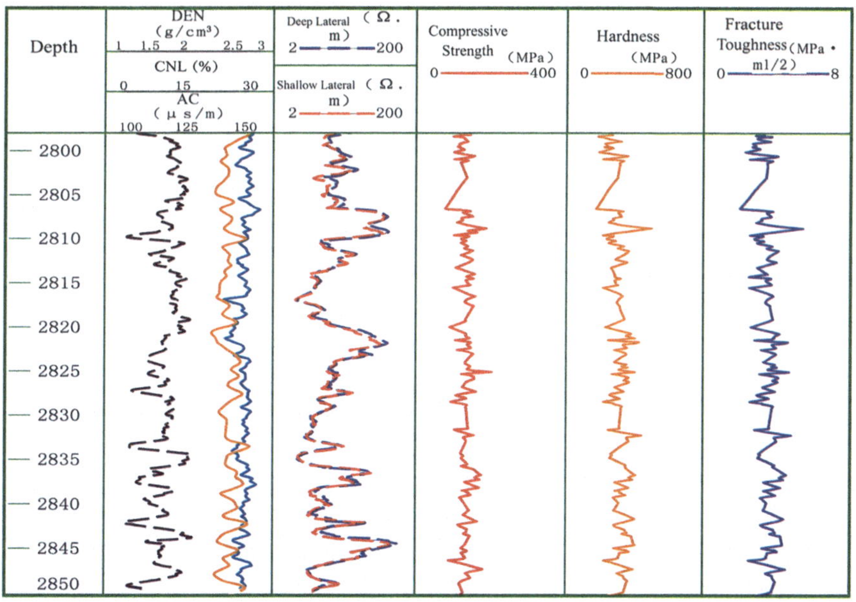

Fig. 7.25 C2-4 scratch results across the whole well

toughness is 0.69 MPa-m1/2, the maximum value is 3.51 MPa-m1/2, and the average value is 1.47 MPa-m1/2 (Fig. 7.26).

7.2.1 Logging Response

Rock mechanics research is commonly based on regional tectonic and stratigraphic frameworks. Sonic, density, gamma-ray, and other logging data are used to construct rock mechanics models. These models have been widely applied in petroleum engineering and provide essential data for development-plan design. They also play an important role in well-pattern optimization, wellbore stability analysis during drilling, and hydraulic fracturing design. It plays a vital role in formulating well network layout, analyzing well wall stability during drilling, and performing hydraulic fracturing.

Based on the available data, a rock mechanics characteristic analysis was conducted on the C41-70 well section of the Songliao Basin, and a comparison was made with the rock mechanics characteristic analysis of the C2-4 well section in the Hero Ridge area. According to the theoretical relationship between the P-wave (compressional wave), S-wave (shear wave), and rock dynamic parameters provided by the propagation equations of P-wave and S-wave, the compressional wave slowness Δt_c and shear wave slowness Δt_s were obtained from the logging data. Using the bulk density ρ_b from density logging, various rock mechanics parameters can be calculated:

$$G_{\text{dyn}} = \frac{\rho_{\text{b}}}{(\Delta t_{\text{s}})^2} \tag{7.1}$$

$$K_{\text{dyn}} = \rho_{\text{b}} \left[\frac{1}{(\Delta t_{\text{c}})^2} \right] - \frac{4}{3} G_{\text{dyn}} \tag{7.2}$$

$$E_{\text{dyn}} = \frac{9G_{\text{dyn}} \times K_{\text{dyn}}}{G_{\text{dyn}} + 3K_{\text{dyn}}} \tag{7.3}$$

$$V_{\text{dyn}} = \frac{3K_{\text{dyn}} - 2G_{\text{dyn}}}{6K_{\text{dyn}} + 2G_{\text{dyn}}} \tag{7.4}$$

$$\sigma_H - \alpha P_P = \frac{V}{1-V}(\sigma_V - \alpha P_P) + \frac{E}{1-V^2}\varepsilon_H + \frac{EV}{1-V^2}\varepsilon_{\text{h}} \tag{7.5}$$

$$\sigma_{\text{h}} - \alpha P_P = \frac{V}{1-V}(\sigma_V - \alpha P_P) + \frac{E}{1-V^2}\varepsilon_{\text{h}} + \frac{EV}{1-V^2}\varepsilon_H \tag{7.6}$$

In the equations, G_{dyn} represents the dynamic shear modulus, K_{dyn} is the dynamic bulk modulus, E_{dyn} is the dynamic Young's modulus, V_{dyn} is the dynamic Poisson's ratio, ρ_b is the bulk density, Δt_s and Δt_c are the shear wave slowness and compressional wave slowness, respectively, σ_H is the maximum horizontal principal stress,

Fig. 7.26 Distribution of mechanical parameters in the whole well interval of laminated shale: (**a**) Compressive strength, (**b**) hardness, and (**c**) fracture toughness

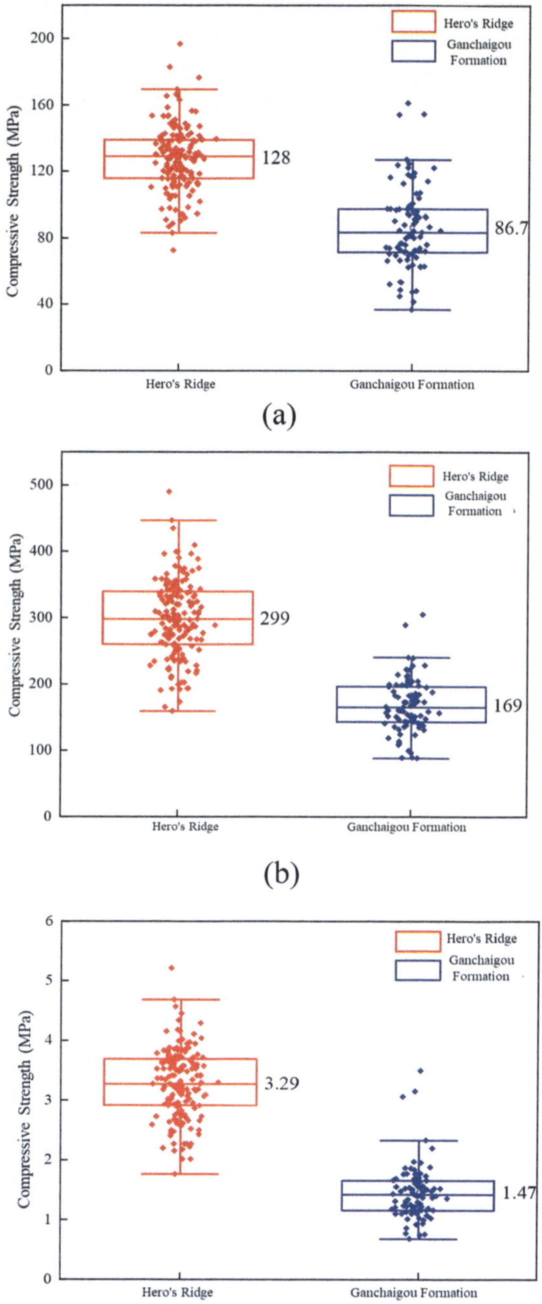

(a)

(b)

(c)

σ_h is the minimum horizontal principal stress, P_P is the pore pressure, E and V are the static Young's modulus and Poisson's ratio, and $\varepsilon_H$and ε_h are the tectonic stress coefficients.

The uniaxial compressive strength (UCS) of rock is typically derived from well logging curves. Several empirical formulas are available for calculating UCS based on logging data. In this study, the rock compressive strength is determined using the rock's elastic modulus (Figs. 7.27 and 7.28).

Using the available information, the rock mechanical parameters of the treated well section were calculated and analyzed, and the calculated rock modulus and rock strength parameters are given in Figs. 7.30 and 7.31. From the results of the treatment, it can be seen that the Young's modulus of the rock of the well section of the Songliao Basin is distributed in the range of about 8–25 GPa, the Poisson's ratio is in the range of about 0.10–0.30, and the compressive strength is in the range of about 50–150 MPa. For shale reservoirs in different regions, there are large differences. The Young's modulus of the rock in the well section of the Hero's Ridge area is distributed around 20–40 GPa, Poisson's ratio is around 0.20–0.40, and the compressive strength is around 100–200 MPa. The horizontal minimum principal stress and maximum principal stress of the striated shale in the Songliao Basin are smaller than those of the shale in the Hero's Ridge area.

The reason for this difference stems from the differences in organic matter, mineral composition, clay content, depth of burial, and so on. From the histograms of acoustic time difference and natural gamma frequency of Songliao Basin and Hero's Ridge shales (Fig. 7.29), it can be concluded that the acoustic time difference of Songliao Basin shales is mostly distributed in the range of 300–340 μs/m, and the natural gamma is distributed in the range of 70–100 API, while the acoustic time difference of Hero's Ridge shales is mostly distributed in the range of 210–230 μs/m, and the natural gamma is distributed in the range of 100–120 API. From the histograms, it can be seen that the natural gamma values of the Songliao Basin shales are not much different from those of the Heroic Ridge shales, but the acoustic time difference is significantly larger than that of the Heroic Ridge shales.

7.2.2 Correlation Analysis

Calibration of logging data is performed by correlating logging data (usually sonic) with laboratory UCS measurements. Although good correlations were obtained, the confidence level for their use in rock strength prediction remains low due to the limited number of UCSs in core tests and the differences in resolution between logging and core measurements.

In order to calibrate the rock strength and logging datasets, the rock strength needs to be adjusted to the logging scale. Comparisons between logging and core measurements can be subject to large errors due to differences in resolution. High-resolution profiles are an ideal tool to characterize the length scale of physical properties, taking into account the spatial structure at different scales. Using scratch

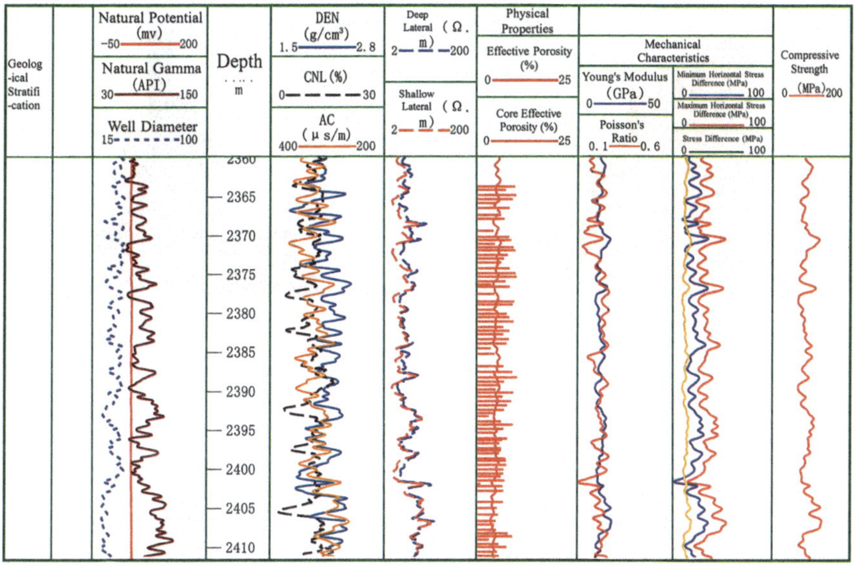

Fig. 7.27 Results of comprehensive interpretation of rock mechanics evaluation in well C41-70 of Qingshankou Formation

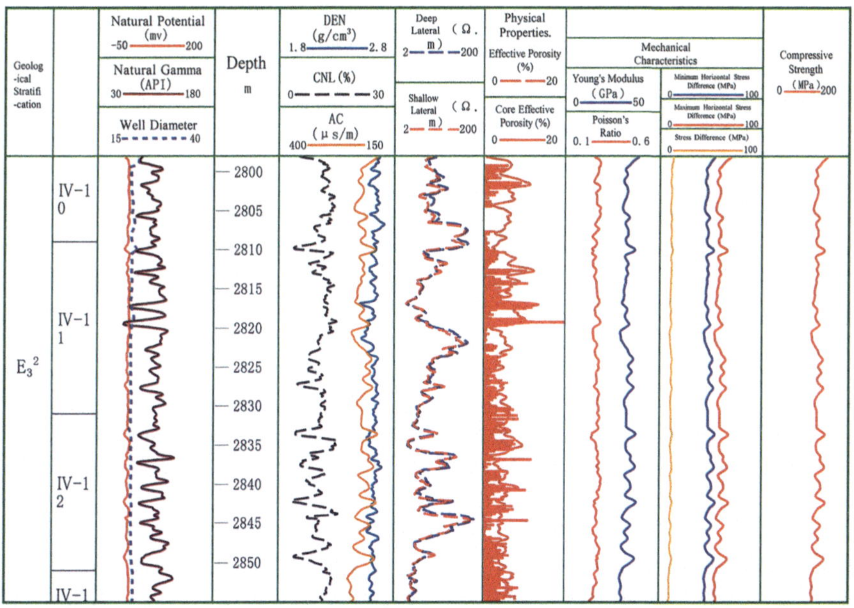

Fig. 7.28 Results of comprehensive interpretation of rock mechanics evaluation in well C2-4 of Ganchaigou formation

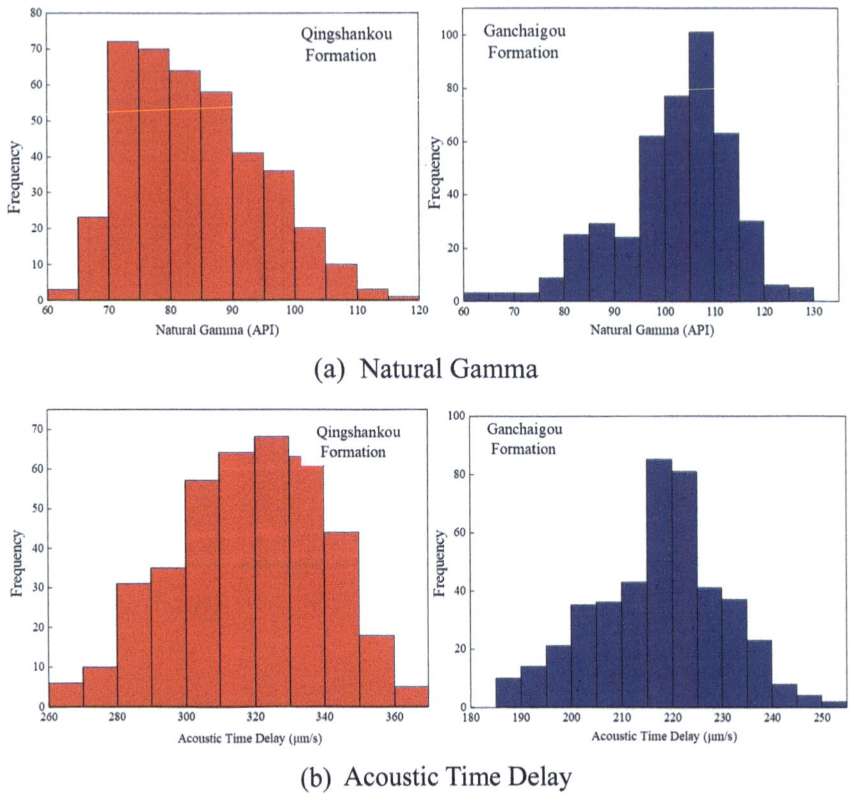

Fig. 7.29 Graph of frequency histogram of well logging response of shale: (**a**) Natural gamma ray and (**b**) acoustic time difference

test data, rock strength data can be directly correlated with logging data. Therefore, scratch experiments were conducted on cores from the entire C41-70 well section in the Songliao Basin and the entire C2-4 well section in the Hero's Ridge. The scratch compressive strength of each shale sample was then averaged in order to obtain the same resolution as well logging measurements.

As shown in Figs. 7.30 and 7.31, the compressive strength data obtained from the scratch experiments in the two wells are compared with the compressive strength derived from the logging curves and the uniaxial compressive strength measured from the core samples. The experimental results show that the rock strengths from the scratch experiments are in good agreement with the rock strengths from the uniaxial experiments. Since the rock strengths obtained from the scratch experiment were measured directly in the core section and correlated well with the values of uniaxial compressive strengths, there is a high degree of confidence in utilizing this information to correlate with the logging curves to predict the mechanical properties.

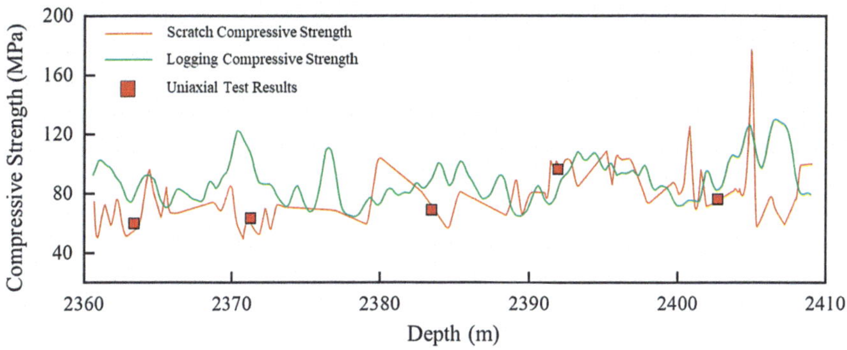

Fig. 7.30 Log compressive strength and uniaxial compressive strength and scratch compressive strength

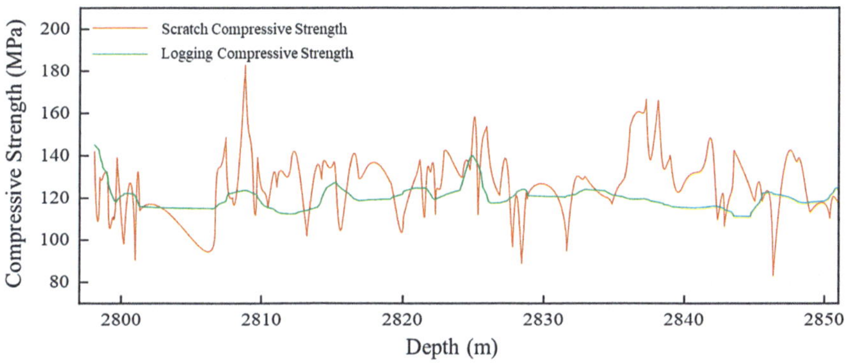

Fig. 7.31 Log compressive strength and uniaxial compressive strength and scratch compressive strength

Identifying the major compositional and structural elements that govern rock strength is becoming increasingly important in rock strength prediction. Key reservoir intrinsic properties such as porosity, grain size, and mineral composition have a significant impact on the geomechanical properties of rocks. Rock strength decreases with increasing clay (shale) content and increases with increasing grain contact. In contrast, the effects of mineral composition and pore type on the strength of shale in the Qingshankou Formation have not been systematically studied in rock mechanics. Therefore, it is necessary to further understand the effect of mineral composition on rock strength.

Principal Component Multiple Regression Prediction

In order to predict the rock strength of uncored well sections, some kind of link is established between the experimental data from scratching and logging. Seismic wave velocity has a linear, polynomial, or logarithmic relationship with rock strength (UCS). McNally established an exponential relationship between UCS and sonic logging based on thousands of compressive strength tests on core samples. The use of sonic data for compressive strength estimation is a traditional method commonly accepted in Australian coal mining. However, this relationship is not applicable to shales with complex mineral compositions and grain layers and their development. Indeed, velocity is related to the elastic nature and density of the rock and it provides a general correlation with rock strength. However, establishing a direct relationship between velocity and strength is not always possible. To overcome this problem, Hatherly proposed an alternative method for estimating rock strength. Compressive strength was compared with porosity and mineral composition, which were considered to have a major influence on rock strength.

Therefore, in this study, continuous rock compressive strength profiles obtained from scratch experiments were compared with rock mineral compositions and porosities inferred from logging interpretations. Figure 7.32 shows the rock strength profiles from the scratch results, along with the logged mineral composition and porosity variations in the available core sections. Principal component analysis (PCA) allows an understanding of which mineral compositions have a dominant influence on the rock strength properties. Table 7.2 lists the eigenvalues of the four principal components along with the variance explained and cumulative explained for the raw data. The first four principal components have a cumulative contribution of 87.665%, indicating that these four principal components contain and explain more than 87.665% of the information. The variance explained rate of principal component 1 is 43.013%, which is greater than the variance contributed rates of the second, third, and fourth principal components, indicating that principal component 1 reflects the most information and the compressive strength is mainly controlled by the first principal component.

For the principal direction (vector F1), it can be seen from the component matrix table in Table 7.2 that the principal component loadings of porosity as well as the volume fractions of quartz X, calcite Y, and dolomite Z account for a large percentage of the total, and thus, it can be assumed that these four can be used as the four key factors and have a significant effect on the change of rock compressive strength (Table 7.3).

The four variables of porosity and the volume fractions of quartz, calcite, and mica were analyzed by multivariate correlation with the compressive strength data, and the following correlations were obtained:

$$\mathrm{UCS} = -0.38 * X + 0.06 * Y - 2.07 * \phi + 0.04 * Z + 113.43 \tag{7.7}$$

The results were calculated by multiple regression analysis and the predicted compressive strength of the rock along the cored layer segments was compared with the measured compressive strength as shown in Fig. 7.33, and it was found that there is a good correlation between the two but there is an error, so further prediction analysis is needed. This can be applied to the uncored layer segments to reduce the uncertainty in rock strength prediction.

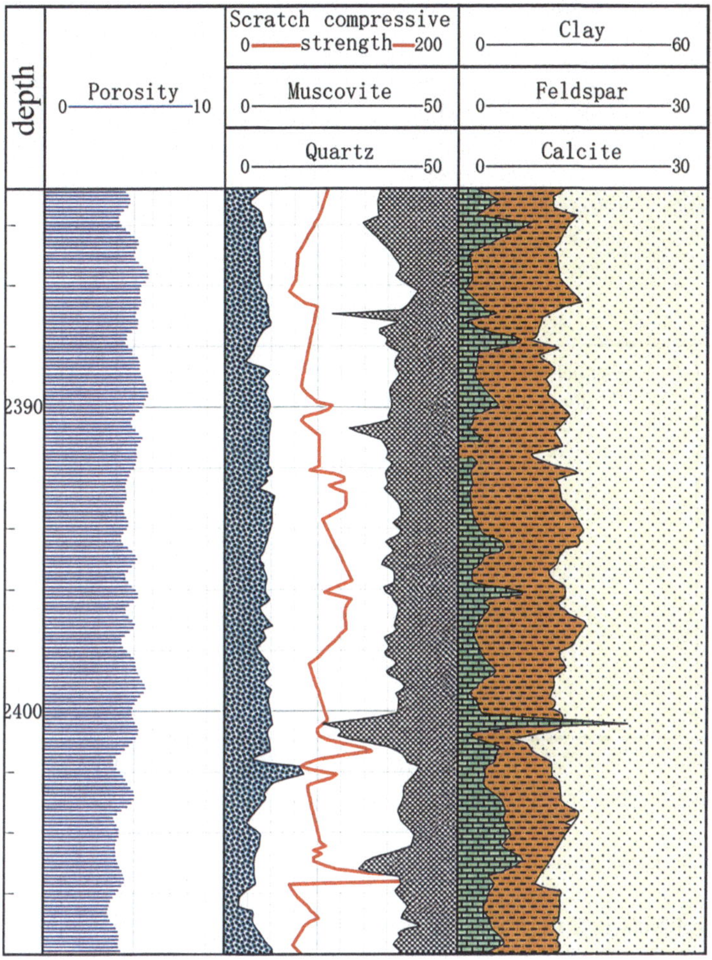

Fig. 7.32 Rock strength profiles, porosity, and mineral composition

Table 7.2 Eigenvalues and variance interpretation rates of each principal component

Ingredient	Characteristic root	Explanation of variance/%	Cumulative variance explained/%
1	2.581	43.013	43.013
2	1.146	19.092	62.105
3	0.831	13.854	75.96
4	0.702	11.706	87.665
5	0.407	6.78	94.445
6	0.337	4.555	100

Table 7.3 Component matrix

Ingredient matrix					
Title	Component 1	Component 2	Component 3	Component 4	Component 5
Quartz	−0.242	0.446	0.523	−0.412	0.837
Feldspar	0.229	0.763	−0.365	−0.06	−0.487
Calcite	0.313	0.154	0.385	0.965	−0.305
Clay minerals	−0.103	−0.126	−0.633	0.144	0.726
Porosity	−0.283	−0.2	0.227	−0.541	−0.877
Mica	0.296	−0.109	0.442	0.083	0.362

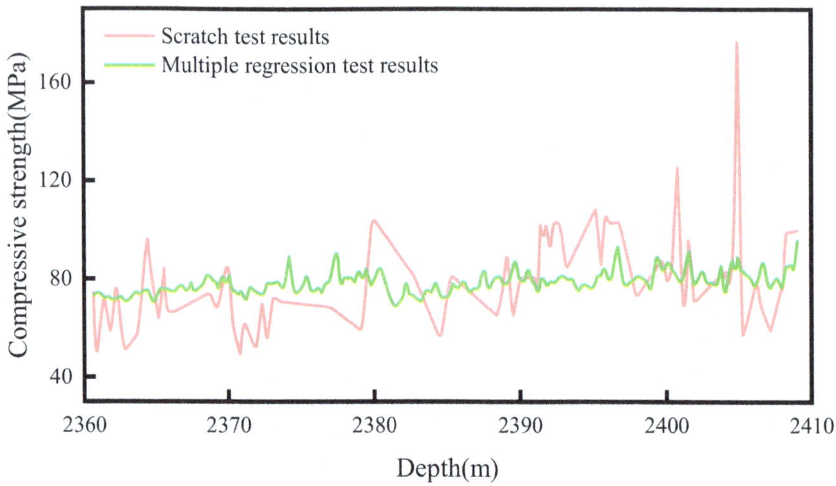

Fig. 7.33 Rock strength measurement vs. rock strength prediction

BP Neural Network Regression Prediction

Through the principal component analysis in the previous section, quartz, dolomite, porosity, and dolomite were determined to be the input layers and compressive strength was the output layer. Prior to the start of the experiment, multiple logging curves needed to be synthesized, as different treatments can lead to different results in terms of gauge interpretation. These numerically different samples have a significant impact on the results of the training. Therefore, it is necessary to normalize the input sample data. Considering the practical aspects of data fitting, normalization was used to map the raw data into the interval [1–3]. This method not only reduces the interference of different measures and values on training but also improves the convergence rate and accuracy of the model to a large extent.

$$X_{norm} = \frac{X - X_{min}}{X_{max} + X_{min}} \tag{7.8}$$

Forward Propagation

First assume that the number of nodes in the input, implicit, and output layers are, respectively, m, l, and n. Set the input vector $X = [X_i]_m$ passes from the input layer to the implicit layer through the weights w_{ij}^1 and thresholds θ_j^1. Utilizing the activation function (sigmoid function) $f(\mu)$, calculate the implicit layer output value $H = [h_j]_l$, where the activation value of each neuron in the hidden layer is

$$s_j^1 = \sum_{i=1}^{m} w_{ij}^1 x_i - \theta_j^1, j = 1, 2, \cdots, l \tag{7.9}$$

activation function $f(\mu)$ is

$$f(\mu) = \frac{1}{1 + e^{-\mu}} \tag{7.10}$$

Bringing the activation function values into the activation function yields

$$h_j = f\left(s_j^1\right) = f\left(\sum_{i=1}^{m} w_{ij}^1 x_i - \theta_j^1\right), j = 1, 2, \cdots, l \tag{7.11}$$

Activation values for each node in the output layer S_k^2 and output values y_k为:

$$s_k^2 = \sum_{j=1}^{l} w_{jk}^2 h_j - \theta_k^2, k = 1, 2, \cdots, n \tag{7.12}$$

$$y_k = f\left(s_k^2\right), k = 1, 2, \cdots, n \tag{7.13}$$

Backward Propagation

If the output signal $Y = [y_k]_n$ and desired output signal $T = [t_k]_n$ inconsistency. Then, it is transferred to the back propagation process. The output error is inverted layer by layer through the implicit layer toward the input layer, and the error is decreased along the gradient direction by adjusting the output layer to implicit layer connection weights and output layer thresholds as well as the implicit layer to input layer connection weights and implicit layer thresholds.

The correction error of the output and implicit layers is

$$d_k^2 = (t_k - y_k) f'\left(s_k^2\right), k = 1, 2, \cdots, n \tag{7.14}$$

$$d_j^1 = \sum_{k=1}^{n} w_{jk}^2 d_k^2 f'\left(s_j^1\right), j = 1, 2, \cdots, l \tag{7.15}$$

Correct the weights and thresholds:

$$\Delta w_{jk}^{2} = \eta d_{k}^{2} h_{j} \tag{7.16}$$

$$\Delta \theta_{k}^{2} = \eta d_{k}^{1} \tag{7.17}$$

$$\Delta w_{ij}^{2} = \eta d_{j}^{1} h_{i} \tag{7.18}$$

$$\Delta \theta_{j}^{1} = \eta d_{j}^{2} \tag{7.19}$$

where f' is the derivative of the activation function and η is the learning step size (learning rate $0 < \eta < 1$).

For group Z input module $X^1, \cdots, X^P, \cdots X^Z \left[X^a = \left(x_1^a, \cdots, x_j^a, \cdots x_m^a \right) \right]$, expected value $T^1, \cdots, T^P, \cdots T^Z \left[T^a = \left(t_1^a, \cdots, t_j^a, \cdots t_m^a \right) \right]$. The actual output value is $Y^1, \cdots, Y^P, \cdots Y^Z \left[Y^a = \left(y_1^a, \cdots, y_j^a, \cdots y_m^a \right) \right]$. Then the error is

$$E = \sum\nolimits_{P}^{Z} E^P = \sum\nolimits_{P=1}^{Z} \sum\nolimits_{k=1}^{n} \frac{1}{2} \left(t_k^p - y_k^p \right)^2 \tag{7.20}$$

For a given accuracy ε, the algorithm ends if $E < \varepsilon$.

The data from the test section 2360–2384 m is used as the training set and the data from the test section 2385–2410 m is used as the test set, and the results are shown in Figs. 7.34 and 7.35. Among the quantitative prediction models developed using the BP neural network method, the error of the BP artificial neural network model is smaller than that of the multiple regression analysis, which indicates that the prediction of compressive strength capacity with the BP neural network model is more reliable.

7.3 Scratch Brittleness Evaluation

The striated shale is characterized by typical high clay content, and the clay minerals in clayey shale are more developed; the clay minerals in feldspathic and mafic shale are relatively less; besides clay minerals, the content of brittle minerals, such as quartz and feldspars, in the shale is also higher, but the high brittle minerals content does not mean that the shale has a better brittleness characteristic. Quartz minerals in shale show obvious differences in their causes of formation and morphology. In feldspathic and mixed shales, the main source of quartz is the clasts of terrestrial origin, which have large grains, show sub-angular layered distribution, and contribute greatly to the compressibility of shales; in clayey shales, quartz mainly consists of authigenic silica, which has fine grains, stellate distribution, and is surrounded by clay minerals, which makes its contribution to the compressibility relatively small, and therefore, there is some difficulty in the spreading and holding process.

Moreover, the horizontal laminar joints in the textured shale are extremely well developed, with hundreds to thousands of joints per meter, which have a significant

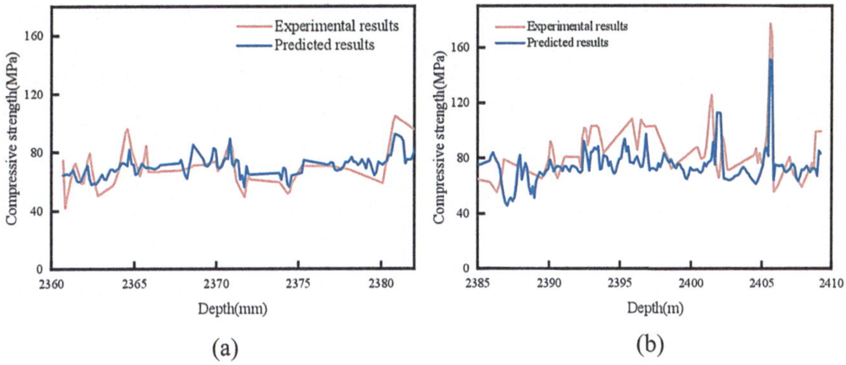

Fig. 7.34 Comparison of experimental and predicted values. (**a**) Comparison of training set prediction results. (**b**) Comparison of test set prediction results

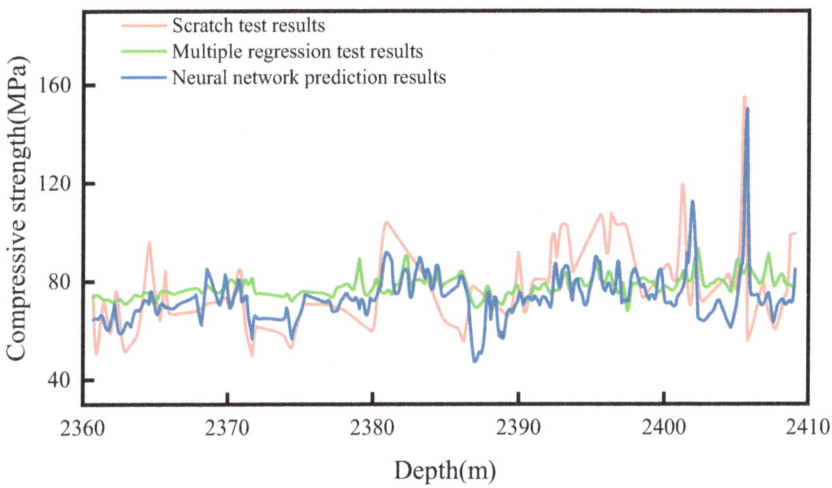

Fig. 7.35 Multiple regression prediction and neural network prediction are compared with experimental results

impact on hydraulic fracture crack expansion. When a hydraulic fracture meets with a shale texture, it may have various morphologies such as penetration, steering, termination or stepwise expansion, which further increases the uncertainty of the fracture network morphology. Therefore, under the control of high content of clay minerals and strong non-homogeneous characteristics in shale, clarifying its mechanical properties and scientifically understanding its compressibility have far-reaching theoretical and practical application values for guiding the selection of targets, the design of fracturing techniques, and the optimization of construction parameters in shale oil engineering. At present, scholars at home and abroad have established a number of compressibility evaluation models based on shale experimental research. Compressibility characterizes the degree of difficulty for tight oil and gas reservoirs to be effectively transformed, and it has become an important evaluation criterion for the development potential of tight unconventional natural gas. Compressibility refers to the ability of a rock reservoir to be effectively fractured and form fractures during hydraulic fracturing. Initially, most scholars used a single indicator of brittleness to evaluate the compressibility of tight oil and gas reservoirs [4, 5], arguing that the higher the brittleness, the easier it is to break up and develop a complex fracture network. Therefore, in this paper, two indicators, the scratch compressive strength and the striation development index, are used to assess the compressibility of shale.

7.3.1 Lamina Development Degree

At present, the degree of development of reservoir texture is mainly described qualitatively by means of thin-section analysis and imaging logging, and most of the data obtained are point data, which are usually used as validation indexes for compressibility evaluation without being included in the parameter assignment for compressibility evaluation.

By comparing the compressive strength curves of different lithologies (Fig. 7.36), it is found that the compressive strength of homogeneous siltstone as a whole fluctuates between 80 and 120 MPa and basically fluctuates within 20 MPa above and below the mean value. The number of times per centimeter that exceeded 20 MPa above and below the mean value of compressive strength was about 0.03 times/cm, which is negligible. And the curve fluctuation of shale samples of other lithologies is obvious. The clayey shale has developed grain layer structure and shows high inhomogeneity, and its compressive strength fluctuates drastically in the range of 0–120 MPa. The number of times per centimeter exceeding 20 MPa above and below the average compressive strength is about 5.45 times/cm. long quartz shale, the compressive strength is larger, but the fluctuation amplitude is reduced compared with the clay shale, along the direction of the scratches, the compressive strength value fluctuates between 40 and 130 MPa, and the number of times per centimeter exceeding 20 MPa above and below the average compressive strength is about 4.21 times/cm. In mélangeous shale, the compressive strength is larger, but the frequency of fluctuation is reduced, the value of compressive strength fluctuates between 50 and 125 MPa along the direction of scratch, and the number of times per centimeter exceeding 20 MPa above and below the average value of compressive strength is about 3.15 times/cm. In the dolomitic shale sample, the 0–15 mm test interval is dominated by dolomite, whereas the 15–40 mm interval consists mainly of shale. Along the scratch direction, the compressive strength in the dolomite-dominated interval fluctuates between 90 and 120 MPa. A similar range of 90–120 MPa is observed in the shale interval, indicating a relatively small strength contrast between the two intervals. In the shale test interval, the compressive strength fluctuates between 40 and 110 MPa along the scratch direction. The frequency of deviations exceeding ± 20 MPa from the average compressive strength is

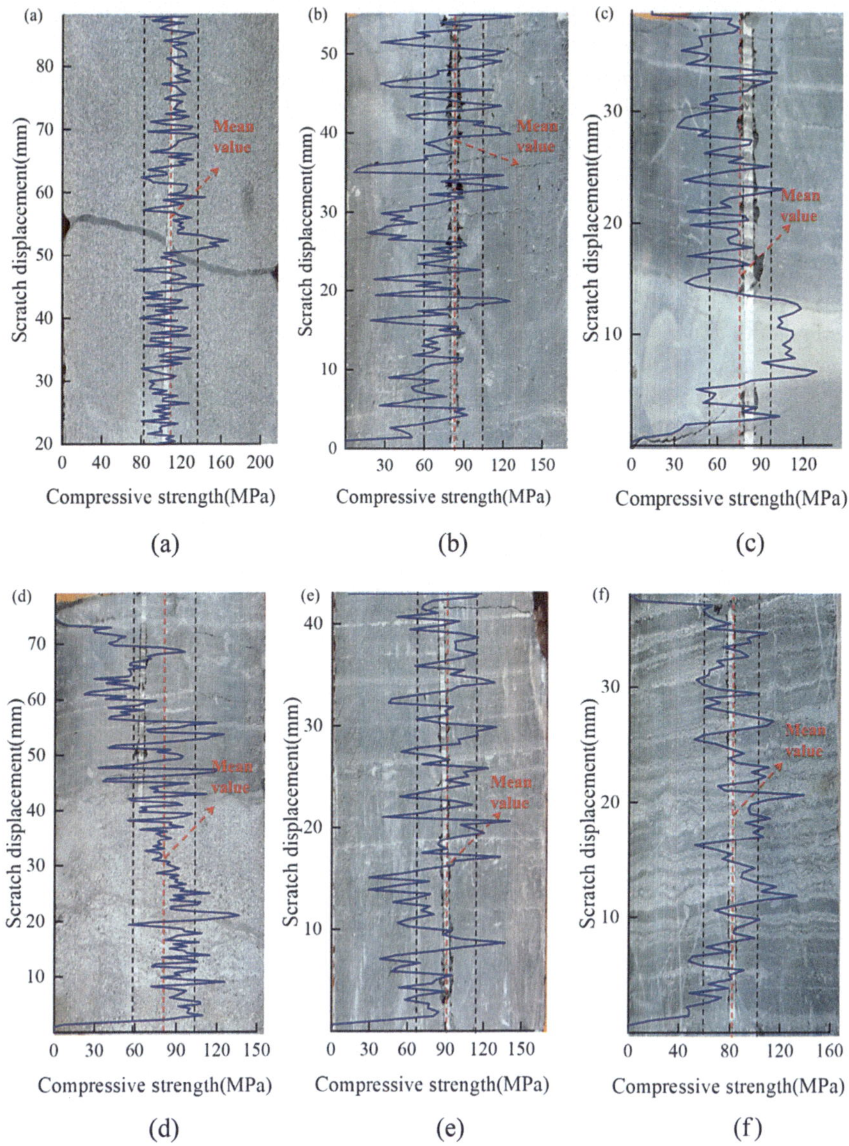

Fig. 7.36 Scratch compressive strength curves of different lithologies: (**a**) Homogeneous siltstone, (**b**) clay shale, (**c**) dolomite, (**d**) sandstone intercalation, (**e**) felsic shale, and (**f**) mixed shale

approximately 4.1 times/cm, indicating strong mechanical heterogeneity. For the sandstone-interbedded sample, the overall compressive strength is relatively higher. The 40–80 mm test interval is composed of shale, where the compressive strength fluctuates between 30 and 120 MPa along the scratch direction. The 0–40 mm test interval is composed of sandstone, where the compressive strength varies between

60 and 120 MPa. The overall frequency of deviations exceeding ± 20 MPa from the average compressive strength is approximately 3.15 times/cm.

Comparing the compressive strength curves of different layers (Fig. 7.37), it is found that the fluctuation of the compressive strength curves is not the same at different depths. The shale grain layer structure of layer Q4 is developed, and the tiny cracks that are not easily distinguishable by naked eyes are also more developed, the compressive strength value of the rock is small, and the shale non-homogeneity is very drastic, and the change of the undulation along the direction of the scratches is obvious. Conversely, the shale laminae in layer Q9 are not developed, the compressive strength value is larger, the shale non-homogeneity is not intense, and the undulation along the direction of the scratches is not obvious.

The number of fluctuations of the scratch curves of shales with different lithologies is obviously different, so this paper proposes to use the number of fluctuations of the scratch compressive strength curves to quantitatively characterize the degree of grain layer development, using the scratch compressive strength curves of homogeneous siltstones as the standard, with the mean value of the scratch compressive strength curves as the baseline, and plus or minus 20 MPa as the upper and lower bounds. The number of fluctuations per centimeter above this range was used to characterize the index of grain layer development.

$$\beta = B\left(\overline{UCS} - 20, \overline{UCS} + 20\right)/L \tag{7.21}$$

where L denotes the distance from the scratch surface of the sample. $B\left(\overline{UCS} - 20, \overline{UCS} + 20\right)$ the number of fluctuations of the curve above this range, $\overline{UCS}$ denotes the average value of the compressive strength, and β denotes the index of the development of the grain layer.

7.3.2 Brittleness Evaluation

For the 2360–2410 m reservoir section of C41-70 well in Songliao Basin, it is divided into four strata: Q1 (2398–2410), Q2 (2384–2398), Q3 (2370–2384), and Q4 (2355–2370). Based on the above method to characterize the formation development index of well C41-70, the formation is classified into four types of reservoirs by making a rendezvous diagram (Fig. 7.38) based on the already obtained formation development index and compressibility strength. Those with good compressibility are class I reservoirs, those with better compressibility are class II reservoirs, those with medium compressibility are class III reservoirs, and those with poor compressibility are class IV reservoirs. In this way, the indoor compressibility evaluation model of grain shale was established. The classification basis of compressibility evaluation is shown in Table 7.4.

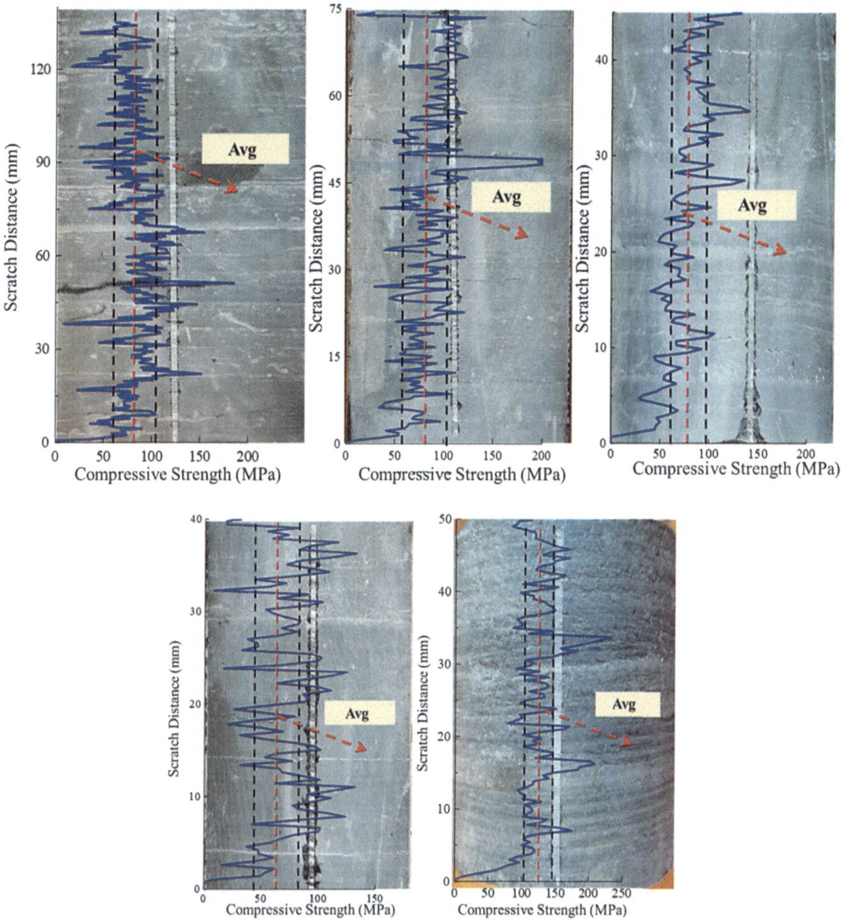

Fig. 7.37 Scratch compressive strength curves at different levels: (**a**) Q1, (**b**) Q2, (**c**) Q3, (**d**) Q4, and (**e**) Q9

The rendezvous plot of striation development index and compressive strength found that the compressibility grade of reservoir C41-70 in Songliao Basin is medium, with a larger percentage of class III and class IV reservoirs.

In addition, scratch experiments were also carried out on the shale in well section C2-4 of the Ganchagou Formation, and a comparison of the compressibility evaluations of the two major regions revealed (Figs. 7.39 and 7.40) that the overall compressive strength of the shale in the Ganchagou Formation is on the large side, with an average compressive strength value of 129.61 MPa, while the grain development index is on the small side. While the average value of compressive strength of shale in well section C41-70 of Qingshankou Formation is 78.97 MPa, and the grain layer is developed. Most of the formations in well C2-4 have good compressibility, and class II reservoirs and class I reservoirs account for a relatively large percentage of the wells.

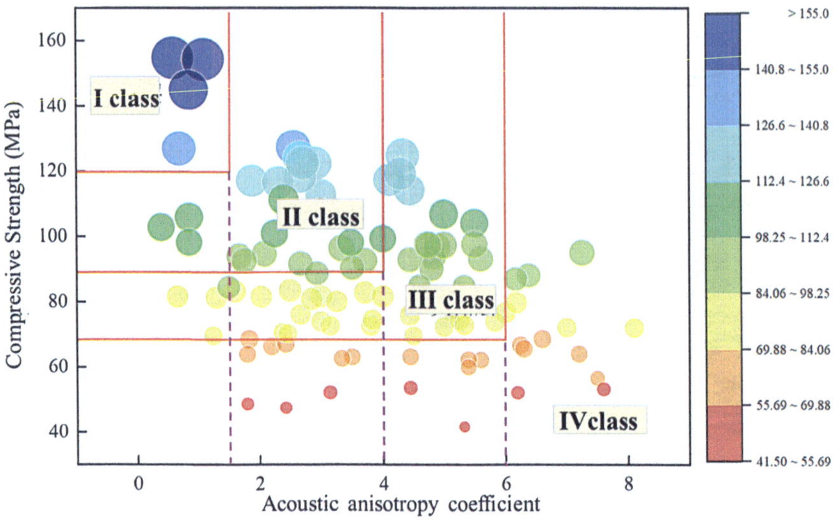

Fig. 7.38 C41-70 well laminae development index and compressive strength cross plot

Table 7.4 Compressibility evaluation level classification criteria

Reservoir class	Evaluation criteria	Class
I class	UCS > 90, $\lambda < 0.475$	Best
II class	75 < UCS < 90, $0.475 < \lambda < 0.535$	Better
III class	65 < UCS < 75, $0.535 < \lambda < 0.60$	Medium
IV class	UCS < 65, $\lambda > 0.60$	Worst

According to Figs. 7.40, 7.41, and 7.42, the average value of compressive strength of striated shale in layers Q8–Q10 of Qingshankou Formation in Songliao Basin is larger, and the striated layer is not developed, which belongs to class I. Q1 belongs to class II reservoir, Q2 and Q3 belong to class III reservoir, and Q4 belongs to class IV reservoir. Most of the formations in well C2-4 of the Dry ChaiGou Formation have good compressibility, and class II reservoirs and class I reservoirs account for a relatively large percentage.

The non-homogeneity of reservoirs is the result of the joint action of tectonic activity, sedimentary process, and diagenetic process, which influences the development results of reservoirs. Regarding the classification of reservoir inhomogeneity, different scholars have their own viewpoints and schemes, among which the anisotropic properties of rocks are considered to be one of the main factors affecting inhomogeneity, and at the same time, there is a close correlation between the development degree of shale grain layer and inhomogeneity. The so-called anisotropy of shale refers to the fact that the physical properties of a substance change with its orientation and show different properties in different directions. The sound velocity characteristic parameters of the core in different directions can be used to describe the anisotropy of rocks. Typically, the anisotropy coefficient of sound

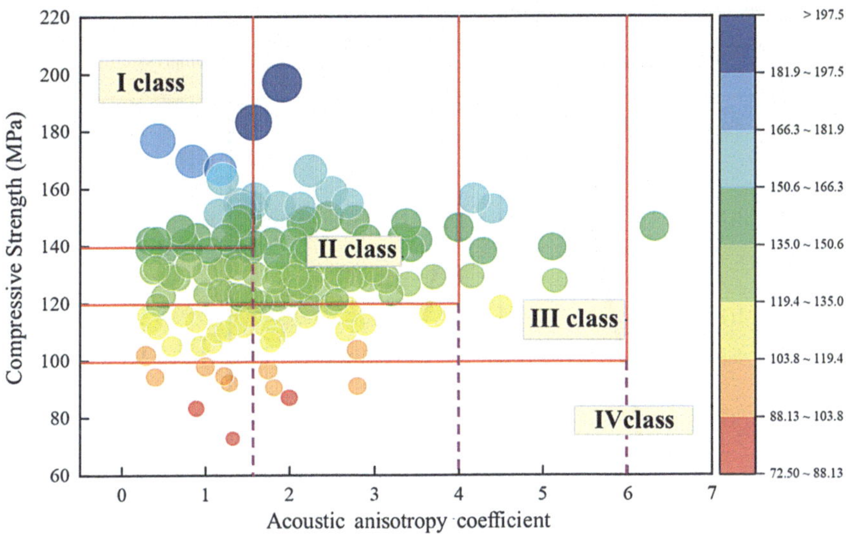

Fig. 7.39 C2-4 well laminae development index and compressive strength cross plot

waves is calculated by using the velocity difference of the core in each direction to quantitatively characterize the anisotropy of the rock. This characterization method provides a basis for assessing sound velocity and energy anisotropy in horizontal wells.

$$\lambda = \frac{V_y}{V_x} \tag{7.22}$$

where λ denotes the acoustic anisotropy coefficient, V_y denotes the longitudinal wave velocity, and V_x denotes the transverse wave velocity.

The mineral composition of the striated shale is complex, with large variations in the content of various minerals in different rock samples and extreme non-homogeneity. The degree of development of the grain layer is large. Figure 7.43 shows that the anisotropy coefficient of acoustic waves with a large grain layer development index is also large, with a strong positive correlation and good agreement with anisotropy. The above has been quantitatively characterized by indoor scratching experiments, through the grain layer development index and acoustic anisotropy coefficient to establish a relationship, and then use the acoustic anisotropy system to characterize the degree of grain layer development.

A clear positive correlation was found through the relationship between the striation development index and the acoustic anisotropy coefficient, and based on the predicted compressive strength and acoustic anisotropy coefficient rendezvous diagram (Fig. 7.44), a model for evaluating the compressibility of the site based on the logging curves was established. Similarly, the reservoirs were divided into four categories, with good compressibility as class I reservoirs, better compressibility as

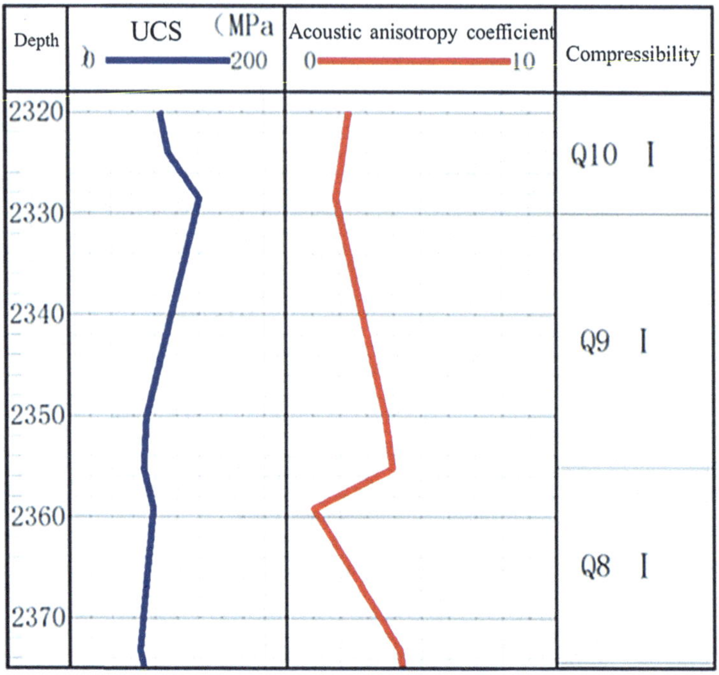

Fig. 7.40 Evaluation of compressibility of the Q8–Q10 layers in the Qingshankou formation

Fig. 7.41 Evaluation of compressibility of the Q1–Q4 layers in the Qingshankou formation

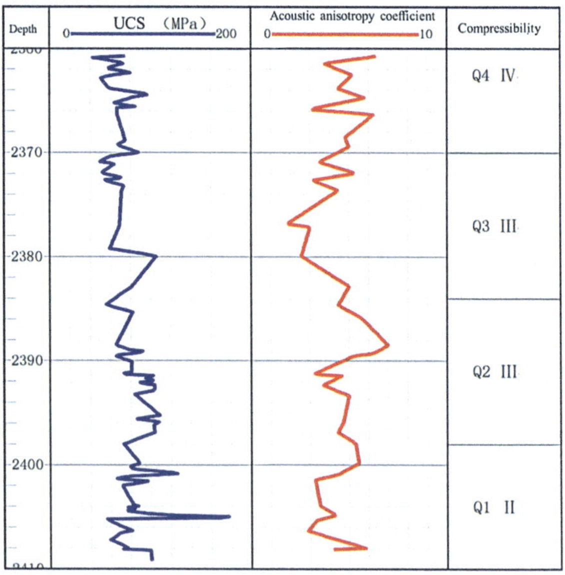

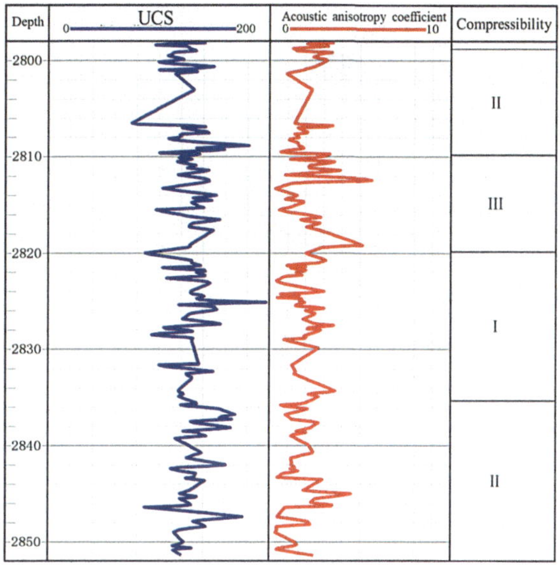

Fig. 7.42 Evaluation of the compressibility of the Ganchaigou formation

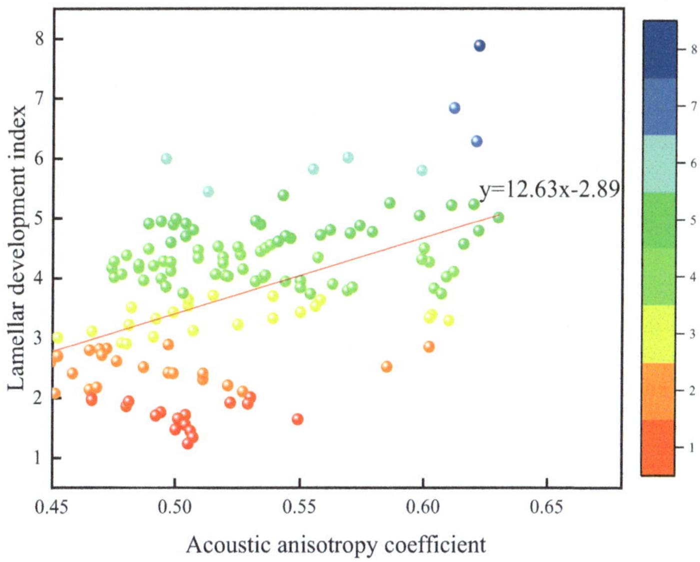

Fig. 7.43 Relation between acoustic anisotropy coefficient and lamina development index

class II reservoirs, medium compressibility as class III reservoirs, and poor compressibility as class IV reservoirs. In this way, the compressibility evaluation model of striated shale is established, and the basis of compressibility evaluation classification is shown in Table 7.5.

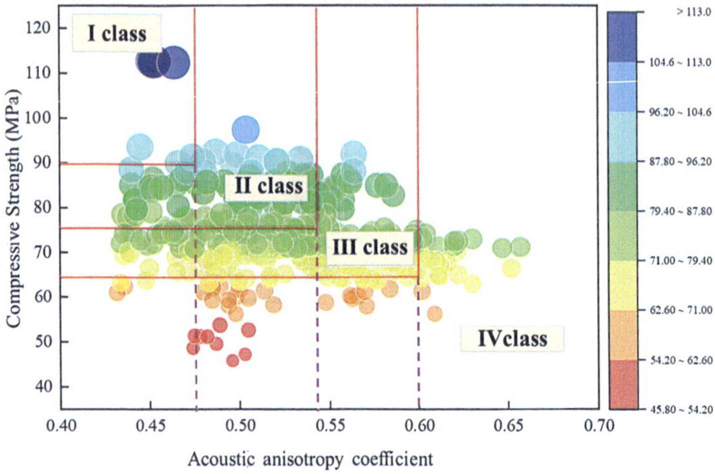

Fig. 7.44 Acoustic anisotropy coefficient versus compressive strength

Table 7.5 Compressibility evaluation level classification criteria

Reservoir class	Evaluation criteria	Class
I class	UCS > 90, $\lambda < 0.475$	Best
II class	75 < UCS < 90, $0.475 < \lambda < 0.535$	Better
III class	65 < UCS < 75, $0.535 < \lambda < 0.60$	Medium
IV class	UCS < 65, $\lambda > 0.60$	Worst

References

1. Guo H, Jia W, Peng P, et al. The composition and its impact on the methane sorption of lacustrine shales from the upper Triassic Yanchang formation, Ordos Basin, China. Mar Pet Geol. 2014;57: 509–20.
2. Jarvie DM, Hill RJ, Ruble TE, et al. Unconventional shale-gas systems: the Mississippian Barnett Shale of north-Central Texas as one model for thermogenic shale-gas assessment. AAPG Bull. 2007;91(4):475–99.
3. Rickman MM, Petre E, et al. A practical use of shale petrophysics for stimulation design optimization: all shale plays are not clones of the Barnett Shale. SPE Annual Technical Conference and Exhibition, Denver, Colorado, USA, 21–24 September 2008.
4. Bybee K. Proper evaluation of shale-gas reservoirs leads to more-effective hydraulic-fracture stimulation. J Pet Technol. 2009;61(7):59–61.
5. Guo H, Jia W, Peng P, et al. The composition and its impact on the methane sorption of lacustrine shales from the Upper Triassic Yanchang formation, Ordos Basin, China. Mar Pet Geol. 2014;57: 509–20.

The manufacturer's authorised representative in the EU is Springer Nature Customer Service Centre GmbH, Europaplatz 3, 69115 Heidelberg, Germany. If you have any concerns regarding our products, please contact ProductSafety@springernature.com

Printed and bound by CPI Group (UK) Ltd, Croydon, CR0 4YY
21/09/2026
02219983-0005